ICSA Book Series in Statistics

The ICSA Book Series in Statistics showcases research from the International Chinese Statistical Association that has an international reach. It publishes books in statistical theory, applications, and statistical education. All books are associated with the ICSA or are authored by invited contributors. Books may be monographs, edited volumes, textbooks and proceedings.

More information about this series at http://www.springer.com/series/13402

Lanju Zhang • Ding-Geng (Din) Chen
Hongmei Jiang • Gang Li • Hui Quan
Editors

Contemporary Biostatistics with Biopharmaceutical Applications

Editors
Lanju Zhang
Data and Statistical Sciences, AbbVie Inc.
North Chicago, IL, USA

Ding-Geng (Din) Chen
School of Social Work
University of North Carolina
Chapel Hill, NC, USA

Hongmei Jiang
Department of Statistics
Northwestern University
Evanston, IL, USA

Gang Li
R&D
Janssen Pharmaceuticals
Raritan, NJ, USA

Hui Quan
Sanofi US
Bridgewater, NJ, USA

ISSN 2199-0980 ISSN 2199-0999 (electronic)
ICSA Book Series in Statistics
ISBN 978-3-030-15312-0 ISBN 978-3-030-15310-6 (eBook)
https://doi.org/10.1007/978-3-030-15310-6

This Springer imprint is published by the registered company Springer Nature Switzerland AG.
The registered company address is: Gewerbestrasse 11, 6330 Cham, Switzerland

Preface

This book is a collection of the most significant research papers presented at the 26th ICSA Applied Statistics Symposium. Held on June 25–28, 2017, at Hilton Chicago Downtown, this symposium attracted more than 800 statisticians in academia, government, and industry around the world. With the theme *Statistics for a new generation: challenges and opportunities*, the symposium also attracted hundreds of students. One hundred and fifty-nine invited, topic contributed, and contributed sessions covered the broadest variety of topics across the full spectrum of all statistical theoretical fronts and applications. After the symposium, speakers were invited to contribute to this book. From all submissions, the editors selected 18 chapters after rigorous peer-reviewed and subsequent revisions.

The book is organized into two balanced parts: Part I, *Biostatistical Methodology*, which includes nine chapters that present the most recent theoretical breakthrough in experimental design, modelling, and analysis, and Part II, *Biopharmaceutical Applications*, which consists of nine chapters that depict various statistical applications in the biopharmaceutical industry. Each chapter is self-contained with relevant references provided at the end of the chapter. The following is a quick glimpse of each chapter:

Part I. In Chap. 1, Mao developed an EM algorithm to estimate tumor onset time in carcinogenicity studies under the condition that cause of death is unknown in a subset of animals. Log-rank test was used to compare treatment groups against controls. The proposed new method was shown to outperform the available methods by simulation. In Chap. 2, Pan and Jiang addressed the high-dimensional variable selection problem for associating the microbial compositions with a phenotype. They employed a log contrast model to bypass the usual step for normalization and developed a new method to identify phenotype-associated species using penalized regression and stability selection. In Chap. 3, Wei proposed the use of contemporary aggregation as a dimension reduction method in high-dimensional multivariate time series and showed that this natural and simple method had forecast accuracy superior to existing methods. In Chap. 4, Wu developed PC-ABT, a novel principal component-based adaptive-weight burden test for gene-based association mapping of quantitative traits. This method efficiently accounted for correlation in multiple genotypic variants and was showed to be more powerful than other multiple variant tests that allowed related individuals. In Chap. 5, Wang et al. proposed an adaptive dynamic Bayesian network model that provided an unprecedented tool to elucidate a comprehensive picture of gene regulatory networks. In particular, an unevenly spaced gene expression record can be accommodated. By analyzing real data sets from a surgical study and through extensive simulation studies, the new model was demonstrated for its usefulness and utility. In Chap. 6, Alghamdi et al. focused on ultrafast functional brain imaging studies and proposed an efficient approach for obtaining high-quality stimulus sequences by taking the uncertainty of the autocorrelation of the response into account. The performance of their proposed approach was demonstrated via case studies. In Chap. 7, Zhang et al. derived a global optimization algorithm that provided a guaranteed ε-global optimum for a sparse mixed membership matrix factorization problem. The algorithm was tested on simulated data and small real gene expression data set and found to always bound the global optimum across random initializations and explore multiple modes efficiently. In Chap. 8, Chuang and Yang proposed a nonnegative robust linear model (NRLM) approach that yielded robust, yet interpretable, mixing rate estimates. In a simulation study, NRLM showed a robust performance for finding the relative abundance of specified components when a large amount of noise was present. More importantly, the approach accurately estimated the absolute level of the specified components in the presence of unspecified ones. Finally, it showed a superior performance when applied to deep deconvolution of blood samples. In Chap. 9, O'Brien and Silcox explored optimal experimental designs for parallelism testing in potency bioassays. They derived theoretical optimal designs and proposed several extensions that took practicality into account. One of the designs, reflection design, was demonstrated to be the most efficient and easy to implement since the researcher could merely sketch the drug/compound logistic curves and read off design at some cutoff lines.

Part II. In Chap. 10, Zhang et al. proposed an optimized two-stage phase III clinical trial design that combined three adaptive techniques to offer the opportunity of dose selection and sample size determination based on the first-stage data with strict type I error rate control and robust power across an effect size interval. In Chap. 11, Gou and Chen proposed a generalized framework for critical boundary refinement when conducting hierarchical hypothesis test in a clinical trial involving multiple interim stages with an improvement on the secondary boundary. The framework had a particular advantage when the primary endpoint data can be assessed earlier than the secondary endpoint data. The framework was also extended to include an adaptive update on the refined boundary when the attained sample sizes were different from what they were originally planned. In Chap. 12, Liu et al. proposed an escalation with overdose control design for phase I oncology trials using dose-limiting toxicity (DLT) with two components, one for immediate toxicity in a binary model and the other for late-onset toxicity in a time to event model. They demonstrated that the proposed dose escalation design can incorporate historical knowledge, protect patients from being assigned to toxic doses, and consider early- and late-onset toxicity while maintaining the escalation timeline. In Chap. 13, Yang et al. proposed a new approach via adding a companion constancy test to the non-inferiority test that consequently protected the validity of a non-inferiority trial under Bayesian framework. In addition, historical data of the active control was borrowed in the analysis with two different approaches. In Chap. 14, Lin et al. introduced a nonparametric model which was robust to event time distribution in response-adaptive designs for survival trials. The operating characteristics of the proposed design and the parametric design were compared by simulation studies, including their robustness properties, with respect to model misspecifications. Both advantages and disadvantages of adaptive randomization were discussed in the summary from a practical perspective of clinical trials as well as illustrations by master protocol case studies. In Chap. 15, Lu et al. provided valuable considerations of the design and analysis of the non-randomized studies using the propensity score methodology. Statistical and regulatory perspectives were highlighted. In Chap. 16, Jiang et al. reviewed key methodological and statistical implications of pragmatic clinical trials (PCTs) in the context of drug development and reimbursement, with emphasis on study design and analyses to maximize external validity. The principles of PCTs challenged some well-established guidelines in randomized clinical trials (RCTs), as open-label and treatment switch in intention-to-treat (ITT) population being the most pronounced ones. They provided valuable suggestions on handling these issues. In Chap. 17, Lipkovich et al. enhanced existing SIDES and SIDEScreen methods for biomarker discovery by incorporating stochastic elements in computing the variable importance, expected treatment effect, and replicability index. The improvement was particularly useful when dealing with relatively small data sets, so as to properly account for the uncertainty of the subgroup selection process. The operating characteristics of the Stochastic SIDEScreen were demonstrated to be improved compared with the corresponding deterministic procedure through simulation. Last, but not the least, in Chap. 18, Pantoja-Galicia and Gene Pennello discussed the implicit or explicit trade-offs between false-positive and false-negative test errors provided by the information from the receiver operating characteristic (ROC) curve. They demonstrated how it can impact the evaluation of the performance of a new medical diagnostic test in comparison with an already established test. They illustrated the idea with the comparability of a new test N with respect to a standard test S in terms of the seriousness of a false-positive error relative to a false-negative error using the information from the ROC curve.

The editors are grateful to many people who contributed to the publication of this book. First, we would like to thank the authors of all chapters for their original research and dedication to share through this book. Second, our sincere appreciations go to all the reviewers for their valuable time and excellent review, which significantly improved the presentations and quality of the book. Third, our gratitude goes to the leadership of the executive committee, organization committees, and numerous volunteers of the 26th ICSA Applied Statistics Symposium. This book would not be possible without such a successful symposium. Last, but not least, we would like to acknowledge the support and guidance of Abitha PradeepCoumar, Susan Westendorf, and Shobha Karuppiah from Springer through the entire process of publishing this book.

North Chicago, IL, USA — Lanju Zhang
Chapel Hill, NC, USA — Ding-Geng (Din) Chen
Evanston, IL, USA — Hongmei Jiang
Raritan, NJ, USA — Gang Li
Bridgewater, NJ, USA — Hui Quan

Contents

Part I
Biostatistical Methodology

Chapter 1
Nonparametric Inference on Tumor Incidence with Partially Identified Cause-of-Death Data

Lu Mao

1.1 Introduction

In drug development, the new product must be evaluated for its safety before being applied to humans in a clinical setting. For ethical considerations, such assessment is usually conducted on animals as a surrogate for human testing. Animal studies thus constitute an indispensable component in the pre-clinical development of novel pharmaceutical agents (Chow and Liu 1998).

Of primary concern about the potential hazards of a new drug is its carcinogenicity, i.e., the ability to induce tumor in the recipient. Carcinogenicity studies are typically carried out in the form of the so-called survival-sacrifice experiments with certain strains of mice or rats. Specifically, healthy rodents are randomized to control or treatment groups and are followed until they die naturally or meet the pre-scheduled time for sacrifice. Following death, the animal is necropsied to determine if a tumor is present. If the tumor under investigation is of a lethal kind, i.e., causing death immediately after formation, then the time to tumor onset is observed insofar as it occurs during study. The resulting data can thus be analyzed using standard survival analysis methods such as Kaplan–Meier curves and logrank tests (Fleming and Harrington 1991). If the tumor is completely non-lethal, so much so that it does not affect the chance for the animal's survival at all, then the time to death can be

L. Mao (✉)
Department of Biostatistics and Medical Informatics, School of Medicine and Public Health, University of Wisconsin-Madison, Madison, WI, USA
e-mail: lmao@biostat.wisc.edu

L. Zhang et al. (eds.), *Contemporary Biostatistics with Biopharmaceutical Applications*, ICSA Book Series in Statistics,
https://doi.org/10.1007/978-3-030-15310-6_1

considered as a random check-up time for tumor status. This setting gives rise to the standard current-status (or case 1 interval-censored) data (see, e.g., Groeneboom and Wellner 1992; Huang 1996; Jewell and van der Laan 2003). However, most tumors reside in between the two extremes, that is, they tend to expedite death of the host, but not in a none-or-all fashion. This ambiguity presents a unique challenge for statistical analysis of the data, as the precise time for the event (tumor onset) is unobserved and the check-up (necropsy) time is correlated with the event in an unknown way.

It is intuitively understandable that one can hope to gain unbiased information about the prevalence of tumor only at time points where necropsy is "objective", i.e., not influenced by the underlying tumor onset time (for more formal discussions on statistical identifiability with survival-sacrifice data, see Clifford (1977), Dewanji and Kalbfleisch (1986), and Malani and Van Ryzin (1988), among others). The role of such "objective" necropsies is structurally fulfilled by pre-planned sacrifices. Consequently, if one wishes to identify tumor prevalence along a continuum of time, one must plan to conduct frequent serial sacrifices over the target time window. This necessitates a large cohort of animals and considerable manpower. To save cost, it was recommended that the cause of natural death be ascertained by a pathologist (Peto 1974; Peto et al. 1980; Archer and Ryan 1989), so that if a death is not caused by the tumor, it can be treated as "accidental" and thus serve in the same role as a planned sacrifice.

Given known cause of death, proper analysis of tumor incidence must factor in the contribution of tumor-caused death because it is correlated with the underlying tumor onset time. This is in contrast with the analysis of the standard current-status data under independent check-ups, in which the distribution of check-up time plays no inferential role. There is a vast literature on statistical methodology for survival-sacrifice data with known cause of death. Most of the existing methods, however, either make parametric assumptions or rely on partitioning continuous time into a small number of intervals under multi-state illness-death models (see an excellent review of various such methods by Ahn and Kodell (1998)). In the nonparametric setting, Kodell et al. (1982) derived readily computable estimators for the marginal distributions of tumor onset and tumor-caused death, under the assumption that tumor prevalence among the surviving animals is monotonic over time. As pointed out by later authors, this assumption need not hold. Dinse and Lagakos (1982) proposed unrestricted nonparametric estimators by iteratively maximizing the likelihood with respect to the two distribution functions. Turnbull and Mitchell (1984) used an EM algorithm to compute the nonparametric maximum likelihood estimators. This type of algorithm is applicable in a wide range of settings (Turnbull 1976) but is known to be very slow due to a large number of unknown parameters (Groeneboom and Jongbloed 2014). A computational efficient approach was suggested by van der Laan et al. (1997), who used the Kaplan–Meier estimator for the distribution of tumor-caused death and a weighted least squares estimator for the distribution of tumor onset. An alternative strategy was employed by Gomes (2001), who proposed to estimate the distribution function for tumor onset by maximizing a pseudo-likelihood with plug-in Kaplan–Meier estimator for the distribution of tumor-caused death. The maximum pseudo-likelihood estimator (MPLE) was computed using a modified iterative convex minorant (ICM; see Groeneboom and Wellner 1992) algorithm, and the validity of the MPLE was proved rigorously later (Gomes 2008).

In practice, however, the pathologist may not be able to adjudicate the causes for all deaths with absolute certainty (Peto et al. 1984). As a result, some cases may be associated with equivocal or unknown causes. To account for the partially missing information on cause of death, Kodell and Chen (1987) proposed an EM algorithm by treating the unknown causes as missing data. However, their method is built upon the approach of Kodell et al. (1982) and may thus yield misleading results when the monotonicity assumption of the latter fails. In addition, as pointed out by Dinse (1987), their E step is improper in that it is not conditioned upon all the information available and may thus result in biased inference.

In this article, we propose fully nonparametric inference procedures for survival-sacrifice data when the cause-of-death information is only partially available. The rest of the paper is organized as follows. In Sect. 1.2, we develop an EM-type algorithm to estimate the marginal distributions of tumor onset and tumor-caused death. The E step consists in properly estimating the conditional probabilities for the cause of death (in closed forms) and the M step amounts to weighted Kaplan–Meier estimators for death and a weighted version of the MPLE for tumor onset (Gomes 2001). We also propose a class of logrank-type tests for comparing tumor incidence across treatment arms. In Sect. 1.3, we conduct simulations to assess the finite-sample performance of the proposed methods. A real survival-sacrifice study on pituitary tumor in rats is analyzed using the proposed methods in Sect. 1.4. We conclude the paper by some discussions in Sect. 1.5.

1.2 Methods

1.2.1 Data Structure

Let T and D denote times to tumor onset and to tumor-caused death, respectively. Note that we always have $T \leq D$. Use D_C to denote time to death from a competing cause, which is assumed to be independent of (T, D). In addition to the natural deaths, the experimental animals are subject to serial and/or terminal sacrifice. Let U^* denote time to sacrifice that is independent of (T, D, D_C). We use a composite notation $U = D_C \wedge U^*$ to denote time to "accidental" death, i.e., sacrificial death or one from a competing cause, where $a \wedge b = \min(a, b)$. Clearly, we have that $U \perp (T, D)$ so that accidental death amounts to a random check-up.

The observed data consist of time to death $X = D \wedge U$ along with the label for the type of death and tumor status at death. In a study where the cause of death is fully ascertained, one observes $\Delta_1 = I(\text{Death from tumor})$, $\Delta_2 = I(\text{Accidental death with tumor})$, and $\Delta_3 = I(\text{Accidental death without tumor})$, where $I(\cdot)$ is the indicator function. In the literature, a tumor resulting in death is called a fatal tumor and a tumor found at accidental death is called an incidental tumor (Peto et al. 1980). Clearly, we have that $\sum_{k=1}^{3} \Delta_k = 1$. When the cause of death is only partially ascertained, it is worth noting that uncertainty can only arise between fatal and incidental tumors, i.e., death in the absence of tumor must not be caused by it and thus must be accidental. To formalize the set-up, let $R = I(\text{Cause of death is ascertained})$ so that one observes Δ_1 and Δ_2 if $R = 1$ and observes $\Delta_{12} = \Delta_1 + \Delta_2$ if $R = 0$. Define $R = 1$ if $\Delta_3 = 1$. Throughout, we make the important assumption that missingness in the cause of death depends only on the death time and not on the underlying (unobserved) cause. That is, we assume that

$$\text{pr}(R = 1 \mid \Delta_{12} = 1, \Delta_1, X) = \text{pr}(R = 1 \mid \Delta_{12} = 1, X) =: \pi(X). \tag{1.1}$$

In missing data literature, (1.1) is essentially a standard missing at random (MAR) assumption. Denote a random sample of $(X, \Delta_1, \Delta_2, \Delta_3, R)$ by $(X_i, \Delta_{1i}, \Delta_{2i}, \Delta_{3i}, R_i)$ $(i = 1, \ldots, n)$, where n is the sample size. Then, the observed data can be represented as

$$(R_i \Delta_{1i}, R_i \Delta_{2i}, R_i, X_i), \quad i = 1, \ldots, n. \tag{1.2}$$

The (hypothetical) full data, where the cause-of-death information is available on all animals, are

$$(\Delta_{1i}, \Delta_{2i}, X_i), \quad i = 1, \ldots, n. \tag{1.3}$$

Denote the distribution function for tumor onset time by $F_T(t) = \text{pr}(T \leq t)$, which is our main target of inference. Of secondary interest is the distribution of time to death caused by tumor, denoted by $F_D(t) = \text{pr}(D \leq t)$. A nuisance quantity that is dispensed with in the case of full data (Gomes 2001) but that will play a role in our case is the distribution function for accidental death $F_U(t) = \text{pr}(U \leq t)$. While F_T and F_D can be realistically assumed to be continuous over $[0, \tau]$, with τ denoting the study termination time, the situation for F_U depends on the study design. The distribution of U^* may be discrete in the case of infrequent sacrifices or even degenerate at τ in the case of a sole terminal sacrifice. However, if deaths from competing causes are frequent enough to warrant treatment of D_C as continuous (see real examples in e.g., Kodell and Nelson 1980; Peto et al. 1984), then F_U will at least have a continuous component on $[0, \tau]$. If that is true, F_T will be identifiable over $[0, \tau]$ and the results to be established later will be applicable regardless of the sacrifice plan.

Under the MAR assumption (1.1), the likelihood for (1.2) is proportional to

$$\begin{aligned} L_n(F_T, F_D, F_U) = \prod_{i=1}^{n} \left[F_D'(X_i)\{1 - F_U(X_i)\}\right]^{R_i \Delta_{1i}} \left[F_U'(X_i)\{F_T(X_i) - F_D(X_i)\}\right]^{R_i \Delta_{2i}} \\ \times \left[F_D'(X_i)\{1 - F_U(X_i)\} + F_U'(X_i)\{F_T(X_i) - F_D(X_i)\}\right]^{1-R_i} \\ \times \left[F_U'(X_i)\{1 - F_T(X_i)\}\right], \end{aligned} \quad (1.4)$$

where $f'(x) = \mathrm{d}f(x)/\mathrm{d}x$ for any function f. Note that $\Delta_{3i} = R_i - R_i\Delta_{1i} - R_i\Delta_{2i}$ is always observed.

1.2.2 Nonparametric Estimation

Due to the entanglement of the three functional parameters, direct maximization of (1.4) is hard even with state-of-the-art tools of order-constrained optimization (Groeneboom and Jongbloed 2014). Instead, we borrow the idea of the MPLE for the full data (Gomes 2001). The MPLE works as follows. The distribution function F_D is first estimated using the Kaplan–Meier estimator based on (Δ_{1i}, X_i) $(i = 1, \ldots, n)$, and is then inserted into the likelihood to form a pseudo-likelihood. Next, F_T is estimated using the ICM algorithm on the pseudo-likelihood. In our case, one might be tempted to imitate this strategy by first estimating F_D based on the reduced data $(R_i, R_i\Delta_{1i}, X_i)$ $(i = 1, \ldots, n)$, possibly using an EM algorithm. However, this approach is not valid because the cause-of-death information is not MAR given the reduced data. Intuitively, because the information is always known on a subset of accidental deaths (namely, those with tumor absent), tumor-caused deaths are over-represented in those of unknown causes. Thus, the true mixing proportions, i.e., the fractions of tumor-caused and accidental deaths among the equivocal cases, depend on the other parameter F_T also. Consequently, F_D cannot be properly estimated alone in the presence of unidentified cause of death.

To circumvent this problem, we devise another kind of EM-type algorithm, whose E step estimates the mixing proportions using the current iterates of F_T, F_D, and F_U jointly and whose M step updates (F_D, F_U) and F_T separately, in a similar way to the MPLE for the full data. Note that the log-likelihood for the full data (1.3) is

$$\begin{aligned} l_{n,\mathrm{F}}(F_T, F_D, F_U) = \sum_{i=1}^{n} \Big[\Delta_{1i} \log F_D'(X_i)\{1 - F_U(X_i)\} + (1 - \Delta_{1i}) \log F_U'(X_i) \\ + \Delta_{2i} \log\{F_T(X_i) - F_D(X_i)\} + \Delta_{3i} \log\{1 - F_T(X_i)\}\Big]. \end{aligned}$$

Denote the jth iterate of the parameters (F_T, F_D, F_U) by $\theta^{(j)} = (F_T^{(j)}, F_D^{(j)}, F_U^{(j)})$. Then, the E step at the $(j+1)$th iteration computes

$$\begin{aligned} Q_n(F_T, F_D, F_U \mid \theta^{(j)}) &= E\left\{l_{n,\mathrm{F}}(F_T, F_D, F_U) \mid \text{Observed data; } \theta^{(j)}\right\} \\ &= \sum_{i=1}^{n}\Big[w_i^{(j)} \log F_D'(X_i)\{1 - F_U(X_i)\} + (1 - w_i^{(j)}) \log F_U'(X_i) \\ &\quad + (1-w_i^{(j)}) \log\{F_T(X_i) - F_D(X_i)\} + \Delta_{3i} \log\{1 - F_T(X_i)\}\Big], \end{aligned}$$

where

$$\begin{aligned} w_i^{(j)} &= R_i \Delta_{1i} + (1 - R_i) E(\Delta_{i1} \mid \Delta_{1i} + \Delta_{2i} = 1, X_i; \theta^{(j)}) \\ &= R_i \Delta_{1i} + \frac{(1 - R_i) F_D^{(j)\prime}(X_i)\{1 - F_U^{(j)}(X_i)\}}{F_D^{(j)\prime}(X_i)\{1 - F_U^{(j)}(X_i)\} + F_U^{(j)\prime}(X_i)\{F_T^{(j)}(X_i) - F_D^{(j)}(X_i)\}}. \end{aligned} \tag{1.5}$$

The second term on the far right hand side of (1.5) estimates the mixing probability of tumor-caused death given an unknown cause. The computation of $F_D^{(j)\prime}$ and $F_U^{(j)\prime}$ in (1.5) is a delicate issue, which we shall remark upon later.

Direct maximization of $Q_n(F_T, F_D, F_U \mid \theta^{(j)})$, a weighted version of the full-data log-likelihood, is still not easy because of the entanglement of F_D and F_T in the term $\log\{F_T(X_i) - F_D(X_i)\}$. It is now that we adopt the idea of the two-step procedure of the MPLE for the full data. First, $F_D^{(j+1)}$ and $F_U^{(j+1)}$ are obtained through maximizing the conditional expectation of the log-likelihood for (Δ_{1i}, X_i) $(i = 1, \ldots, n)$ given the observed data (1.2) and $\theta^{(j)}$. One can easily show that this leads to the weighted Kaplan–Meier estimators

$$F_D^{(j+1)}(t) = 1 - \prod\nolimits_{0 \le s \le t}\left\{1 - \frac{\sum_{i=1}^{n} w_i^{(j)} I(X_i = s)}{\sum_{i=1}^{n} I(X_i \ge s)}\right\},$$

$$F_U^{(j+1)}(t) = 1 - \prod\nolimits_{0 \le s \le t}\left\{1 - \frac{\sum_{i=1}^{n} (1 - w_i^{(j)}) I(X_i = s)}{\sum_{i=1}^{n} I(X_i \ge s)}\right\}.$$

So, the functions $F_D^{(j)}$ and $F_U^{(j)}$ used in the E step (1.5) are step functions taking jumps on certain subsets of the X_i. Write $F\{t\} = F(t) - F(t-)$ for $F = F_D, F_U$. Then, we can estimate the $F^{(j)\prime}$ in (1.5) by

$$F^{(j)\prime}(t) = \mathrm{avg}\{F^{(j)}\{X_i\} : |X_i - t| \le \kappa_n, i = 1, \ldots, n\}, \tag{1.6}$$

where $\mathrm{avg}(A)$ is the average of the elements in set A and $\kappa_n > 0$ is a pre-set bin width.

Remark 1.1 One might question the need for aggregating neighboring points in the estimation of $F^{(j)\prime}(X_i)$ instead of using $F^{(j)\prime}\{X_i\}$ itself. The latter is the strategy adopted in the EM algorithms for certain regression models for competing risks data with unknown failure causes (e.g., Mao and Lin 2017; Mao et al. 2017). The peculiarity here is that, in a fully nonparametric setting such as ours, if there is no tie at X_i and the cause information is missing, estimating the mixing probability is essentially the same as estimating the success probability based on an unobserved outcome of a single Bernoulli trial. It is clear that the latter setting offers no hope of getting a consistent estimate for the parameter of interest. In a regression setting, on the other hand, the mixing probabilities are intrinsically constrained by the model and thus do not have this singularity problem. The strategy taken in (1.6) overcomes the difficulty in the nonparametric setting under the mild condition that the true mixing proportions are reasonably smooth over time.

Next, we insert $F_D = F_D^{(j+1)}$ back into $Q_n(F_T, F_D, F_U \mid \theta^{(j)})$, and, shedding all terms unrelated to F_T, obtain an objective function for F_T:

$$Q_{1n}(F_T \mid F_D^{(j+1)}, \theta^{(j)}) = \sum_{R_i \Delta_{1i}=0} \left[(1 - w_i^{(j)}) \log\{F_T(X_i) - F_D^{(j+1)}(X_i)\} + \Delta_{3i} \log\{1 - F_T(X_i)\}\right].$$

So, we compute

$$F_T^{(j)} = \arg\max_{F_T} Q_{1n}(F_T \mid F_D^{(j+1)}, \theta^{(j)}). \tag{1.7}$$

The function $Q_{1n}(F_T \mid F_D^{(j+1)}, \theta^{(j)})$ takes the form of a weighted log-pseudo-likelihood of the full data. Therefore, the optimization problem (1.7) can be tackled by the ICM algorithm developed by Gomes (2001) for that purpose. Specifically, Q_{1n} is maximized subject to the (natural) constraints that F_T is a distribution function and that $F_T(t) \geq F_D(t)$ for all $t \in [0, \tau]$. The ICM is an order-constrained analog of the Newton-Raphson algorithm and converges at a similar (quadratic) rate (see, e.g., Groeneboom and Jongbloed 2014). Because the outer loop of the EM, with its linear convergence rate, moves on much slower than does the inner loop of the ICM, it is advisable to replace the full maximization of (1.7) in the M step with a single-step update in the interest of computational efficiency.

Thus, we iterate the E and M steps until the difference between two successive iterates becomes very small, i.e.,

$$\sum_{i=1}^{n} |F_T^{(j+1)}(X_i) - F_T^{(j)}(X_i)| + |F_D^{(j+1)}(X_i) - F_D^{(j)}(X_i)| + |F_U^{(j+1)}(X_i) - F_U^{(j)}(X_i)| < \varepsilon_0,$$

where $\varepsilon_0 > 0$ is a very small number. Denote the final state of (F_T, F_D) at convergence by $(\widehat{F}_T, \widehat{F}_D)$. With suitable regularity conditions on the joint distribution of (T, D) and with a positivity condition such as $\pi(X) \geq \delta$ almost surely for some $\delta > 0$, one may expect that the proposed estimators are consistent, with $\widehat{F}_D$ converging to the truth at the $n^{-1/2}$ rate and $\widehat{F}_T$ at the $n^{-1/3}$ rate. That $\widehat{F}_T$ converges slower than the standard $n^{-1/2}$ rate is due to the fact that, unlike D, the information about T is always incomplete as in standard interval censoring problems (Groeneboom and Wellner 1992; Huang and Wellner 1997).

1.2.3 A Class of K-Sample Logrank-Type Tests

In practice, it is often of interest to compare different treatment arms (e.g., control vs varying dosages of a drug) on the tumor incidence. We propose a class of K-sample tests that is based on the estimated F_T for each sample.

Use F_{Tk} to denote the cumulative distribution function of T for the kth sample $(k = 1, \ldots, K)$. Then, we estimate F_{Tk} using the methods described in Sect. 1.2.2 based on the data in the kth sample. Denote the estimate by $\widehat{F}_{Tk}$. Likewise, we construct an estimate $\widehat{F}_{T0}$ for F_{T0}, the cumulative distribution function for T under the null hypothesis of no treatment effect, based on the pooled sample. Then, the corresponding estimates for the cumulative hazard functions are

$$\widehat{\Lambda}_k(t) = \int_0^t \frac{\mathrm{d}\widehat{F}_{Tk}(s)}{1 - \widehat{F}_{Tk}(s-)}, \qquad k = 0, 1, \ldots, K-1.$$

Using the idea of Gray (1988) in mimicking the logrank test for standard right-censored data (Fleming and Harrington 1991, Ch 3), we construct test statistics of the form

$$z_k = \int_0^\tau W_k(s)\{\mathrm{d}\widehat{\Lambda}_k(t) - \mathrm{d}\widehat{\Lambda}_0(t)\}, \qquad k = 1, \ldots, K-1, \tag{1.8}$$

where $W_k(s)$ is a properly chosen weight function. We assume that $W_k \to w_k$, where w_k is a continuous function. A popular choice for the weight functions is the Harrington-Fleming class (1982)

$$W_k(s) = \{1 - \widehat{F}_{T0}(s)\}^\rho, \qquad \rho \geq 0, \tag{1.9}$$

where ρ controls the amount of emphasis placed on early tumorigenesis.

Write $z = (z_1, \ldots, z_{K-1})^{\mathrm{T}}$. Although the $\widehat{\Lambda}_k(t)$ have non-standard convergence rates, z is expected to be asymptotically multivariate normal (with mean zero under the null hypothesis) because its components are smooth functionals of the $\widehat{\Lambda}_k(t)$ (see, e.g., Groeneboom and Wellner 1992, Ch I.3). Thus, the test statistic $S_n^2 = ||z||^2$ should be asymptotically distributed as a multiple of the chi-square with $K - 1$ degrees of freedom. In practice, the null distribution of S_n^2 may be assessed by permutation, and an asymptotic level-α test can be constructed by rejecting the null if the observed S_n^2 is greater than the $100(1 - \alpha)$th percentile of its permutation distribution.

Let $\lambda_k(t)$ denote the hazard function of T for the kth sample. Following the reasoning of Fleming and Harrington (1991, Ch 3) and Gray (1988), it can be shown that, under suitable regularity conditions, the tests are consistent against the following class of hazard-order alternatives: there exist k and $k' \in \{1, \ldots, K\}$ such that

$$\lambda_k(t) \leq \lambda_{k'}(t), \qquad \text{for all } t \in [0, \tau],$$

with strict inequality for some t.

1.3 Simulation Studies

We performed simulation studies to assess the finite-sample performance of the proposed methods. Let T follow a Weibull distribution with cumulative hazard function $\Lambda_T(t) = t^{1.25}$. Let $D - T$ be an independently and identically distributed random variable as T. Let $U = D_C \wedge \tau$, where $D_C \sim \text{Expn}(0.5)$ and $\tau = 4$. We used the following logistic regression model to generate missing cause of death data:

$$\text{pr}(R = 0 \mid \Delta_1, X, \Delta_{12} = 1) = \frac{\exp(0.1Z - 1)}{\exp(0.1Z - 1) + 1}.$$

Under this set-up, about 40% of the subjects are known to experience fatal tumor ($\Delta_1 = 1$), about 15% are known to experience incidental tumor ($\Delta_2 = 1$), about 20% die with tumor from unknown causes ($\Delta_{12} = 1$), and the remaining about 25% die tumor-free. Thus, among those that die with tumor, information on the cause of death is missing on about 26.7% of them.

We set the sample size to be $n = 400, 800$, and used the bin widths $\kappa_n = \tau n^{-1/2} = 0.2, 0.14$, respectively. The convergence criterion was $\varepsilon_0 = 10^{-4}$. For each simulated dataset, we used the proposed method to estimate F_T and F_D. We compared the estimates to those from three *ad hoc* approaches based on the naive use of MPLE of Gomes (2001) for full data after removal or imputation of the missing data. The first is the complete-case analysis, where analysis is restricted to observations with known causes of death. The second is to impute all the unknown causes to tumor-caused. The third is to impute all the unknown causes to accidental. For each sample size, we calculated the average of the estimates based on 1000 replicates and plotted the results along with the corresponding true values.

Figure 1.1 displays the true and estimated curves for F_T. The curves computed by the proposed method agree with the true values quite well. There is slight under-estimation near time zero, but the problem seems to be mitigated by increasing the sample size. On the other hand, all the *ad hoc* methods exhibit severe biases. We examine and explain the direction of bias for each method one by one. Understandably, the complete case analysis under-estimates the tumor incidence because the discarded cases are all tumor-bearing ones. The two imputation methods have opposite directions of bias, over-estimation for imputing as accidental and under-estimation for imputing as tumor-caused. This can be explained as follows. Naturally, time to tumor-caused death is correlated with time to tumor onset. As a result, tumor onset is expected to have occurred closer to the death time if death is tumor-caused than if it is accidental (one may think of the extreme case of a lethal tumor). Thus, fatal and incidental tumors contribute differently to the estimated tumor incidence. Another angle to look at the relative contribution is through the likelihood function. An incidental tumor contributes $F_T - F_D$ to the likelihood, while a fatal tumor contributes only F'_D and thus has an impact on F_T only through the inequality $F_T \geq F_D$. In a word, an incidental tumor contributes more to tumor incidence than does a fatal tumor.

The survival curves for tumor-caused death, i.e., $1 - F_D$, are plotted in Fig. 1.2. Again, the agreement between estimates by the proposed method and the true values are satisfactory, even somewhat better than the estimation of F_T. This is not surprising because we expect a faster rate of convergence for $\widehat{F}_D$ (see Sect. 1.2.2). Interestingly, the complete-case analysis does not appear to incur much bias. After all, the discarded observations are a mix of the two causes, so that no categorical preference is given to one or the other. The naive imputation methods still lead to substantial biases, the directions of which are in this case straightforward to explain.

Next, we assessed the impact of the choice for the bin width κ_n on the estimation of F_T. We used the same setting as the first set of simulations and varied κ_n over $\tau n^{-1/5}$, $\tau n^{-1/3}$, $\tau n^{-1/2}$, and $\tau n^{-2/3}$. We evaluated the estimates by the proposed method for F_T at different time points under each scenario based on 1000 replications. The results are summarized in Table 1.1. The bias and standard error for all values of κ_n look fairly similar, suggesting a certain degree of robustness of the aggregation approach with regard to the choice of the bin width in the E step.

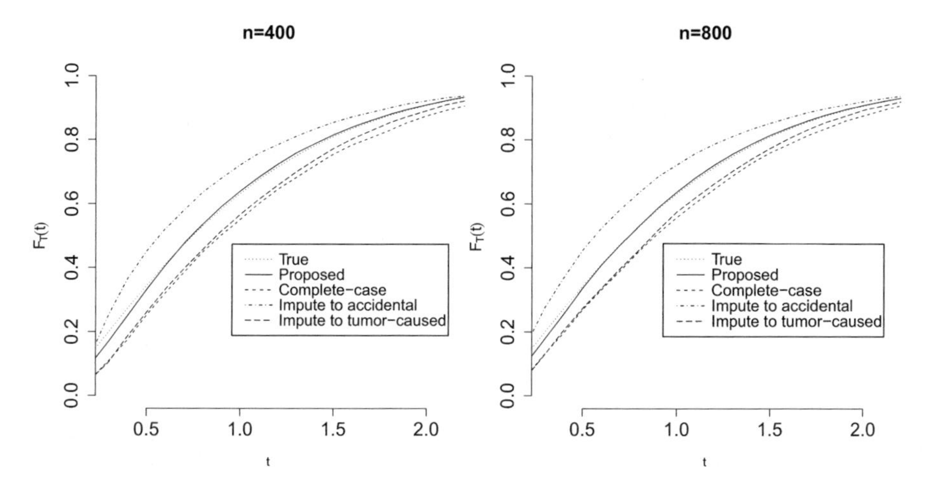

Fig. 1.1 True and estimated cumulative distribution functions for tumor-onset time using the proposed method, complete-case analysis, imputation to accidental death, and imputation to tumor-caused death. Each scenario is replicated 1000 times

1.4 A Pituitary Tumor Study

A large-scale animal study was conducted by the British Industrial Biological Research Association to evaluate the carcinogenic effects of nitrosamine in drinking water (Peto et al. 1984). The experiments involved 5000 rodents and assessed different nitrosamines in different dosages. Here we consider a subset of the study data. The data consist of observations on 384 inbred Colworth rats. Among them, half were randomized to the control group, whose drinking water contained no added substances; the other half were randomized to the treatment group, where different daily doses (0.033, 0.066, or 0.132 ppm) of N-nitrosodimethylamine (NDMA) were administered to the drinking water. Death time, status about onset of pituitary tumor, and the (possibly unknown) cause of death were recorded. There were 10 rats whose tumor status was not ascertained and are thus excluded. The final dataset contains $n = 374$ rats, with 185 in the control group and 189 in the treatment group. The maximum length of follow-up is $\tau = 1234$ days. The dataset contains 281 unique time points for death. So, the traditional approach of grouping the data into a small number of time intervals (e.g., Kodell et al. 1994) could potentially lose a lot of information.

As a first step, we provide some descriptive statistics, along the lines of Peto et al. (1980), about the cause of death and tumor status by treatment in Table 1.2. In the columns of the table, "Fatal" indicates fatal tumor, i.e., one that results in

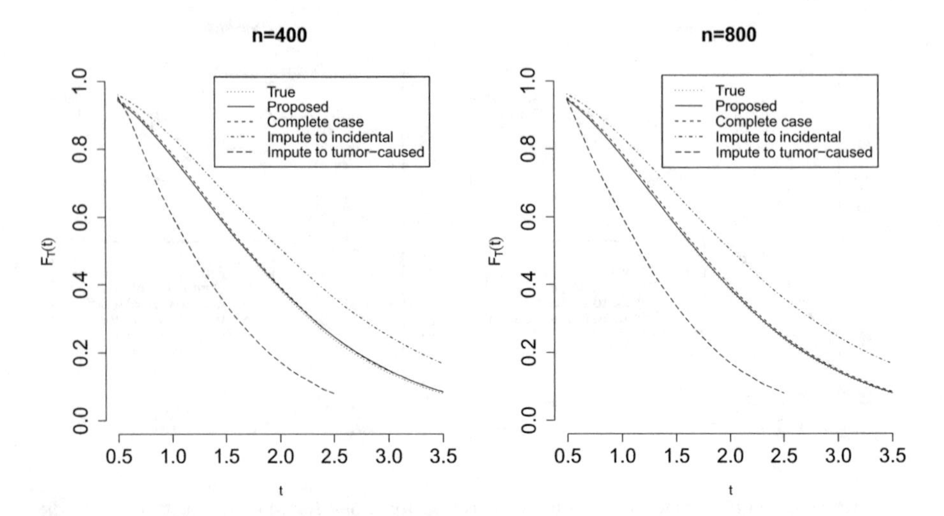

Fig. 1.2 True and estimated survival curves for tumor-caused death (in the absence of accidental death) using the proposed method, complete-case analysis, imputation to accidental death, and imputation to tumor-caused death. Each scenario is replicated 1000 times

Table 1.1 Estimation of $F_T(t)$ based on different choices of κ_n

		$F_T(0.5) = 0.343$	$F_T(1.0) = 0.632$	$F_T(1.5) = 0.810$	$F_T(2.0) = 0.907$
n	κ_n	Bias (SE)	Bias (SE)	Bias (SE)	Bias (SE)
400	1.20	−0.018 (0.108)	0.012 (0.084)	0.014 (0.057)	0.004 (0.039)
	0.54	−0.029 (0.113)	0.011 (0.085)	0.013 (0.057)	0.003 (0.039)
	0.20	−0.025 (0.111)	0.010 (0.084)	0.014 (0.059)	0.004 (0.039)
	0.07	−0.022 (0.116)	0.006 (0.085)	0.012 (0.058)	0.006 (0.039)
800	1.05	−0.012 (0.081)	0.008 (0.065)	0.006 (0.046)	0.005 (0.029)
	0.43	−0.008 (0.085)	0.009 (0.066)	0.008 (0.045)	0.004 (0.031)
	0.14	−0.009 (0.083)	0.010 (0.070)	0.006 (0.044)	0.006 (0.030)
	0.05	−0.011 (0.085)	0.007 (0.067)	0.007 (0.046)	0.003 (0.031)

Note: Bias and SE are the empirical bias and standard error, respectively, of the estimator. Each scenario is replicated 1000 times

a tumor-caused death; "Incidental" indicates incidental tumor, i.e., one found at accidental death; "Unknown" indicates unknown cause of death with tumor; "No tumor" indicates no tumor found. By inspection of the crude numbers, we do not find material difference between the two groups in terms of tumor incidence or death rate.

Table 1.2 Descriptive statistics for the pituitary tumor dataset

	Fatal	Incidental	Unknown	No tumor	Total
Control	25	23	2	135	185
Treatment	28	24	3	134	189
Overall	53	47	5	269	374

We provide a more formal analysis by plotting the group-specific nonparametric estimates for the cumulative incidence of pituitary tumor and the survival rate for tumor-caused death using the proposed methods with $\kappa_n = \tau n^{-1/2} = 63.8$ and $\varepsilon_0 = 10^{-4}$. The graphs are shown in Fig. 1.3. Again, neither function shows clear advantage or disadvantage of the treatment compared with the control. Next, we use the Harrington-Fleming logrank-type test described in Sect. 1.2.3 with $\rho = 1$ to test the group difference in tumor incidence. The test statistic is calculated to be $S_n^2 = 0.040$ with a p-value of 0.22 with reference to a null distribution based on 1000 permutations. Thus, we conclude that there is no significant tumor-inducing effect of NDMA on the experimental rats. This conclusion is in agreement with the analysis results based on the totality of the data using various discrete-time methods (see, e.g., Gart 1986, Ch 5).

1.5 Concluding Remarks

We have extended the nonparametric estimation procedure of Gomes (2001) to accommodate missing information on the cause of death. It is worth noting that missingness can only occur with tumor-bearing animals. This implies that the missing mechanism is never completely at random. Consequently, a naive complete-case analysis is bound to incur bias, as demonstrated in our numerical studies. In this sense, our methods provide the much needed tools for proper analysis of such data in practice.

The proposed estimation procedure makes use of the ICM algorithm of Gomes (2001) in the M step and thus avoids treating all incomplete observations of tumor onset time as missing data. Therefore, our method is likely to be computationally much more efficient than any adaptation of the EM algorithm of Turnbull and Mitchell (1984). Our approach also improves upon the EM algorithm of Kodell and Chen (1987) in the following ways. First, we remove the unnecessary (and fallible) assumption that $\{1 - F_T(t)\}/\{1 - F_D(t)\}$ is monotone in t; second, we compute the mixing proportion properly in the E step by conditioning on all the observed data (see a detailed critique by Dinse (1987)).

Testing for difference in tumor incidence across treatment groups is presumably the main goal for a carcinogenicity study. Most testing procedures proposed in the literature require discretizing time into a few intervals so as to maintain the nominal type I error rate (Kodell et al. 1994). In addition to loss of infor-

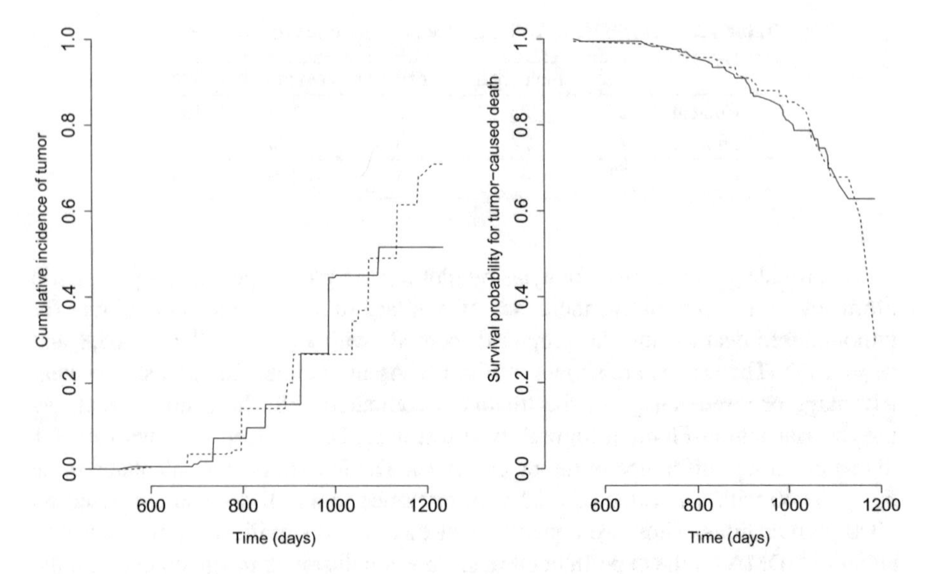

Fig. 1.3 Analysis of the pituitary tumor study. Left, estimated cumulative distribution functions for pituitary tumor onset; right, estimated survival functions for tumor-caused death. Solid, treatment; dashed, control

mation, the choice of the time intervals introduces subjectivity in the analysis. Our proposed class of logrank-type tests is superior to the traditional methods in that it makes fuller use of the data and is a completely automated procedure.

Using our approach for studies where a sizable portion (say >60%) of natural deaths have unknown causes is not recommended. For one thing, a large number of unknown mixing probabilities slow down the EM algorithm considerably. More importantly, as observed by Kodell and Chen (1987), it is imprudent to rely on a small number of known cases to infer the status of others. In such scenarios, methods designed for data without cause-of-death information (e.g., McKnight and Crowley 1984; Dewanji and Kalbfleisch 1986) would apply.

Supplementary Materials

An R package implementing the proposed methodology is posted on the author's website https://biostat.wisc.edu/~lmao/.

References

Ahn, H., Kodell, R.L.: Analysis of long-term carcinogenicity studies. In: Chow, S.C., Liu, J.P. (eds.) Design and Analysis of Animal Studies in Pharmaceutical Development, pp. 259–290. CRC Press, Boca Raton (1998)

Archer, L.E., Ryan, L.M.: Accounting for misclassification in the cause-of-death test for carcinogenicity. J. Am. Stat. Assoc. **84**, 787–791 (1989)

Chow, S.C., Liu, J.P. (eds.): Design and Analysis of Animal Studies in Pharmaceutical Development. CRC Press, Boca Raton (1998)

Clifford, P.: Nonidentifiability in stochastic models of illness and death. Proc. Natl. Acad. Sci. **74**, 1338–1340 (1977)

Dewanji, A., Kalbfleisch, J.D.: Nonparametric methods for survival/sacrifice experiments. Biometrics **42**, 325–341 (1986)

Dinse, G.E.: Discussion of "Handling cause of death in equivocal cases using the em algorithm" by RL Kodell and JJ Chen. Commun. Stat. Theory Methods, **16**, 2587–2592 (1987)

Dinse, G.E., Lagakos, S.W.: Nonparametric estimation of lifetime and disease onset distributions from incomplete observations. Biometrics **38**, 921–932 (1982)

Fleming, T.R., Harrington, D.P.: Counting Processes and Survival Analysis. Wiley, New York (1991)

Gart, J.J.: Statistical Methods in Cancer Research: The Design and Analysis of Long-Term Animal Experiments. Oxford University Press, Oxford (1986)

Gomes, A.E.: Characterization of the NPMPLE of the disease onset distribution function for a survival-sacrifice model. Braz. J. Probab. Stat. **15**, 135–145 (2001)

Gomes, A.E.: Consistency of the non-parametric maximum pseudo-likelihood estimator of the disease onset distribution function for a survival-sacrifice model. J. Nonparametr. Stat. **20**, 39–46 (2008)

Gray, R.J.: A class of K-sample tests for comparing the cumulative incidence of a competing risk. Ann. Stat. **16**, 1141–1154 (1988)

Groeneboom, P., Jongbloed, G.: Nonparametric Estimation Under Shape Constraints. Cambridge University Press, Cambridge (2014)

Groeneboom, P., Wellner, J.A.: Information Bounds and Nonparametric Maximum Likelihood Estimation. Springer, Berlin (1992)

Harrington, D.P., Fleming, T.R.: A class of rank test procedures for censored survival data. Biometrika **69**, 553–566 (1982)

Huang, J.: Efficient estimation for the proportional hazards model with interval censoring. Ann. Stat. **24**, 540–568 (1996)

Huang, J., Wellner, J.A.: Interval censored survival data: a review of recent progress. In: Lin, D.Y., Fleming, T.R. (eds.) Proceedings of the First Seattle Symposium in Biostatistics: Survival Analysis, pp. 123–169. Springer, New York (1997)

Jewell, N.P., van der Laan, M.: Current status data: review, recent developments and open problems. In: Handbook of Statistics, vol. 23, pp. 625–642. Elsevier, Amsterdam (2003)

Kodell, R.L., Chen, J.J.: Handling cause of death in equivocal cases using the EM algorithm. Commun. Stat. Theory Methods **16**, 2565–2585 (1987)

Kodell, R.L., Nelson, C.J.: An illness-death model for the study of the carcinogenic process using survival/sacrifice data. Biometrics **36**, 267–277 (1980)

Kodell, R.L., Shaw, G.W., Johnson, A.M.: Nonparametric joint estimators for disease resistance and survival functions in survival/sacrifice experiments. Biometrics. 43–58 (1982)

Kodell, R.L., Chen, J.J., Moore, G.E.: Comparing distributions of time to onset of disease in animal tumorigenicity experiments. Commun. Stat. Theory Methods **23**, 959–980 (1994)

Malani, H.M., Van Ryzin, J.: Comparison of two treatments in animal carcinogenicity experiments. J. Am. Stat. Assoc. **83**, 1171–1177 (1988)

Mao, L., Lin, D.Y.: Efficient estimation of semiparametric transformation models for the cumulative incidence of competing risks. J. R. Stat. Soc. Ser. B **79**, 573–587 (2017)

Mao, L., Lin, D.Y., Zeng, D.: Semiparametric regression analysis of interval-censored competing risks data. Biometrics **73**, 857–865 (2017)

McKnight, B., Crowley, J.: Tests for differences in tumor incidence based on animal carcinogenesis experiments. J. Am. Stat. Assoc. **79**, 639–648 (1984)

Peto, R.: Guidelines on the analysis of tumour rates and death rates in experimental animals. Br. J. Cancer **29**, 101–105 (1974)

Peto, R., Pike, M.C., Day, N.E., Gray, R.G., Lee, P.N., Parish, S., Peto, J., Richards, S., Wahrendorf, J.: Guidelines for simple, sensitive significance tests for carcinogenic effects in long-term animal experiments. IARC Monogr. Eval. Carcinog. Risk Chem. Hum. **Supplement**, 311–426 (1980)

Peto, R., Gray, R., Brantom, P., Grasso, P.: Nitrosamine carcinogenesis in 5120 rodents: chronic administration of sixteen different concentrations of NDEA, NDMA, NPYR and NPIP in the water of 4440 inbred rats, with parallel studies on NDEA alone of the effect of age of starting (3, 6 or 20 weeks) and of species (rats, mice or hamsters). IARC Sci. Publ. **57**, 627–665 (1984)

Turnbull, B.W.: The empirical distribution function with arbitrarily grouped, censored and truncated data. J. R. Stat. Soc. Ser. B **38**, 290–295 (1976)

Turnbull, B.W., Mitchell, T.J.: Nonparametric estimation of the distribution of time to onset for specific diseases in survival/sacrifice experiments. Biometrics **40**, 41–50 (1984)

van der Laan, M.J., Jewell, N.P., Peterson, D.R.: Efficient estimation of the lifetime and disease onset distribution. Biometrika **84**, 539–554 (1997)

Chapter 2
Variable Selection for High Dimensional Metagenomic Data

Pan Wang and Hongmei Jiang

2.1 Introduction

The advent of next-generation sequencing technologies has greatly promoted the development of metagenomics in the past 10 years. Different from traditional and classical genomics which studies one individual organism in pure culture, metagenomics studies the genetic material recovered directly from the environment such as soil, water, and human gut. Therefore metagenomics is also referred as environmental genomics and community genomics. Using direct sequencing of the genetic materials collected from an environmental sample, metagenomics allows researchers to study the collection of multiple microorganisms, especially species that are difficult or even impossible to culture in the laboratory.

Metagenomics has been widely used in different fields including biological and medical research. Numerous pieces of evidence have shown that microbes living on and inside our body are associated with the occurrence and progression of different diseases such as inflammatory bowel disease, obesity, and various types of cancer (Furnari et al. 2012, Ley et al. 2006, Qin et al. 2012, Turnbaugh et al. 2009, to name a few examples). Recent studies have also linked strong evidence that gut microbiota modulates the efficacy and effects of treatments such as therapeutics and diets (Krautkramer et al. 2016; Liu et al. 2017; Matson et al. 2018; Yu et al. 2017). Therefore, comprehensive characterization of the microbes, and their interactions with each other, the host, and the environmental factors, will help develop strategies for diagnosis, treatment, and even prevention of some diseases.

Depending on the sequencing technology, thousands or millions of sequence reads are generated in each metagenomic sample. Sequence reads are mapped and

---

P. Wang · H. Jiang (✉)
Department of Statistics, Northwestern University, Evanston, IL, USA
e-mail: panwang2012@u.northwestern.edu; hongmei@northwestern.edu

L. Zhang et al. (eds.), *Contemporary Biostatistics with Biopharmaceutical Applications*, ICSA Book Series in Statistics,
https://doi.org/10.1007/978-3-030-15310-6_2

assigned to reference genomes or grouped into operational taxonomic units (OTU) or taxa in the taxonomy tree. After taxonomic assignment, we get the number of reads assigned to each taxon or OTU. Because the samples are usually taken from the natural environment, and there are noises from library preparation and sequencing, it would be impossible to make sure that the total number of sequencing reads is the same across different samples in one study. In fact, real data shows that the total number of sequencing reads varies significantly across samples. So it would be inappropriate to use the count data directly in a standard regression analysis. In general, rarefying or normalization has to be done before performing a downstream analysis.

In this chapter, we develop statistical methods to identify OTUs associated with a phenotype such as body mass index and disease status. We employ a log contrast model for metagenomic count data bypassing the need for rarefying or normalization. We propose a new method to identify phenotype associated species or OTUs using penalized regression and stability selection. In the log contrast model, one of the OTUs will serve as the reference OTU. We propose an averaging approach to avoid finding a particular reference OTU. The proposed method can also be applied to variable selection for regression analysis with compositional covariates. In fact, the log contrast model for count data is equivalent to the model for scaling-based normalized data. We present the proposed method in details in Sect. 2.2 and perform simulation studies to compare the performance of different approaches in Sect. 2.3 and apply them to real data in Sect. 2.4. We present some conclusions and discussions in Sect. 2.5.

2.2 Methods

2.2.1 Metagenomic Data

In the general process of metagenomic sequencing studies, a sample is usually taken from a natural community such as soil and seawater, or a host-associated community such as the human gut. All or partial DNA is extracted directly from the microbes contained in the sample, and then sequenced by sequencers (such as Sanger, Roche 454 or Illumina Sequencing). The resulted dataset contains thousands or millions of mixed sequence reads from the multiple genomes. Although there is a reduced cost for next-generation sequencing technologies, targeted sequencing such as 16S ribosomal RNA (rRNA) sequencing has been widely used to identify and quantify microbes present in a sample. Based on the data from high-throughput sequencing of marker genes, assignment of the sequencing reads to taxonomy can be conducted using some pipelines, for example, QIIME (Caporaso et al. 2010). After taxonomic assignment, we get the OTU count table with rows representing OTUs and columns representing samples and entries being the number of reads assigned to an OTU in a given example. Several factors have a significant impact on the estimated count of each taxon in the experimental process, for example, the sequencing depth or library size, the copy number of the genome, and the possible biases from the step of DNA extraction (Bragg and Tyson 2014).

With the different total number of reads per sample which could vary from thousands to tens of thousands, applying the traditional and classical statistics methods directly on the count data is not appropriate. Normalization or rarefying is usually conducted for the OTU count data before performing a downstream statistical analysis. Rarefying is a commonly used method by microbial biologists by drawing without replacement from each sample such that all samples in one study have the same total number of reads or counts. Samples with a total number of reads below the chosen threshold will be excluded from downstream analysis. One popularly used normalization method is to compute the relative abundance (or proportion) by dividing the OTU count by the total number of reads in the corresponding sample. For the OTU relative abundance data, the sum of each sample (column) is equal to 1 which can be considered as compositional data. It is well known that many traditional and classical statistical methods cannot be applied to the compositional data directly (Aitchison 1986). Other normalization methods include 75th percentile and cumulative sum scaling (Paulson et al. 2013). Some normalization methods developed for RNA sequencing have also been applied to metagenomic data. For a summary of different normalization methods, please refer to Weiss et al. (2017). However, there is no clear consensus on what is the best normalization method for metagenomic data.

2.2.2 Linear Log-Contrast Model

Here we focus on regression analysis with microbial covariates. As mentioned earlier, the OTU counts data cannot be directly used. For the normalized compositional covariates, traditional methods are not appropriate either. When we explain the meaning of one regression coefficient, we usually say that it is the average change of the response variable when the corresponding explanatory variable increases by one unit while holding other explanatory variables as constant. However, for compositional covariates, because of the sum being one we cannot make one component change while holding other components constant. Therefore, transformations such as log-ratio transformation usually are applied to the compositional data. We will use the linear log-contrast model proposed by Aitchison (1982) for regression analysis. It can be used to both raw OTU counts and relative abundance data after scaling normalization. The linear log-contrast model was proposed by Aitchison (1982) for compositional data analysis with application in geology where chemical, mineral and fossil compositions of rock and sediment specimens are studied. Metagenomic data is different from geological data due to its large number of OTUs or species and a small number of samples. That is, we are facing a "large p and small n" problem. We would like to identify some phenotype associated OTUs.

Suppose there are p OTUs. Let $x = (x_1, x_2, \ldots . x_p)^T \in S^p$ be a vector of OTU counts. Using additive log-ratio transformation we first choose one variable as the reference, and all the other variables are transferred to the log-ratio with respect to the reference (Aitchison 1982). For example, if x_p is selected as the reference variable, the additive log-ratio transformed x is defined as $z^p = (log(x_1/x_p), log(x_2/x_p), \ldots, log(x_{p-1}/x_p))^T$. The additive log-ratio transformation, although easy to conduct, has an apparent problem that the choice of reference variable will strongly influence the result of statistical analysis. For instance, if we permute the original sequence of $[x_1, x_2, \ldots . x_p]$ and choose x_r with $r \neq p$ as the reference variable, we may get a new model that varies a lot from the original one. To address this drawback, Aitchison (1982) discussed another transformation model called the centered log-ratio transformation by replacing the reference variable with the geometric mean. However, when doing the variable selection, the selected variables are a combination of $(x_1, x_2, \ldots, x_p)$. This issue would be harmful to both the reduction of dimension and explanation of statistical results. So the additive log-ratio transformation would be a better choice for this purpose.

Suppose there are n samples in the p-dimensional space, and define each sample i as $X_i = (x_{i1}, x_{i2}, \ldots, x_{ip})^T$. Suppose the rth OTU is used as the reference variable where $r \in \{1, 2, \ldots, p\}$. Then the linear regression model is

$$y_i = \Sigma_{j=1, j\neq r}^{p} log(\frac{x_{ij}}{x_{ir}})\beta_{jr} + \epsilon_i, \tag{M1}$$

where x_{ij} represents the (relative) abundance of OTU j in sample i, y_i is the response for sample i, β_{jr} is the regression coefficient for additive log-ratio transferred variable j using the rth variable as reference. This model (M1) can be converted into an equivalent version of linear regression model that

$$y_i = \Sigma_{j=1}^{p} log(x_{ij})\beta_j + \epsilon_i, \ \Sigma_{j=1}^{p}\beta_j = 0. \tag{M2}$$

It is evident that (M2) does not contain a reference variable, and there is no bias in choosing the reference variable if (M2) can be developed directly. Lin et al. (2014) studied the estimation of (M2), where they applied a Lagrangian method to deal with the restriction of the sum of coefficients, and applied Lasso (Tibshirani 1996) penalization function for variable selection. Although this method has its advantage, there still exists limitation that people need to derive new algorithm using the coordinate descent method with various penalization function, and it might be difficult when dealing with more complicated penalization term, for example, the minimax concave penalty (MCP) (Zhang 2012). With the constraint on the coefficients, people are not able to apply the current existing fast speed computing packages directly. To overcome this, one can study (M1) directly. One advantage of this model (M1) is that it is more straightforward to conduct and one can reduce the computing time by applying fast speed variable selection algorithms and computing packages which have been developed. Another advantage is that model (M1) is equivalent for OTU counts data and scaling-based normalized data.

It will produce the same results even if different scaling normalization method is used because $\frac{x_{ij}}{x_{ir}} = \frac{x_{ij}/s_i}{x_{ir}/s_i}$ where s_i is the scaling factor which is the total number of reads in sample i if relative proportion is desired for normalization, or a flexible sample distribution-dependent threshold if cumulative sum scaling is used for normalization.

It is clear that the result from (M1) mainly depends on the choice of the reference variable. If an OTU unassociated with the response is selected as the reference, it may increase the number of falsely selected OTUs. We propose an averaging approach to reduce the chance of choosing a non-related OTU as the reference variable. Currently, in our study, we applied the Elastic Net (Zou and Hastie 2005) for both linear and logistic regression. However, our proposed method is a general framework that can be extended to many different penalization functions.

2.2.3 *Proposed Method*

We examine the association between a response variable and a set of microbial covariates and propose an easy-to-implement approach for the high dimensional variable selection problem. Our method uses the additive log-ratio linear contrast model and tries to get rid of the influence of the reference variable. Our design has no requirement on the penalization terms, which means that we can implement any fast computing software for this model. The primary challenge of directly applying existing high dimensional variable selection techniques is that the sparse model we get depends mainly on which variable is chosen as the reference variable. To reduce this effect, in our stated method, for each variable, we take the average of estimated selection probability using different reference variable x_r. Our new approach has the advantage that we can get rid of the notorious problem of choosing a proper reference variable while being able to control the number of selected essential features by adjusting the threshold probability.

The idea of stability selection is also applied to gain accuracy. It is a general subsampling technique proposed by Meinshausen and Bühlmann (2010). In stability selection, we apply our favorite algorithm of variable selection to a large number of half samples of the original dataset and choose the variables with high selection frequency on the subsamples. In our proposed method we apply a complementary pairs stability selection (CPSS) model introduced in Shah and Samworth (2013).

The following shows the details of the procedure.

(1) Fix one OTU, $x_r \in \{x_1, \ldots, x_p\}$ and consider it as the reference variable for log-ratio transformation. Then we construct the following linear regression model

$$y_i = \Sigma_{j=1, j \neq r}^{p} log(\frac{x_{ij}}{x_{ir}})\beta_{jr} + \epsilon_i.$$

(2) For each OTU variable x_j, based on the model in step (1), we compute its selection probability. This can be done in two different ways.

(a) For the first approach, we apply penalized regression for variable selection. Define L as the cost function under different penalization term,

$$L(\beta, \lambda) = \frac{1}{2n} \Sigma_{i=1}^{n} [y_i - (\Sigma_{j=1, j \neq r}^{p} log(\frac{x_{ij}}{x_{ir}}) \beta_{jr})]^2 + \lambda P(\beta^{-r}).$$

where $\lambda > 0$ is considered as the tuning parameter and $\beta^{-r} = (\beta_{1r}, \beta_{2r}, \ldots, \beta_{pr})$ is the $p - 1$ dimensional vector containing the regression coefficients. The penalty function $P(\cdot)$ could be $\Sigma_{j \neq r}^{p} |\beta_{jr}|$ as specified in LASSO (Tibshirani 1996), or $\lambda_1 \Sigma_{j \neq r} |\beta_{jr}| + \lambda_2 \Sigma_{j \neq r} \beta_{jr}^2$ for Elastic Net, or any other penalty functions such as minimax concave penalty. After applying the penalized regression, we assign the selection probability for variable x_k with x_r being the reference variable as

$$\tilde{\pi}_r(x_k) := 1 \text{ if } \hat{\beta}_{kr} \neq 0,$$

$$\tilde{\pi}_r(x_k) := 0 \text{ if } \hat{\beta}_{kr} = 0.$$

Since x_r is the reference variable, we always assign $\tilde{\pi}_r(x_r) = 1$.

(b) For the second approach, the performance of conventional variable selection algorithm can be further improved by the stability selection. Here we use the CPSS approach proposed in Shah and Samworth (2013). We randomly divide the whole sample into two parts of equal size for B times, and get a set of subsample pairs, $\{(A_{2t-1}^r, A_{2t}^r) : t = 1, \ldots, B, \; A_{2t-1}^r \cap A_{2t}^r = \phi\}$. For each sample of A_{2t-1}^r and A_{2t}^r, apply a variable selection procedure separately and get $\hat{S}(A_{2t-1}^r)$ and $\hat{S}(A_{2t}^r)$ which contain the index of variables which have been selected as associated with the response variable. The selection probability of each variable can then be estimated by

$$\tilde{\pi}_r(x_k) := 1/(2B) \sum_{t=1}^{B} \left(\mathbb{I}_{\{k \in \hat{S}(A_{2t-1}^r)\}} + \mathbb{I}_{\{k \in \hat{S}(A_{2t}^r)\}} \right).$$

We also assign $\tilde{\pi}_r(x_r) = 1$ when x_r is used as the reference variable.

(3) Repeat steps 1 and 2 for each $r \in \{1, 2, \cdots, p\}$, and compute $\tilde{\pi}_r(x_k)$, the estimation of the selection probability for variable x_k with respect to the reference variable x_r (Table 2.1). Then we calculate the selection probability of variable $x_k (k = 1, 2, \cdots, p)$ by taking the average, that is,

$$\bar{\pi}(k) = \frac{1}{p} \sum_{r=1}^{p} \tilde{\pi}_r(x_k).$$

(4) The selected signal index set is

$$\hat{S}_\tau = \{k : \bar{\pi}(k) \geq \tau\}$$

Here, τ is the cutoff probability considered as a tuning parameter. According to Meinshausen and Bühlmann (2010), the results should not be sensitive to the choice of τ when $\tau \in [0.6, 0.9]$. However, this property does not keep as we take the step of averaging. We set $\tau = 0.5$ so that the variables with selection probability better than random guessing is accepted.

Table 2.1 Variable selection probability $\tilde{\pi}_r(x_k)$ for variable x_k using x_r as the reference variable, and the average selection probability $\bar{\pi}(k)$ for variable x_k

	Reference variable				Average
	x_1	x_2	...	x_p	
x_1	$\tilde{\pi}_1(x_1)$	$\tilde{\pi}_2(x_1)$	...	$\tilde{\pi}_p(x_1)$	$\bar{\pi}(1) := \frac{1}{p}\sum_{r=1}^{p}\tilde{\pi}_r(x_1)$
x_2	$\tilde{\pi}_1(x_2)$	$\tilde{\pi}_2(x_2)$	...	$\tilde{\pi}_p(x_2)$	$\bar{\pi}(2) := \frac{1}{p}\sum_{r=1}^{p}\tilde{\pi}_r(x_2)$
...	...	...	...	...	...
x_p	$\tilde{\pi}_1(x_p)$	$\tilde{\pi}_2(x_p)$	...	$\tilde{\pi}_p(x_p)$	$\bar{\pi}(p) := \frac{1}{p}\sum_{r=1}^{p}\tilde{\pi}_r(x_p)$

2.3 Simulation Studies

2.3.1 Simulation Setting for Linear Regression Model

To evaluate the performance of the proposed method, we perform comprehensive simulation studies. Suppose the sample size is n, and the number of variables is p, we generate the independent variable and the response variable with the following steps.

(1) Generate a $n \times p$ matrix $W = (w_{ij})$, which is concerned as the logarithm of the original count data. W follows a multivariate normal distribution $N_p(\theta, \Sigma)$. Let $\theta = (\theta_j)$, where $\theta_j = log(0.5p)$ for $j = 1, \ldots, 5$ and $\theta_j = 0$ otherwise. Let $\Sigma = (\rho^{|i_\rho - j_\rho|})$ with $\rho = 0.5$, and i_ρ and j_ρ are the row and column number of each element in Σ, respectively.
(2) Compute the proportion or percentage of each composition in sample i. Define a proportion matrix $X = (x_{ij})_{n\times p}$, where x_{ij} is the percentage of the jth OTU taken in sample i, and is calculated by $x_{ij} = exp(w_{ij})/\sum_{k=1}^{p} exp(w_{ik})$.
(3) Define a matrix $Z = log(X) = (log(x_{ij}))_{n\times p}$ to contain the log proportions.
(4) Let $\beta_p = (1, -0.8, 0.6, 0, 0, -1.5, -1.5, 2.2, 0, \ldots, 0)^T$. Note β_p is a p-dimensional coefficient vector that only the first eight elements contain

non-zero values. People can readily calculate that the constraint that $\sum_{j=1}^{p} \beta_j = 0$, and this property holds with different dimensions. Combining the value of vector β with the setting in (1), it is obvious that only half of the significant variables have $\theta_j > 0$ while the rest of the half have $\theta_j = 0$. Therefore in this simulation setting, only half of the important variables are highly abundant, and that adds more generality of our simulation results.

(5) Generate a n-dimensional vector $\epsilon = (\epsilon_1, \epsilon_2, \ldots, \epsilon_n)^T$, where the $\epsilon_i' s$ are independent and identically distributed as $N(0, \sigma^2)$ with $\sigma = 0.5$. ϵ is considered as the random error in linear regression model. When generating the response variable for logistic regression model, we gain randomness from Bernoulli distribution.

(6) Generate a n-dimensional vector $Y = (y_1, y_2, \ldots, y_n)^T$. With linear model, let

$$y_i = \Sigma_{j=1}^{p} \beta_j \log(x_{ij}) + \epsilon_i.$$

In the case of logistic regression, let

$$logit(P(y_i = 1|X_i)) = \Sigma_{j=1}^{p} \beta_j \log(x_{ij}),$$

and generate each y_i following Bernoulli distribution $B(1, P(y_i = 1|X_i))$.

2.3.2 *Simulation Results*

There are two essential parts of our proposed method, using stability selection strategy to estimate selection probability and taking an average of different reference variables. So in the simulation study, we compare our proposed approach to those taking out one or two parts. All the simulation results are based on variable selection using Elastic Net penalty. To be specific, the four options we compare are: (1) "SS avg": stability selection is used to estimate the selection probability for regularized regression using each of the p variables as the reference variable. Average selection probability is taken as the variable's selection probability. (2) "SS x_p": Stability selection is used to estimate the selection probability for regularized regression with x_p as the reference variable. (3) "avg": For each reference variable, using the regularization method without stability selection to select important variables. Then for each variable, average selection probability is taken as the variable's selection probability. (4) "x_p": Variable x_p is used as the reference variable, and directly applying the regularization method we chose for variable selection without stability selection or averaging.

We compare the above four approaches on the simulated datasets, and examine the performance of each method by the following criteria:

- $\#FP$: number of variables which are not important but are falsely selected.
- $\#FN$: Number of variables which are important but are falsely dropped.

- True positive rate (TPR):

$$TPR = \frac{\#TP}{\#P} = \frac{\#TP}{\#TP + \#FN}$$

  where #P is the total number of important variables, #TP is the number of truly important variables detected.
- True negative rate (TNR):

$$TNR = \frac{\#TN}{\#N} = \frac{\#TN}{\#TN + \#FP}$$

  where #N is the total number of variables with zero coefficients, #TN is the number of truly non-important variables dropped.
- l_2 PE: $\Sigma_{i=1}^{n}(\tilde{y}_i - \hat{y}_i)^2$, the prediction error for linear regression model based on testing data set generated with the same sample size and dimension as the training set. Here, $\tilde{y}$ is the value of response in the testing set and $\hat{y}_i$ is the predicted response value based on the estimated model from the training set.
- PE: $\Sigma_{i=1}^{n}|\tilde{y}_i - \hat{y}_i|$, the prediction error for logistic regression model based on testing data set generated with the same sample size and dimension as the training set. The testing set is generated using the same simulation setting as the training set but using different seeds.

From Table 2.2, it can be seen that taking an average of selection probability will sharply reduce the number of falsely selected variables and slightly mitigate the amount of incorrectly dropped variables, for both the cases with stability selection and without stability selection. It means that taking an average can help us reduce the effect of choosing a reference variable in the additive log-ratio transformation. We find that the idea of sub-sampling like stability selection works well to control the number of falsely selected variables. There is a noticeable improvement in the new method observed in this simulation setting. Regarding the l_2 prediction error, in most cases, taking average can help improve the prediction accuracy. We also observe that in most cases, the prediction error is lower without stability selection. One of the possible reasons is the conservativeness of the stability selection which is likely to give a probability smaller than one compared to the methods without this procedure.

From Table 2.3, for most cases of the logistic regression model, the number of variables falsely selected is reduced with the proposed averaging approach, and sharply reduced with stability selection. False negative rates are almost the same with or without taking an average and will increase with the application of stability selection strategy. For most of the combinations of n and p, the prediction error from our proposed method is not worse than others. An unusual case from the simulated table of logistic regression is when $n = 100$ and $p = 30$, that we fail to observe a drop in the false positive rate with stability selection compared to other cases. So in the logistic regression model, when the sample size is large enough compared to the dimension of the dataset, our proposed plan might fail to give a benefit over the conventional techniques.

Table 2.2 Comparisons of four variable selection methods for linear regression model with metagenomic covariates: "SS avg" uses both stability selection and proposed averaging approach, "SS x_p" uses stability selection and x_p as the reference variable, "avg" uses the proposed averaging approach without stability selection, "x_p" uses x_p as the reference variable without stability selection

	n	p	#P	#N	#FP	#FN	TPR	TNR	l_2 PE
SS avg	50	30	6	24	1.84	0.39	0.94	0.92	0.61
SS x_p	50	30	6	24	3.11	0.46	0.92	0.87	0.65
avg	50	30	6	24	4.59	0.12	0.98	0.81	0.42
x_p	50	30	6	24	6.55	0.12	0.98	0.73	0.41
SS avg	50	60	6	54	1.28	1.4	0.77	0.98	1.60
SS x_p	50	60	6	54	2.51	1.61	0.73	0.95	1.85
avg	50	60	6	54	7.01	0.24	0.96	0.87	0.52
x_p	50	60	6	54	8.76	0.24	0.96	0.84	0.59
SS avg	50	100	6	94	0.68	2.87	0.52	0.99	3.56
SS x_p	50	100	6	94	1.93	3.04	0.49	0.98	3.73
avg	50	100	6	94	7.80	0.52	0.91	0.92	0.86
x_p	50	100	6	94	10.51	0.58	0.90	0.89	1.02
SS avg	50	200	6	194	0.38	4.09	0.32	1.00	5.59
SS x_p	50	200	6	194	1.49	4.16	0.31	0.99	5.68
avg	50	200	6	194	9.74	1.12	0.81	0.95	1.94
x_p	50	200	6	194	13.08	1.16	0.81	0.93	2.11
SS avg	100	30	6	24	2.04	0	1.00	0.92	0.27
SS x_p	100	30	6	24	2.98	0	1.00	0.88	0.27
avg	100	30	6	24	4.55	0	1.00	0.81	0.30
x_p	100	30	6	24	6.81	0	1.00	0.72	0.30
SS avg	100	60	6	54	1.17	0.08	0.99	0.98	0.32
SS x_p	100	60	6	54	2.44	0.10	0.98	0.95	0.35
avg	100	60	6	54	4.21	0.03	1.00	0.92	0.32
x_p	100	60	6	54	6.61	0.01	1.00	0.88	0.32
SS avg	100	100	6	94	0.89	0.16	0.97	0.99	0.36
SS x_p	100	100	6	94	2.18	0.23	0.96	0.98	0.42
avg	100	100	6	94	3.71	0.02	1.00	0.96	0.30
x_p	100	100	6	94	5.97	0.03	1.00	0.94	0.31
SS avg	100	200	6	194	0.89	0.72	0.88	1.00	0.89
SS x_p	100	200	6	194	2.01	0.76	0.87	0.99	0.89
avg	100	200	6	194	4.28	0.16	0.97	0.98	0.42
x_p	100	200	6	194	6.26	0.16	0.97	0.97	0.44

Here, n is the sample size, p is the total number of variables, $\#P$ is the total number of true positives (variables with non-zero coefficients), $\#N$ is the total number of true negatives (variables with zero coefficients), $\#FP$ is the number of false positives, $\#FN$ is the number of false positives, TPR is the true positive rate, TNR is the true negative rate, and l_2 PE is the prediction error

Table 2.3 Simulation results for logistic regression with metagenomic covariates (the notations are the same as in Table 2.2)

	n	p	#P	#N	#FP	#FN	TPR	TNR	PE
SS avg	50	30	6	24	3.70	2.02	0.66	0.85	0.33
SS x_p	50	30	6	24	5.00	2.30	0.62	0.79	0.34
avg	50	30	6	24	14.52	0.76	0.87	0.39	0.34
x_p	50	30	6	24	15.42	0.92	0.85	0.36	0.35
SS avg	50	60	6	54	1.81	3.67	0.39	0.97	0.39
SS x_p	50	60	6	54	3.23	3.66	0.39	0.94	0.40
avg	50	60	6	54	14.71	1.68	0.72	0.72	0.37
x_p	50	60	6	54	18.27	1.70	0.72	0.66	0.38
SS avg	50	100	6	94	1.49	4.33	0.28	0.98	0.42
SS x_p	50	100	6	94	2.75	4.22	0.30	0.97	0.41
avg	50	100	6	94	17.69	2.33	0.61	0.81	0.40
x_p	50	100	6	94	22.78	2.33	0.61	0.76	0.43
SS avg	50	200	6	194	0.97	4.83	0.20	1.00	0.43
SS x_p	50	200	6	194	2.45	4.81	0.20	0.99	0.44
avg	50	200	6	194	20.25	3.04	0.49	0.90	0.42
x_p	50	200	6	194	22.74	3.14	0.48	0.88	0.44
SS avg	100	30	6	24	18.77	0.14	0.98	0.22	0.27
SS x_p	100	30	6	24	18.12	0.20	0.97	0.24	0.28
avg	100	30	6	24	13.70	0.25	0.96	0.42	0.27
x_p	100	30	6	24	15.44	0.38	0.94	0.36	0.27
SS avg	100	60	6	54	5.96	1.27	0.79	0.89	0.28
SS x_p	100	60	6	54	7.70	1.39	0.77	0.86	0.28
avg	100	60	6	54	21.47	0.56	0.91	0.59	0.30
x_p	100	60	6	54	23.89	0.73	0.88	0.56	0.32
SS avg	100	100	6	94	4.31	2.02	0.66	0.95	0.30
SS x_p	100	100	6	94	5.86	2.07	0.66	0.94	0.30
avg	100	100	6	94	24.54	0.84	0.86	0.74	0.33
x_p	100	100	6	94	26.18	1.06	0.82	0.72	0.32
SS avg	100	200	6	194	2.44	3.20	0.47	0.99	0.35
SS x_p	100	200	6	194	3.79	3.25	0.46	0.98	0.35
avg	100	200	6	194	25.09	1.45	0.76	0.87	0.36
x_p	100	200	6	194	36.82	1.37	0.77	0.81	0.37

2.4 Real Data Analysis

We apply our proposed model to identify the significant features among the OTUs in a mouse skin study from Srinivas et al. (2013) as an example of logistic regression analysis. In their research, they studied the host gene-microbiota interactions contributing to disease risk in a mouse model of epidermolysis bullosa acquisita and treat bacterial species abundances as covariates with the disease. The dataset includes a total of 183 immunized and 78 non-immunized mice. In the group of vaccinated mice, there are two disease status, immunized-healthy (Healthy) and immunized-diseased (EBA) groups. In this part, we focus on the two groups of immunized mice and compare the performance of different methods in finding the association between disease status and OTUs. There are 131 core OTUs in total, and the sample size is 183. To get an estimation of prediction error, let the size of the training set be 120, and that of the testing set be 63.

In Table 2.4, the result of real data analysis shows that taking average will not increase the prediction error. However, the methods without stability selection give better prediction accuracy at the expense of the much more selected variables. This finding coincides with our discovery in the simulation study that taking the average of selection probability will not increase the prediction error, however, sometimes the conservativeness of stability selection might harm the detection of actual significant variables. Especially when there are zeros in the sample, stability selection may be too conservative due to splitting the whole data into two half samples.

Table 2.4 Real data analysis for mouse skin data

	#Selected OTUs	PE
SS avg	3.02	0.37
SS x_p	4.16	0.37
avg	13.91	0.35
x_p	15.12	0.35

2.5 Discussion and Future Work

We propose a general framework for the problem of variable selection for the high dimensional microbiome data associated with a phenotype such as body mass index or disease status. Since the metagenomic count data carries the relative information of the OTUs in samples with different total number of reads, the additive log-ratio transformation can be used. Although the log contrast model is originally proposed for compositional data, it can be applied to the count data too. However, regarding the variable selection, it suffers from the bias of choosing a single reference variable. Our proposed method provides a general framework to let people apply the existing and fast speed algorithm and computing packages directly and reduces the effect of reference variable by the idea of taking the average selection probability. The possible bias of the algorithm can be further reduced by a sub-sampling based method call stability selection. We consider the variable selection problem of both binary and continuous response model and evaluate their performance with different sample size and dimension through the simulation study and real data analysis.

We observe in the simulation study that when the sample size is comparably small and when the number of species or OTUs increases dramatically, the estimated selection probability of each variable may suffer from a severe drop to a value far below our inclusion criteria. In this case, taking the average without stability selection might be a better choice. In the simulation study, we only compare our proposed method with the cases that there is no prior information on which variable being truly important. If a substantially important variable is used as the reference variable, one might get similar results compared to our proposed method and the computing time will be largely reduced. However, currently, to our knowledge, there is no reliable strategy to identify a truly important variable. Our proposed method provides an easy-to-implement approach.

It is well known that the metagenomic data is not only high dimensional but also sparse with excess zeros. With zero counts, we cannot take the log-ratio transformation. For the real data analysis, we use the core OTUs which have been observed in the majority samples, and zero counts are replaced with $1/10$ of the smallest non-zero value in the dataset. Stability selection may be too conservative when there are excess zeros. The sensitivity of variable selection to the replacement of zeros with a relatively small number needs to be further studied. We will also consider other strategies, for example estimating the missing values based on the information from other samples.

The models and methods we propose can be readily applied to the analysis of sub-compositions in the study of metagenomics to find the OTUs in each unit of a high taxonomic rank that associate with the phenotype of interest. To identify the significant OTUs in the sub-composition analysis, we may apply a greedy search algorithm considering each unit of the higher rank taxa as a whole group to be accepted or rejected, and study the importance of sub-composition within each unit separately.

References

Aitchison, J.: The statistical analysis of compositional data. J. R. Stat. Soc. Ser. B (Methodol.) **44**(2), 139–177 (1982)

Aitchison J.: The Statistical Analysis of Compositional Data. Chapman & Hall, London. Reprinted in 2003, with additional material, by The Blackburn Press (1986)

Bragg, L., Tyson, G.W.: Metagenomics using next-generation sequencing. In: Paulsen, I., Holmes, A. (eds.) Environmental Microbiology. Methods in Molecular Biology (Methods and Protocols), vol. 1096. Humana Press, Totowa, NJ (2014)

Caporaso, J.G., et al.: QIIME allows analysis of high-throughput community sequencing data. Nat. Methods **7**(5), 335–336 (2010)

Furnari, M.E., Savarino, L.B., Moscatelli, A., Gemignani, L., Giannini, E.G., Zentilin, P.: Reassessment of the role of methane production between irritable bowel syndrome and functional constipation. J. Gastroenterol. Liver Dis. **21**, 157–163 (2012)

Krautkramer, K.A., Kreznar J.H., et al.: Diet-microbiota interactions mediate global epigenetic programming in multiple host tissues. Mol. Cell **64**(5), 982–992 (2016)

Ley, R.E., Turnbaugh, P.J., Klein, S., Gordon, J.I.: Microbial ecology: human gut microbes associated with obesity. Nature **444**, 1022–1023 (2006)

Lin, W., Shi, P., Feng, R., Li, H.: Variable selection in regression with compositional covariates. Biometrika **101**(4), 785–797 (2014)

Liu, R., Hong, J., et al.: Gut microbiome and serum metabolome alterations in obesity and after weight-loss intervention. Nat. Medicine **23**(7), 859–868 (2017)
Matson, V., et al.: The commensal microbiome is associated with anti-PD-1 efficacy in metastatic melanoma patients. Science **359**(6371), 104–108 (2018)
Meinshausen, N., Bühlmann, P.: Stability selection (with discussion). J. R. Stat. Soc. Ser. B (Methodol.) **72**, 417–473 (2010)
Paulson, J.N., et al.: Differential abundance analysis for microbial marker-gene surveys. Nat. Methods **10**(12), 1200–1202 (2013)
Qin, J.J., et al.: A metagenome-wide association study of gut microbiota in type 2 diabetes. Nature **490**(7418), 55–60 (2012)
Shah, R.D., Samworth, R.J.: Variable selection with error control: another look at stability selection. J. R. Stat. Soc. Ser. B (Methodol.) **75**(1), 55–80 (2013)
Srinivas, G., et al.: Genome-wide mapping of gene-microbiota interactions in susceptibility to autoimmune skin blistering. Nat. Commun. **4**, 2462 (2013)
Tibshirani, R.J.: Regression shrinkage and selection via the lasso. J. R. Stat. Soc. Ser. B (Methodol.) **58**, 267–288 (1996)
Turnbaugh, P.J., et al.: A core gut microbiome in obese and lean twins. Nature **457**, 480–484 (2009)
Weiss, S., et al.: Normalization and microbial differential abundance strategies depend upon data characteristics. Microbiome **5**, 27 (2017)
Yu, T., Guo, F., et al.: Fusobacterium nucleatum Promotes Chemoresistance to Colorectal cancer by modulating autophagy. Cell **170**(3), 548–563.e16 (2017)
Zhang, CH.: Nearly unbiased variable selection under minimax concave penalty. Ann. Stat. **38**(2), 894–942 (2012)
Zou, H., Hastie, T.: Regularization and variable selection via the elastic net. J. R. Stat. Soc. Ser. B (Methodol.) **67**(2), 301–320 (2005)

Chapter 3
Dimension Reduction in High Dimensional Multivariate Time Series Analysis

William W. S. Wei

3.1 Introduction

Multivariate time series are of interest in many fields such as economics, finance, epidemiology, physical science, geoscience, and many others. When modeling multivariate time series, the vector autoregressive (VAR) and vector autoregressive moving average (VARMA) models are possibly the most widely used models, because of their capability to represent the dynamic relationships among variables in a system and their usefulness in forecasting unknown future values. These models are described in many time series textbooks including Hannan (1970), Hamilton (1994), Reinsel (1997), Wei (2006), Lütkepohl (2007), Tsay (2013), Box et al. (2015), and many others.

Let $\mathbf{Z}_t = [Z_{1,t}, Z_{2,t}, \cdots, Z_{m,t}]'$, $t = 0, \pm 1, \pm 2, \ldots$, be a m-dimensional jointly stationary real-valued vector process so that $E(Z_{i,t}) = \mu_i$ is constant for each $i = 1, 2, \ldots, m$ and the cross-covariance between $Z_{i,t}$ and $Z_{j,s}$ for all $i = 1, 2, \ldots, m$ and $j = 1, 2, \ldots, m$, are functions only of the time difference $(s - t)$. A useful class of vector time series models is the following vector autoregressive moving average model of order p and q, shorten as VARMA(p, q),

$$\mathbf{\Phi}_p(B)\mathbf{Z}_t = \mathbf{\Theta}_q(B)\mathbf{a}_t, \tag{3.1}$$

where we assume the series is mean adjusted for simplicity, $\mathbf{\Phi}_p(B) = \mathbf{\Phi}_0 - \mathbf{\Phi}_1 B - \cdots - \mathbf{\Phi}_p B^p$ and $\mathbf{\Theta}_q(B) = \mathbf{\Theta}_0 - \mathbf{\Theta}_1 B - \cdots - \mathbf{\Theta}_q B^q$ are autoregressive and moving average matrix polynomials of order p and q, respectively, $\mathbf{\Phi}_i$ and $\mathbf{\Theta}_j$ are nonsingular $m \times m$ matrices, and $\mathbf{a}_t$ is a sequence of m-dimensional white noise

W. W. S. Wei (✉)
Department of Statistical Science, Temple University, Philadelphia, PA, USA
e-mail: wwei@temple.edu; https://astro.temple.edu/~wwei/

L. Zhang et al. (eds.), *Contemporary Biostatistics with Biopharmaceutical Applications*, ICSA Book Series in Statistics,
https://doi.org/10.1007/978-3-030-15310-6_3

processes with mean zero vector and positive definite variance-covariance matrix $\boldsymbol{\Sigma}$. Since one can always invert $\boldsymbol{\Phi}_0$ and $\boldsymbol{\Theta}_0$, and combine them into $\boldsymbol{\Sigma}$, with no loss of generality, we will assume in the following discussion that $\boldsymbol{\Phi}_0 = \boldsymbol{\Theta}_0 = \mathbf{I}$, the $m \times m$ identity matrix. Since any VARMA model can be approximated by a vector AR model, in practice, one often use the following vector autoregressive model of order p, shorten as VAR(p),

$$
\begin{aligned}
&\left(\mathbf{I} - \boldsymbol{\Phi}_1 B - \cdots - \boldsymbol{\Phi}_p B^p\right) \mathbf{Z}_t = \mathbf{a}_t, \\
&\text{or} \\
&\mathbf{Z}_t = \boldsymbol{\Phi}_1 \mathbf{Z}_{t-1} - \cdots - \boldsymbol{\Phi}_p \mathbf{Z}_{t-p} + \mathbf{a}_t
\end{aligned}
\tag{3.2}
$$

where the zeros of $|\mathbf{I} - \boldsymbol{\Phi}_1 B - \cdots - \boldsymbol{\Phi}_p B^p|$ lie outside of the unit circle or, equivalently, the roots of $|\lambda^p \mathbf{I} - \lambda^{p-1} \boldsymbol{\Phi}_1 - \cdots - \boldsymbol{\Phi}_p| = 0$ are all inside of the unit circle. In the following discussion, we will use VAR(p) model for our illustrations.

With the development of computer and internet, we have data exploration. For a m-dimensional multivariate time series, m being hundred and thousand is very common. Simply consider $m = 100$, a simple VAR(2) model has $2(100 \times 100) = 20{,}000$ parameters. For observations obtained yearly, we cannot estimate the model parameters even with a hundred-year data.

To solve the problem, after introducing some existing methods, we will suggest the use of aggregation as a dimension reduction method, which is very natural and simple to use. We will compare our proposed method with other existing methods in terms of forecast accuracy through both simulations and empirical examples. The presentation is organized as follows. Section 3.2 introduces and discusses several existing methods to handle high-dimensional time series. The proposed procedure is introduced in Sect. 3.3. Monte Carlo simulations and empirical data analysis are presented in Sects. 3.4 and 3.5, respectively. Lastly, concluding remarks are given in Sect. 3.6.

3.2 Existing Methods

In this section, we briefly review the existing methods that handle time series modeling in high-dimensional setting, including various regularization methods (Sect. 3.2.1), the space-time AR model if the data ate collected from different locations (Sect. 3.2.2), model-based clustering (Sect. 3.2.3), and factor model (Sect. 3.2.4).

3.2.1 Regularization Methods

Let $\mathbf{Z}_t = [Z_{1,t}, Z_{2,t}, \cdots, Z_{m,t}]'$, $t = 1, 2, \ldots, N$, be a m-dimensional time series with N observations. It is well known that the least squares method can be used to fit the VAR(p) model by minimizing

$$\sum_{t=1}^{N} \left\| \mathbf{Z}_t - \sum_{k=1}^{p} \mathbf{\Phi}_k \mathbf{Z}_{t-k} \right\|_2, \tag{3.3}$$

where $\| \|_2$ is Euclidean (L^2) norm of a vector. More compactly, in practice, with data $\mathbf{Z}_t = [Z_{1,\,t}, Z_{2,\,t}, \cdots, Z_{m,\,t}]'$, $t = 1, 2, \ldots, N$, we can present the VAR(p) model in Eq. (3.2) in the matrix form,

$$\underset{N\times m}{\mathbf{Y}} = \underset{(N\times mp)}{\mathbf{X}} \underset{(mp\times m)}{\mathbf{\Phi}} + \underset{(N\times m)}{\boldsymbol{\xi}}, \tag{3.4}$$

where

$$\mathbf{Y} = \begin{bmatrix} \mathbf{Z}_1' \\ \mathbf{Z}_2' \\ \vdots \\ \mathbf{Z}_N' \end{bmatrix}, \mathbf{X} = \begin{bmatrix} \mathbf{X}_1' \\ \mathbf{X}_2' \\ \vdots \\ \mathbf{X}_N' \end{bmatrix}, \mathbf{\Phi} = \begin{bmatrix} \mathbf{\Phi}_1' \\ \mathbf{\Phi}_2' \\ \vdots \\ \mathbf{\Phi}_p' \end{bmatrix}, \boldsymbol{\xi} = \begin{bmatrix} \mathbf{a}_1' \\ \mathbf{a}_2' \\ \vdots \\ \mathbf{a}_N' \end{bmatrix},$$

$$\mathbf{X}_t' = \left[\mathbf{Z}_{t-1}', \mathbf{Z}_{t-2}', \cdots, \mathbf{Z}_{t-p}' \right].$$

So, minimizing (3.3) is equivalent to

$$\underset{\mathbf{\Phi}}{\arg\min} \|\mathbf{Y} - \mathbf{X\Phi}\|_F, \tag{3.5}$$

where $\| \|_F$ is Frobenius norm of a matrix.

For VAR model in high-dimensional setting, many regularization methods have been developed, which assume sparse structures on coefficient matrices $\mathbf{\Phi}_k$ and use regularization procedure to estimate parameters. These methods include the Lasso (Least Absolute Shrinkage and Selection Operator) method, the lag-weighted lasso method, and the hierarchical vector autoregression method, among others.

3.2.1.1 The Lasso Method

One of the most commonly used regularization methods is Lasso method proposed by Tibshirani (1996) and extended to vector time series setting by Hsu et al. (2008). Formally, the estimation procedure for the VAR model is through

$$\underset{\mathbf{\Phi}}{\arg\min} \left\{ \|\mathbf{Y} - \mathbf{X\Phi}\|_F + \lambda \|vec\,(\mathbf{\Phi})\|_1 \right\}, \tag{3.6}$$

where the second term is the regularization through L_1 penalty with λ being its control parameter. λ can be determined by cross-validation. The lasso method does not impose any special assumption on the relationship of lag orders and tends to over select the lag order p of the VAR model. This leads to the development of some modified methods.

3.2.1.2 The Lag-Weighted Lasso Method

Song and Bickel (2011) proposed a method that incorporates lag-weighted lasso (lasso and group lasso structures) approach for the high-dimensional VAR model. They placed group lasso penalties introduced by Yuan and Lin (2006) on the off-diagonal terms, $\mathbf{\Phi}(j, -j) = \{\phi_{ji}\}_{i \neq j}$, and lasso penalties on the diagonal terms, $\mathbf{\Phi}_k(jj)$. The regularization for $\mathbf{\Phi}_k$ is

$$\sum_{j=1}^{m} \| \mathbf{\Phi}_k (j, -j) \, \mathbf{W}(-j) \|_2 + \lambda \sum_{j=1}^{m} w_j \mid \mathbf{\Phi}_k(jj) \mid, \tag{3.7}$$

where $\mathbf{W}(-j) = diag\,(w_1, \cdots, w_{j-1}, w_{j+1}, \cdots, w_m)$, a $(m - 1) \times (m - 1)$ diagonal matrix with w_j being the positive real-valued weight associated with the jth jvariable for $1 \leq j \leq m$, which is chosen to be the standard deviation of $Z_{j,\,t}$. λ is the control parameter that controls the extent to which other lags are less informative than its own lags. The first term of (3.7) is group lasso penalty, second term is lasso penalty, and they impose regularization on other lags and its own lags respectively. Let $\alpha > 1$ and $(k)^{\alpha}$ be the other control parameter for different regularization for different lags, the estimation procedure is based on

$$\underset{\mathbf{\Phi}_1, \ldots, \mathbf{\Phi}_p}{\arg\min} \left\{ \| \mathbf{Y} - \mathbf{X}\mathbf{\Phi} \|_F + \sum_{k}^{p} k^{\alpha} \left[\sum_{j=1}^{m} \| \mathbf{\Phi}_k (j, -j) \, \mathbf{W}(-j) \|_2 + \lambda \sum_{j=1}^{m} w_j \| \mathbf{\Phi}_k(jj) \|_1 \right] \right\}. \tag{3.8}$$

3.2.1.3 The Hierarchical Vector Autoregression (HVAR) Method

More recently, Nicholson et al. (2018) proposed the hierarchical vector autoregression method for high-dimensional time series. Particularly, they assume various predefined sparse assumptions on the coefficient matrices of the VAR model. Let $\mathbf{\Phi}_k(i)$ be the ith row of the coefficient matrix $\mathbf{\Phi}_k$ and $\mathbf{\Phi}_k(ij)$ be the ijth element of the coefficient matrix $\mathbf{\Phi}_k$. To express their model, we denote

$$\mathbf{\Phi}_{k:p} = \left[\mathbf{\Phi}_k, \cdots, \mathbf{\Phi}_p \right]' \in \mathbf{R}^{m(p-k+1) \times m},$$

$$\mathbf{\Phi}_{k:p}(i) = \left[\mathbf{\Phi}_k(i), \cdots, \mathbf{\Phi}_p(i) \right]' \in \mathbf{R}^{m(p-k+1) \times 1},$$

and

$$\mathbf{\Phi}_{k:p}(ij) = \left[\mathbf{\Phi}_k(ij), \cdots, \mathbf{\Phi}_p(ij) \right]' \in \mathbf{R}^{(p-k+1) \times 1}.$$

Consider the $m \times m$ matrix of elementwise coefficient lags L as

$$L_{ij} = \max\left\{k : \mathbf{\Phi}_k(ij) \neq 0\right\}, \tag{3.9}$$

where we let $L_{ij} = \mathbf{0}$ if $\mathbf{\Phi}_k(ij) = 0$ for all $k = 1, \cdots, p$. Thus, each L_{ij} denotes the maximal coefficient lag for the ijth component, meaning L_{ij} is the smallest k such that $\mathbf{\Phi}_{k+1:p}(ij) = \mathbf{0}$.

The method includes three types of sparse structures for coefficient matrices of the VAR model. They are (a) the component wise structure, which allows each of the m marginal equations from Eq. (3.2) to have its own maximal of lag orders, but it requires all components within each equation must share the same maximal lag orders, such that $L_{ij} = L_i$ for $i = 1, \cdots, m$ (b) the own-other structure, which assumes a series' own lag are more informative than lags from other series and emphasizes on the importance of diagonal elements of the coefficient matrices $\mathbf{\Phi}_k$, such that $L_{ij} = L_i^{other}$ for $i \neq j$ and $L_{ii} \in \left\{L_i^{other}, L_i^{other} + 1\right\}$, for $i = 1, \cdots, m$, and (c) the elementwise structure, which places no stipulated relationship.

The parameter estimation is based on a convex optimization algorithm. For the component wise structure, the parameters are estimated through

$$\underset{\mathbf{\Phi}}{\arg\min}\left\{\frac{1}{2}\|\mathbf{Y} - \mathbf{X}\mathbf{\Phi}\|_F + \lambda \sum_{i=1}^{m}\sum_{k=1}^{p}\left\|\mathbf{\Phi}_{k:p}(i)\right\|_2\right\}, \tag{3.10}$$

where again λ is the control parameter controlling sparsity such that bigger λ means $\hat{\mathbf{\Phi}}_{k:p}(i) = 0$ for more i and for smaller k. This means that if $\hat{\mathbf{\Phi}}_{k:p}(i) = \mathbf{0}$, then $\hat{\mathbf{\Phi}}_{k':p}(i) = \mathbf{0}$, for all $k' > k$. For the own-other structure, the objective function is

$$\underset{\mathbf{\Phi}}{\arg\min}\left\{\frac{1}{2}\|\mathbf{Y} - \mathbf{X}\mathbf{\Phi}\|_F + \lambda \sum_{i=1}^{m}\sum_{k=1}^{p}\left[\left\|\mathbf{\Phi}_{k:p}(i)\right\|_2 + \|\mathbf{D}\|_2\right]\right\}, \tag{3.11}$$

where $\mathbf{D}$ is a vector concatenating $\mathbf{\Phi}_k(i, -i) = \{\mathbf{\Phi}_k(ij) : j \neq i\}_{(m-1)\times 1}$ and $\mathbf{\Phi}_{(k+1):p}(i)$. The additional second penalty allows coefficient matrices to be sparse such that the influence of component i itself may be nonzero at lag lk even though the influence of other components is zero at that lag. This indicates that for all $k' > k$, $\hat{\mathbf{\Phi}}_k(i) = \mathbf{0}$ implies $\hat{\mathbf{\Phi}}_{k'}(i) = \mathbf{0}$, and $\hat{\mathbf{\Phi}}_k(ii) = \mathbf{0}$ implies $\hat{\mathbf{\Phi}}_{k+1}(i, -i) = \mathbf{0}$. Finally, for (c) the elementwise structure, the objective function is given by

$$\underset{\mathbf{\Phi}}{\arg\min}\left\{\frac{1}{2}\|\mathbf{Y} - \mathbf{X}\mathbf{\Phi}\|_F + \lambda \sum_{i=1}^{m}\sum_{j=1}^{m}\sum_{k=1}^{p}\left\|\mathbf{\Phi}_{k:p}(ij)\right\|_2\right\}. \tag{3.12}$$

This structure in (3.12) indicates that each of the components of coefficient matrix can have its own maximum lags. Thus, this is the most flexible structure proposed by Nicholson et al. (2018) which would perform well if L_{ij} differs for all i and j, however, would be suboptimal if $L_{ij} = L_i$. The Hierarchical Vector Autoregression Method can be fitted by using R package *BigVAR* in the CRAN.

3.2.2 The Space-Time AR (STAR) Model

Similar to the regularization methods that control the values of parameters, when modeling time series associated with spaces or locations, it is very likely that many elements of $\boldsymbol{\Phi}_k$ are not significantly different from zero for pairs of locations that are spatially far away and uncorrelated given information from other locations. Thus, a model incorporating spatial information is not only helpful for parameter estimation, but also for dimension reduction and forecasting.

For a zero-mean stationary spatial time series, the space-time autoregressive moving average **STARMA** $\left(p_{a_1,\ldots,a_p}, q_{m_1,\ldots,m_q}\right)$ model is defined by

$$\mathbf{Z}_t = \sum_{k=1}^{p}\sum_{\ell=0}^{a_k}\phi_{k,\ell}\mathbf{W}^{(\ell)}\mathbf{Z}_{t-k} + \mathbf{a}_t - \sum_{k=1}^{q}\sum_{\ell=0}^{m_k}\theta_{k,\ell}\mathbf{W}^{(\ell)}\mathbf{a}_{t-k}, \tag{3.13}$$

where the zeros of $\det\left(\mathbf{I} - \sum_{k=1}^{p}\sum_{\ell=0}^{a_k}\phi_{k,\ell}\mathbf{W}^{(l)}B^k\right) = 0$ lie outside the unit circle, $\mathbf{a}_t$ is a Gaussian vector white noise process with zero mean vector $\mathbf{0}$, and covariance matrix structure

$$E\left[\mathbf{a}_t\mathbf{a}_{t+k}'\right] = \begin{cases}\boldsymbol{\Sigma}, \text{ if } k = 0,\\ \mathbf{0}, \text{ if } k \neq 0,\end{cases} \tag{3.14}$$

and $\boldsymbol{\Sigma}$ is a $m \times m$ symmetric positive definite matrix. The **STARMA** $\left(p_{a_1,\ldots,a_p}, q_{m_1,\ldots,m_q}\right)$ model becomes a space-time autoregressive **STAR** $\left(p_{a_1,\ldots,a_p}\right)$ model when $q = 0$. The **STAR** models were first introduced by Cliff and Ord (1975) and further extended to **STARMA** models by Pfeifer and Deutsch (1980a, b, c). Since a stationary model can be approximated by an autoregressive model, because of its easier interpretation, the most widely used **STARMA** models in practice are **STAR** $\left(p_{a_1,\ldots,a_p}\right)$ models,

$$\mathbf{Z}_t = \sum_{k=1}^{p}\sum_{\ell=0}^{a_k}\phi_{k,\ell}\mathbf{W}^{(\ell)}\mathbf{Z}_{t-k} + \mathbf{a}_t, \tag{3.15}$$

where $\mathbf{Z}_t$ is a zero-mean stationary spatial time series or a proper differenced and transformed series of a nonstationary spatial time series.

The spatial information is introduced to the model by weighting matrices $\mathbf{W}^{(\ell)} = \left[w_{(i,j)}^{(\ell)}\right]$. Suppose that there are total m locations and we let $\mathbf{Z}_t = [Z_{1,\,t},\ Z_{2,\,t},\ \cdots,\ Z_{m,\,t}]'$ be the vector of times series of these m locations. Based the spatial orders, with respect to the time series at location *i*, we will assign weights related to this location, $w_{i,j}^{(\ell)}$, such that they are nonzero only when the location *j* is the ℓth order neighbors of location *i*, and the sum of these weights is equal to 1. In other word, with respect to location *i*, we have $\sum_{j=1}^{m} w_{i,j}^{(\ell)} = 1$, where

$$w_{i,j}^{(\ell)} = \begin{cases} (0,\, 1]\,, \text{if location } j \text{ is the } \ell^{\text{th}}\text{order neighbor of location } i, \\ \quad 0, \;\; \text{otherwise.} \end{cases}$$

Combing these weights, $w_{i,j}^{(\ell)}$, for all *m* locations, we have the spatial weight matrix for the neighborhood, $\mathbf{W}^{(\ell)} = \left[w_{(i,j)}^{(\ell)}\right]$, which is a $m \times m$ matrix with $w_{(i,j)}^{(\ell)}$ being nonzero if and only if locations *i* and *j* are in the same ℓ^{th} order neighbor and each row summing to 1. The weight can be chosen to reflect physical properties such as border length or distance of neighboring locations. One can also assign equal weights to all the locations of the same spatial order. Clearly, $\mathbf{W}^{(0)} = \mathbf{I}$, an identity matrix, because each location is its own zeroth order neighbor.

It should be noted that the space-time autoregressive moving average (**STARMA**) model is a special case of VARMA model,

$$\mathbf{\Phi}_p(B)\mathbf{Z}_t = \mathbf{\Theta}_q(B)\mathbf{a}_t, \tag{3.16}$$

where $\mathbf{\Phi}_p(B) = \mathbf{I} - \sum_{k=1}^{p}\sum_{\ell=0}^{a_k} \phi_{k,\ell}\mathbf{W}^{(\ell)}B^k$, and $\mathbf{\Theta}_q(B) = \mathbf{I} - \sum_{k=1}^{q}\sum_{\ell=0}^{m_k} \theta_{k,\ell}\mathbf{W}^{(\ell)}B^k$.

3.2.3 *Model-Based Clustering*

Clustering or cluster analysis has been used by researchers to group data into some homogeneous groups for a long time. It was possibly originated in anthropology and psychology. There are many methods in clustering, for example, hierarchical clustering and k-means algorithm. The earlier works include Tryon (1939), Cattell (1943), Ward (1963), Macqueen (1967), McLachlan and Basford (1988), and others. These methods were extended to the model-based cluster approach with an associated probability distribution by researchers including Banfield and Raftery (1993), Fraley and Raftery (2002), Wang and Zhou (2008), Scrucca (2010), and others. More recently, Wang et al. (2013) introduce a robust model-based clustering method for forecasting high dimensional time series, and in this section, we will use their approach for an illustration. Let p_h be the probability a time series belongs to cluster *h*. The method first groups multiple time series into *H* mutually exclusive clusters, $\sum_{h=1}^{H} p_h = 1$, and assumes that each mean adjusted time series in a given cluster follows the same AR(*p*) model. Thus, for *i*th time series that is in cluster *h*, we have,

$$Z_{i,t} = \sum_{k=1}^{p} \phi_k^{(h)} Z_{i,t-k} + \sigma_h \varepsilon_{i,t}, \text{ for } t = p+1, \ldots, N, \tag{3.17}$$

where $h = 1, 2, \ldots, H$, the $\varepsilon_{i,\,t}$ are *i. i. d.* $N(0, 1)$ random variables, independent across time and series.

Let $\boldsymbol{\theta}_h = \left(\phi_1^{(h)}, \phi_2^{(h)}, \ldots, \phi_p^{(h)}, \sigma^{(h)}\right)$ be the vector of all parameters in cluster h, and $\boldsymbol{\Theta} = (\boldsymbol{\theta}_1, \cdots, \boldsymbol{\theta}_H, \boldsymbol{\eta})$, where $\boldsymbol{\eta} = (p_1, \cdots, p_H)$. The estimation procedure is accomplished through the Bayesian Markov Chain and Monte Carlo method.

3.2.4 Factor Analysis

In previous sections, we mainly have reviewed methods that are based on the VAR model but with different model constraints and estimation procedures. However, there exist many other models for multivariate time series analysis, such as transfer function model (Box et al. 2015), state space model (Kalman 1960), and canonical correlation analysis (Box and Tiao 1977). More recently, Stock and Watson (2002a, b) introduced factor model for dimension reduction and forecasting. Matteson and Tsay (2011) proposed the dynamic orthogonal component analysis. In this review section, we will concentrate on the factor model.

The factor model is also called diffusion index approach and can be written as

$$\mathbf{Z}_t = \mathbf{L}\mathbf{F}_t + \boldsymbol{\varepsilon}_t, \tag{3.18}$$

where $\mathbf{F}_t = (F_{1,\,t}, F_{2,\,t}, \ \ldots \ , F_{k,\,t})'$ is a $(k \times 1)$ vector of factors at time t, $\mathbf{L} = [\ell_{ij}]$ is a $(m \times k)$ loading matrix, ℓ_{ij} is the loading of the ith variable on the jth factor, $i = 1, 2, \ \ldots \ , m, j = 1, 2, \ \ldots \ , k$, and $\boldsymbol{\varepsilon}_t = (\varepsilon_{1,\,t}, \ \ldots \ , \varepsilon_{m,\,t})'$ is a $(m \times 1)$ vector of noises with $E(\boldsymbol{\varepsilon}_t) = \mathbf{0}$, and $Cov(\boldsymbol{\varepsilon}_t) = \boldsymbol{\Sigma}$. Let $Z_{i,\,t\,+\,\ell}$ be ith component of $\mathbf{Z}_{t\,+\,\ell}$, once values of factors are obtained, we can build a forecast equation for the $\ell -$ step ahead forecast, such that

$$Z_{i,t+\ell} = \boldsymbol{\beta}'\mathbf{F}_t + \varepsilon_{i,t+\ell}, \tag{3.19}$$

where $\boldsymbol{\beta} = (\beta_1, \ \ldots \ , \beta_k)'$ denotes the coefficient vector and $\varepsilon_{i,\,t\,+\,\ell}$ is a sequence of uncorrelated zero-mean random variables. Note that the Eq. (3.19) can be further extended to:

$$Z_{i,t+\ell} = \boldsymbol{\beta}'\mathbf{F}_t + \boldsymbol{\alpha}'\mathbf{X}_{i,t} + \varepsilon_{i,t+\ell}, \tag{3.20}$$

where $\mathbf{X}_{i,\,t}$ is a $m \times 1$ vector of lagged values of $Z_{i,\,t\,+\,\ell}$ and/or other observed variables. We follow the approach proposed by Eq. (3.7) of Bai and Ng (2002) plus the penalty term $k[(m + T)/mT] \log [mT/(m + T)]$ to select the number of factors in our simulation studies and empirical examples. Other methods or penalties in Bai and Ng (2002) can be used. However, this is beyond the scope of this presentation.

3.3 Proposed Method for High Dimension Reduction

In many applications, a large number of individual time series may follow similar pattern so that we could aggregate them together. By doing so, we can reduce the dimension of the multivariate time series into a manageable and meaningful size. Specifically, we will concentrate on the VAR model described in Sect. 3.2, and propose aggregation as our method of dimension reduction.

Given a vector time series, assume that after model identification, it follows the VAR(p) model,

$$\mathbf{Z}_t = \sum_{k=1}^{p} \mathbf{\Phi}_k \mathbf{Z}_{t-k} + \mathbf{a}_t, \tag{3.21}$$

where $\mathbf{Z}_t$ is mean adjusted stationary m-dimensional original time series. Let

$$\mathbf{Y}_t = \mathbf{A}\mathbf{Z}_t, \tag{3.22}$$

where $\mathbf{A}$ is a $s \times m$ aggregation matrix with $s < m$, and $\mathbf{Y}_t = [Y_{1,\,t}, \cdots, Y_{s,\,t}]'$. Presently, the elements in $\mathbf{A}$ are assumed to be binary, such that its (ij) element is 1 when $Z_{j,\,t}$ is included in the aggregate $Y_{i,\,t}$, and is 0 otherwise. In other word, the elements of row i in $\mathbf{A}$ construct $Y_{i,\,t}$ as the sum of designated elements of $\mathbf{Z}_t$. We will call $\mathbf{Y}_t$ the aggregate series and $\mathbf{Z}_t$ the non-aggregate series.

It can be shown that the aggregate series $\mathbf{Y}_t$ will also follow a VAR(p) model. However, in practice, we normally use the same model identification procedure to fit a VAR(P) model for some P such that

$$\mathbf{Y}_t = \sum_{k=1}^{P} \mathbf{\Phi}_k^{(a)} \mathbf{Y}_{t-k} + \boldsymbol{\xi}_t, \tag{3.23}$$

where $\mathbf{\Phi}_k^{(a)}$ for $k = 1, \cdots, P$ are $s \times s$ coefficient matrices, and $\boldsymbol{\xi}_t$ follows s-dimensional i.i.d. normal distribution with mean zero and covariance $\mathbf{\Sigma}^{(a)}$. The order P can be selected by existing methods such as AIC, BIC, and sequential likelihood ratio test (A detail review of order selection methods can be found in Lütkepohl (2007)).

By using the aggregation, we reduce the dimension of the time series from m to s. Suppose we are interested in the $\ell -$ step ahead forecast $\hat{\mathbf{Y}}_t(\ell)$ for the aggregate variable $\mathbf{Y}_{t+\ell}$. There are two ways to forecast: (1) forecasting from the non-aggregate data first and then aggregating its forecasts. Mathematically, it can be represented as

$$\hat{\mathbf{Y}}_t(\ell) = \mathbf{A}\hat{\mathbf{Z}}_t(\ell), \tag{3.24}$$

where $\hat{\mathbf{Z}}_t(\ell)$ is the $\ell-$ step ahead forecasts from the model (3.21); (2) modeling and forecasting directly from the aggregates from aggregate model (3.23). Our proposed method takes the second procedure to reduce the dimension when modeling the data.

For the VARMA and VAR models, many results (Rose 1977; Tiao and Guttman 1980; Wei and Abraham 1981; Kohn 1982; Lütkepohl 1984) have shown that it is preferable to forecast the original time series first and then aggregate the forecasts (method 1), rather than forecast the aggregate time series directly (method 2). They also established the conditions for those two methods to be equivalent, which can be summarized below:

Theorem 3.1 *Consider a m-dimensional non-aggregate VARMA(p, q)model,* $\mathbf{Z}_t = \sum_{i=1}^{p}\mathbf{\Phi}_i\mathbf{Z}_{t-i} + \mathbf{a}_t - \sum_{j=1}^{q}\mathbf{\Theta}_j\mathbf{a}_t$, *and its s–dimensional aggregate,* $\mathbf{Y}_t = \mathbf{A}\mathbf{Z}_t$, *modelled with a VARMA(p, q) model,* $\mathbf{Y_t} = \sum_{i=1}^{p}\mathbf{\Phi}_i^{(a)}\mathbf{Y}_{t-i} + \boldsymbol{\xi}_t - \sum_{j=1}^{q}\mathbf{\Theta}_j^{(a)}\boldsymbol{\xi}_t$. *The condition for the forecasts of* $\mathbf{Y}_t$ *from aggregate model to be equivalent to the aggregate of forecasts from the non-aggregate model are:* $\mathbf{A}\mathbf{\Phi}_i = \mathbf{\Phi}_i^{(a)}\mathbf{A}$, *for* $i = 1, \ldots, p$ *and* $\mathbf{A}\mathbf{\Theta}_j = \mathbf{\Theta}_j^{(a)}\mathbf{A}$, *for* $j = 1, \ldots, q$, *where* $\boldsymbol{\xi}_t = \mathbf{A}\mathbf{a}_t$.

For the STARMA and STAR model, Gehman (2015) proved similar results that given a non-aggregate data that follows a STARMA model and modeling its aggregate data as the same order as the non-aggregate data, the mean squared forecast error is always larger when using aggregate model under the assumption that parameters are known.

Results shown above are based on the assumptions that all parameters are known. When parameters are unknown, Lütkepohl (1984) showed that forecasts from the aggregate data might outperform forecasts from the non-aggregate data, since parameter estimates could be noisy. This argument is more obvious in the high dimension setting since so many parameters need to be estimated, which provides a reasoning that forecasting from the aggregate data could be better in some situations.

3.4 Simulation Studies

In this section, we evaluate the performance of different methods in forecasting aggregates via Monte Carlo simulations. We consider three scenarios that were all simulated from the $m = 50$-dimensional VAR(*1*) model

$$\mathbf{Z}_t = \mathbf{\Phi}_1\mathbf{Z}_{t-1} + \mathbf{a}_t, \tag{3.25}$$

where $\mathbf{\Phi}_1$ is coefficient matrix and $\mathbf{a}_t$ is vector white noise, which is simulated from 50–dimensional normal random variable with zero mean and identity covariance matrix. The number of observation used for in-sample modeling and estimation are set to be $N = 100,\ \ 500$. Additional 5 out-of-sample observations are used to compute MSFE. To compare the performances, we simulate 200 realizations for each scenario. We consider two aggregation schemes. First, two-region aggregation, indicating that we aggregate first 25 time series and last 25 time series. Thus, the resulting aggregated time series is bivariate. Second, total aggregate, meaning that the aggregation matrix A is a row vector with elements all equal to one. We chose mean squared forecast error (*MSFE*) as the evaluation metric and define $MSFE(\ell)$ as the ℓ-step-ahead forecasts mean squared error of forecasts, such that

$$MSFE\,(\ell) = \frac{1}{200s} \sum_{k=1}^{s} \sum_{i=1}^{200} \left[Y_{k,t+\ell}^{(i)} - Y_{k,t}^{(i)}\,(\ell) \right]^2.$$

Methods compared in this section include: (1) the VAR model based on non-aggregate data and estimated through least square; (2) the univariate AR model for each time series with the lag orders selected by AIC, denoted by AR; (3) the Lasso method; (4) the lag-weighted Lasso method; (5) the HVAR method with component wise structure, denoted by HVAR-C; (6) the HVAR method with own-other structure, denoted by HVAR-OO; (7) the HVAR method with elementwise structure, denoted by HVAR-E; (8) the factor model with one lag; (9) the model-based cluster method with maximum four clusters; and (10) the proposed method.

3.4.1 Scenario 1

In scenario 1, we assume $\mathbf{\Phi}_1$ to be a diagonal matrix with the diagonal elements generated from uniform distribution $U(0.2, 0.4)$. This is a very simple case in which there is no interdependence between each individual time series, and the AR coefficients for each series are similar. Thus, a simple model based on the univariate AR model for each time series would possibly produce reasonable fitting and forecasts.

Table 3.1 displays the *MSFEs* and corresponding standard deviations of two-region aggregation. The smallest *MSFE* in each category are bolded to aid presentation. It appears that the VAR method has much larger *MSFE* compared to all other methods when $N = 100$. This is due to large parameter estimation errors when N is relative small. As the sample size N increases to 500, the *MSFE* of the VAR method approaches to other methods. Although, all methods except VAR method perform similarly in terms of *MSFE*, the proposed method produces the smallest *MSFE* in most cases. Further, it seems that all regularization methods produce similar *MSFEs*.

Table 3.1 Scenario 1: based on 200 repetitions, mean square forecast errors (MSFE) and standard deviations of two-region aggregation are reported

Sample	$\in$	VAR	AR	Lasso	Lag-weighted	HVAR-C	HVAR-OO	HVAR-E	Factor	Cluster	Proposed
$T = 100$	1	43.50 (61.25)	22.98 (35.27)	24.19 (39.41)	24.02 (36.41)	25.05 (35.76)	23.40 (33.92)	24.22 (33.93)	26.97 (36.53)	23.31 (36.17)	**22.25** (36.32)
	2	52.64 (81.01)	34.46 (56.19)	35.29 (55.94)	35.31 (56.12)	35.23 (56.98)	35.45 (57.13)	34.85 (57.13)	34.95 (56.53)	34.78 (56.03)	**34.05** (53.47)
	3	24.76 (36.65)	**22.91** (34.63)	23.28 (35.04)	23.19 (34.87)	22.96 (34.80)	23.04 (34.88)	23.08 (34.89)	29.78 (39.80)	23.31 (34.54)	23.07 (35.01)
$T = 500$	1	29.49 (43.46)	28.18 (43.36)	28.80 (43.06)	28.45 (42.15)	28.15 (41.95)	27.76 (42.10)	27.87 (42.10)	28.55 (41.50)	28.75 (40.96)	**27.68** (40.96)
	2	28.18 (42.25)	27.16 (41.48)	27.49 (41.30)	27.22 (41.21)	28.19 (43.11)	27.36 (42.04)	27.28 (41.90)	26.21 (39.14)	27.67 (41.75)	**26.08** (41.10)
	3	**27.79** (33.64)	28.11 (34.78)	27.94 (34.05)	27.95 (34.25)	27.97 (34.24)	27.89 (34.18)	27.95 (34.31)	28.32 (38.01)	28.38 (34.90)	27.93 (34.36)

Smallest MSFEs are in boldface

Table 3.2 presents the *MSFEs* and their standard deviations of total aggregation. The results in Table 3.2 are almost consistent with results in Table 3.1. The proposed method is still among one of the best method.

3.4.2 Scenario 2

In the scenario 2, the coefficient matrix $\mathbf{\Phi}_1$ is generated from "band" matrix pattern shown in Fig. 3.1 where black points are corresponding to non-zero entries and white areas are corresponding to zero entries. The non-zero diagonal entries of $\mathbf{\Phi}_1$ are fixed to be 0.3 and the non-zero off-diagonal elements are fixed to be 0.1. This coefficient structure indicates that each time series depend largely on its own past, and weakly depend on other series that are close. Tables 3.3 and 3.4 presents the *MSFEs* of two different aggregation schemes. Again, the *MSFEs* and standard deviations of VAR method are much larger than all other methods when sample size $N = 100$. For $N = 100$, the proposed method outperforms all other methods. For $N = 500$, the proposed method outperforms all other methods when $\ell = 1$, and the factor model performs the best when $\ell = 2, 5$. Among all regularization methods, the HVAR-OO produces relative smaller *MSFEs* when $N = 100$. This is because HVAR-OO assumes the diagonal elements to be more informative which are close to the true coefficient matrix structure. HVAR-E has smaller *MSFEs* than other regularization methods when $N = 500$. This is due to its flexible structure assumption.

3.4.3 Scenario 3

In the scenario 3, the coefficient matrix $\mathbf{\Phi}_1$ is generated from "cluster" matrix pattern. We set the diagonal elements of $\mathbf{\Phi}_1$ to be 0.3. Then, we randomly select $2m$ elements from off-diagonal elements of $\mathbf{\Phi}_1$ and assign each of them with value 0.1 (see Fig. 3.2). Tables 3.5 and 3.6 show the *MSFEs* and the corresponding standard deviations. In this more complicated simulation setting, the AR model and model-based cluster method have very large *MSFEs* for both sample sizes we consider. This is because they largely ignore the interdependences between each time series. For both $N = 100$ and 500, the proposed method and the factor model are the top 2 methods in terms of *MSFEs*.

3.5 Empirical Examples

In this section, two real data examples are considered, including the macroeconomic time series data and the sexually transmitted disease time series data.

Table 3.2 Scenario 2: based on 200 repetitions, mean square forecast errors (MSFE) and standard deviations of total aggregation are reported

Sample	€	VAR	AR	Lasso	Lag-weighted	HVAR-C	HVAR-OO	HVAR-E	Factor	Cluster	Proposed
$T = 100$	1	98.65 (147.22)	49.63 (70.68)	50.20 (68.91)	50.10 (69.23)	50.17 (73.43)	48.81 (71.76)	49.33 (68.28)	55.29 (76.65)	49.68 (70.37)	**48.67** (70.09)
	2	94.76 (129.28)	**61.17** (84.39)	63.42 (86.69)	62.31 (84.23)	62.28 (87.52)	61.35 (84.53)	61.28 (84.32)	66.84 (97.42)	61.20 (84.12)	61.63 (86.52)
	3	43.98 (63.44)	40.98 (59.97)	42.13 (61.32)	41.89 (60.93)	41.42 (60.17)	41.76 (60.61)	41.66 (60.59)	45.23 (63.84)	**40.72** (56.53)	41.31 (60.42)
$T = 500$	1	57.89 (82.42)	54.79 (74.66)	57.24 (80.69)	57.32 (80.14)	57.89 (82.14)	57.50 (76.02)	55.57 (75.32)	59.07 (81.88)	54.25 (75.17)	**53.98** (73.77)
	2	57.07 (86.77)	56.77 (86.31)	57.06 (87.73)	56.92 (87.58)	57.58 (89.71)	56.96 (87.19)	56.99 (87.07)	58.42 (87.97)	56.28 (85.90)	**56.16** (86.77)
	3	58.33 (70.67)	58.79 (70.76)	58.36 (70.66)	58.21 (70.37)	58.41 (70.79)	58.44 (70.49)	58.46 (70.47)	**54.07** (63.81)	58.72 (70.65)	58.97 (71.59)

Smallest MSFEs are in boldface

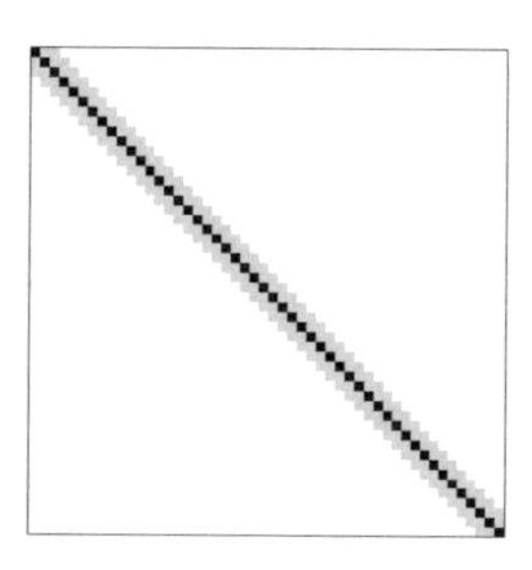

Fig. 3.1 Pattern of $\mathbf{\Phi}_1$ in scenario 2

3.5.1 The Macroeconomic Time Series

We compare *MSFE*s of different methods and assess the effectiveness of the proposed method through the collection of time series of US macroeconomic indicators. The data is collected from Stock and Watson (2009) and Koop (2013). The full data list contains 168 quarterly macroeconomic variables from Quarter 1 of 1959 to Quarter 4 of 2007, representing information about many aspects of the U.S. economy. We retrieve 61 time series from the full dataset, leading to a dataset with $m = 61$ and $N = 196$. Time series are transformed to stationary using the suggestion of Stock and Watson (2009). Those 61 time series can be aggregated into three main macroeconomic measures by its nature: gross domestic product (GDP), industrial production index (IPS), and constant elasticity of substitution (CES). Details of the dataset is given in the Appendix.

The main interest of this section is on accurately forecasting three aggregate variables: GDP, IPS, and CES, since they are important measures of the U.S economy. In this application, we used data from Quarter 1 of 1959 to Quarter 3 of 1992 for model fitting, and then compute the rolling out of sample one-step-ahead forecasts, starting from Quarter 4 of 1992 to Quarter 4 of 2007. The *MSFE*s of univariate AR, VAR, Lasso, Lag-weighted lasso, HVAR-C, HVAR-OO, HVAR-E, factor model with 20 factors, model-based cluster clustering, and the proposed method are compared in this application (Table 3.7).

Three regularization methods, including HVAR-C, HVAR-OO, and HVAR-E, perform the best as their *MSFE*s are below 0.7. The proposed method performs close to those three methods and have smaller *MSFE*s than all other methods. The benchmark univariate AR method outperforms VAR, factor model, and model-based cluster clustering, but does not perform as well as the proposed aggregation method.

3.5.2 The Sexually Transmitted Disease Data

In this section, we provide an illustration using a spatial time series. The data set contains yearly sexually transmitted disease (STD) morbidity rates reported to National Center for HIV/AIDS, viral Hepatitis, STD, and TB Prevention (NCHH-

Table 3.3 Scenario 2: based on 200 repetitions, mean square forecast errors (MSFE) and standard deviations of two-region aggregation are reported

Sample	$\in$	VAR	AR	Lasso	Lag-weighted	HVAR-C	HVAR-OO	HVAR-E	Factor	Cluster	Proposed
$T = 100$	1	60.07 (88.08)	35.38 (47.16)	40.67 (55.97)	40.53 (54.37)	35.09 (49.40)	31.61 (42.04)	33.03 (44.24)	33.49 (41.79)	33.29 (47.60)	**25.68** (34.05)
	2	96.52 (131.14)	64.89 (83.75)	67.18 (92.72)	65.01 (89.13)	61.28 (83.62)	55.88 (75.92)	56.71 (79.64)	57.22 (66.61)	61.19 (80.12)	**40.89** (62.56)
	3	139.34 (203.14)	88.50 (130.39)	91.18 (127.07)	89.18 (121.25)	93.26 (124.12)	85.64 (126.97)	86.97 (124.33)	94.67 (126.11)	85.40 (119.40)	**75.77** (114.31)
$T = 500$	1	26.62 (43.12)	37.17 (41.48)	28.61 (45.08)	27.02 (41.10)	29.31 (46.53)	28.36 (45.12)	26.67 (44.63)	33.87 (42.08)	38.27 (55.61)	**23.07** (39.34)
	2	47.43 (86.33)	66.17 (100.05)	52.04 (89.29)	51.90 (82.21)	51.31 (95.19)	50.57 (90.69)	48.34 (84.90)	51.86 (73.49)	67.27 (104.11)	**42.40** (73.76)
	3	82.48 (108.93)	100.91 (123.56)	84.78 (107.83)	84.21 (101.30)	87.72 (112.24)	86.46 (109.56)	85.38 (106.01)	**74.07** (104.64)	102.03 (121.23)	78.53 (97.67)

Smallest MSFEs are in boldface

Table 3.4 Scenario 2: based on 200 repetitions, mean square forecast errors (MSFE) and standard deviations of total aggregation are reported

Sample	€	VAR	AR	Lasso	Lag-weighted	HVAR-C	HVAR-OO	HVAR-E	Factor	Cluster	Proposed
$T = 100$	1	121.27 (166.05)	67.72 (88.77)	79.50 (105.12)	76.31 (102.18)	73.50 (107.03)	63.52 (89.22)	66.45 (91.97)	78.79 (100.34)	77.03 (102.87)	**53.19** (70.24)
	2	214.58 (286.97)	132.14 (173.65)	128.11 (172.10)	118.21 (161.25)	129.32 (176.69)	117.72 (161.75)	121.51 (170.92)	118.62 (161.40)	130.98 (163.21)	**95.11** (121.50)
	3	297.62 (426.84)	182.33 (266.46)	172.08 (233.07)	169.94 (233.18)	185.36 (256.29)	170.15 (245.52)	166.59 (234.41)	192.95 (247.35)	188.52 (231.23)	**150.99** (202.51)
$T = 500$	1	54.00 (69.08)	69.94 (90.14)	54.99 (74.65)	52.28 (70.53)	55.17 (74.23)	54.05 (70.61)	52.72 (91.97)	60.59 (91.83)	62.12 (99.46)	**49.16** (63.31)
	2	94.76 (125.59)	116.69 (157.20)	93.82 (130.48)	91.82 (122.12)	95.19 (131.10)	98.82 (125.08)	89.16 (122.00)	99.46 (139.41)	118.62 (158.12)	**82.04** (118.87)
	3	186.63 (254.26)	214.26 (302.01)	187.51 (247.53)	186.21 (232.72)	192.41 (252.89)	189.95 (253.20)	187.12 (248.16)	**166.28** (215.91)	216.60 (312.10)	172.06 (232.37)

Smallest MSFEs are in boldface

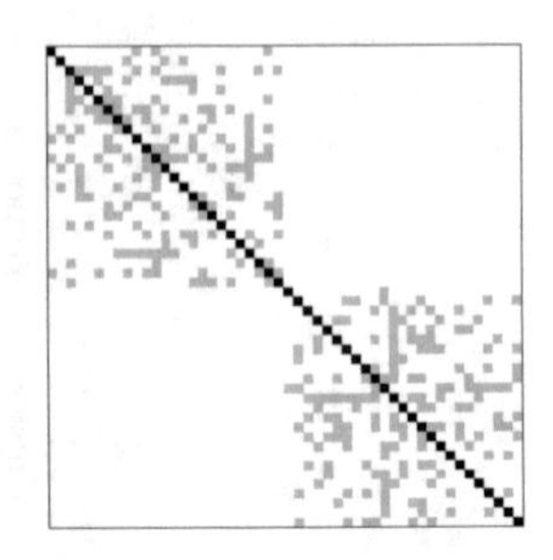

Fig. 3.2 Pattern of $\mathbf{\Phi}_1$ in scenario 3

STP), Center for HIV, and Centers for Disease Control and Prevention (CDC) from 1984 to 2014. The dataset was retrieved from CDC's website (www.cdc.gov/std/stats/) and includes 50 states plus D.C. The rates per 100,000 persons are calculated as the incidence of STD reports, divided by the population, and multiple by 100,000. For illustration, the time series of individual state are shown in Fig. 3.3. In modeling sexually transmitted disease data, researchers are interested in forecasting aggregate data based on nine Morbidity and Mortality Weekly Report (MMWR) regions or four Sexually Transmitted Disease (STD) regions (see Fig. 3.4).

For the analysis, we standardized each time series and remove data from following states, Montana, North Dakota, South Dakota, Vermont, Wyoming, Alaska, and Hawaii, due to missing data. Hence, the dimension of data is $m = 44$ and $N = 29$. We used the first 24 observations for model fitting, and the rest of observations for evaluating the forecasting performance. The *MSFE*s averaged across the lags are reported. Methods considered include: univariate AR, VAR, Lasso, Lag-weighted lasso, HVAR-C, HVAR-OO, HVAR-E, factor model with 10 factors, model-based cluster clustering, and the proposed method. In addition, we also add the STAR model for comparison in this application, as it is one of most naturally considered models for spatial time series analysis, which can also be reviewed as a dimension reduction method.

The *MSFE*s when forecasting sexual transmitted disease at 9 MMWR regions are presented in Table 3.8. VAR fails to estimate the parameters. STAR($2_{2,1}$) performs the best in this case, following by factor model, model-based cluster clustering, proposed method, and univariate AR. All regularization methods have much larger *MSFE*s.

Table 3.9 displays the *MSFE*s when forecasting sexual transmitted disease at 4 STD regions. Again, VAR fails to estimate the parameters due to the number of parameters to estimate is bigger than the number of observations. The proposed method performs the best among all methods. STAR($2_{2,1}$) model has second smallest *MSFE*s. Factor model and Model-based cluster clustering also perform reasonably well. The *MSFE* for the univariate AR is in the middle. Again, all regularization methods do not perform well.

Table 3.5 Scenario 3: based on 200 repetitions, mean square forecast errors (MSFE) and standard deviations of two-region aggregation are reported

Sample	ϵ	VAR	AR	Lasso	Lag-weighted	HVAR-C	HVAR-OO	HVAR-E	Factor	Cluster	Proposed
$T = 100$	1	58.59 (79.16)	105.66 (152.43)	59.34 (93.34)	58.13 (92.61)	42.43 (62.31)	41.65 (62.71)	47.79 (69.91)	83.27 (81.52)	102.13 (131.09)	**26.91** (36.94)
	2	90.44 (107.71)	172.88 (266.76)	92.25 (147.69)	89.76 (151.00)	71.34 (102.08)	72.69 (106.57)	87.25 (126.10)	122.54 (196.01)	168.10 (231.22)	**50.61** (71.77)
	3	200.81 (312.03)	369.70 (539.85)	221.68 (332.73)	198.28 (293.89)	196.56 (279.96)	198.79 (280.07)	231.06 (331.01)	233.49 (376.56)	310.36 (423.91)	**151.58** (211.32)
$T = 500$	1	21.98 (31.26)	63.40 (86.55)	25.65 (32.79)	25.39 (31.08)	24.09 (32.72)	23.28 (30.24)	23.31 (30.23)	31.04 (44.46)	64.82 83.28)	**21.40** (26.74)
	2	49.08 (61.87)	110.83 (161.49)	56.89 (79.29)	56.89 (79.29)	56.19 (78.34)	52.42 (70.28)	51.88 (69.55)	61.46 (90.61)	112.21 (167.51)	**47.20** (67.96)
	3	114.33 (160.87)	202.02 (314.87)	125.73 (178.88)	125.24 (175.12)	120.38 (175.33)	121.67 (179.78)	123.55 (177.49)	**97.45** (132.56)	205.16 (310.29)	116.06 (157.49)

Smallest MSFEs are in boldface

Table 3.6 Scenario 3: based on 200 repetitions, mean square forecast errors (MSFE) and standard deviations of total aggregation are reported

Sample	€	VAR	AR	Lasso	Lag-weighted	HVAR-C	HVAR-OO	HVAR-E	Factor	Cluster	Proposed
$T = 100$	1	107.19 (149.76)	182.26 (257.29)	116.37 (170.56)	103.15 (149.18)	84.45 (122.70)	80.48 (118.01)	93.58 (127.87)	133.16 (222.40)	188.26 (217.39)	**54.95** (74.07)
	2	186.46 (260.73)	408.39 (294.77)	179.01 (246.78)	170.24 (250.17)	140.49 (175.49)	140.87 (174.82)	166.53 (210.67)	196.96 (333.81)	292.90 (402.15)	**112.13** (146.59)
	3	346.11 (595.51)	544.34 (761.25)	356.22 (532.16)	341.03 (541.36)	319.08 (440.95)	326.22 (452.28)	369.65 (522.60)	338.71 (556.76)	751.73 (548.12)	**308.98** (453.83)
$T = 500$	1	49.95 (67.17)	208.07 (132.74)	54.48 (84.02)	53.28 (80.97)	53.19 (70.37)	53.09 (72.87)	53.52 (73.30)	86.99 (124.47)	133.21 (202.15)	**48.03** (64.02)
	2	96.22 (137.05)	217.57 (357.48)	111.54 (174.73)	110.27 (173.67)	101.45 (149.36)	101.99 (155.19)	103.41 (157.41)	113.79 (155.56)	216.27 (357.20)	**93.04** (136.73)
	3	206.39 (267.12)	368.43 (522.56)	234.19 (312.03)	230.59 (309.10)	221.89 (290.89)	226.44 (308.09)	224.87 (296.66)	**171.55** (275.44)	364.12 (502.16)	219.96 (285.35)

Smallest MSFEs are in boldface

Table 3.7 MSFEs of forecasting three aggregate macroeconomics variables (GDP, IPS, and CES)

	MSFE
Univariate AR	0.838
VAR	1.537
Lasso	0.744
Lag-weighted lasso	0.752
HVAR-C	0.683
HVAR-OO	0.667
HVAR-E	0.699
Factor model	0.883
Model-based cluster clustering	1.465
The proposed method	0.715

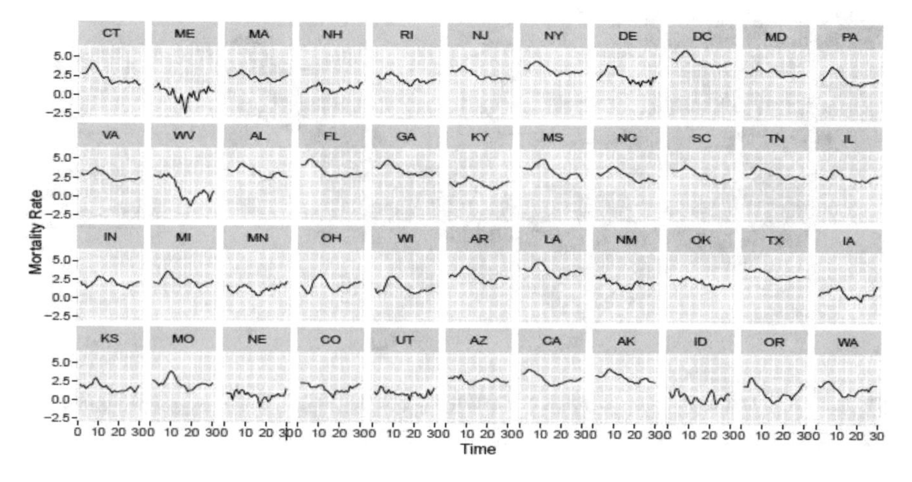

Fig. 3.3 Yearly sexually transmitted disease (STD) time series for each state

Supplementary Material
Supplementary Material is available and includes details of parameter estimation results for two empirical data analyses examples in Sect. 3.5.

3.6 Concluding Remarks

Big data and high dimensional problem are all over the place in the time of fast computer and internet. We propose aggregation as a dimension reduction method. It is very natural and simple to use, and as supported by both simulation and empirical examples in term of its performance in forecasting, it is a useful and good method for dimension reduction.

The aggregation matrix $\mathbf{A}$ and its associated s can be in many different specifications. Even based on practical considerations, we can specify different forms

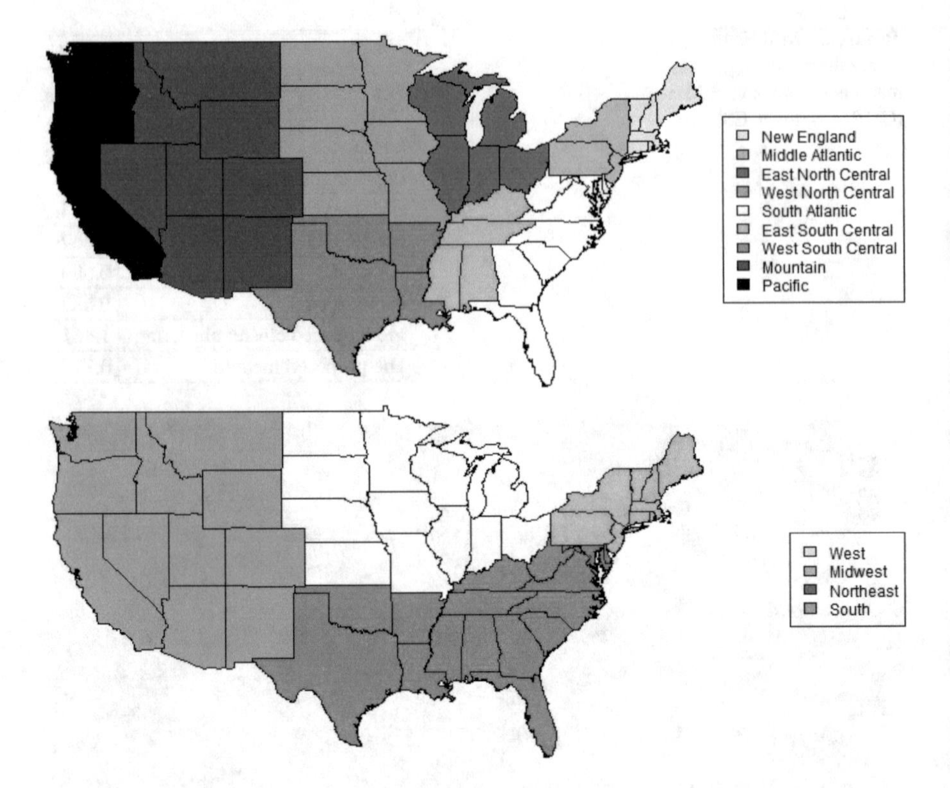

Fig. 3.4 Top: U.S. states grouped into 9 MMWR regions; bottom: U.S. states grouped into 4STD regions

of **A** and *s*. By choosing $\mathbf{s} = 1$, **A** becomes a row vector, and any m-dimensional multivariate time series $\mathbf{Z}_t$ will aggregate to become a univariate time series Y_t. For most of time, the result of aggregation is meaningful. For example, in sales data, we can specify them in terms of regions or kinds (categories). In term of the housing sales of the 3144 US counties, we can aggregate them into the housing sales of 50 states, into the housing sales of the four regions (East, West, North, and South), and further into the total housing sales of the whole country. We can also specify **A** and *s* based on data-driven considerations, which we will continue to investigate in future study.

Acknowledgments The author wants to thank his PhD student, Zeda Li, who helped him develop software code for the analyses of many data sets in the presentation.

Table 3.8 MSFEs in forecasting sexual transmitted disease rate in the 9 MMWR regions

	MSFE
Univariate AR	4.54
VAR	NA
Lasso	5.63
Lag-weighted lasso	5.05
HVAR-C	5.79
HVAR-OO	5.65
HVAR-E	5.56
STAR($2_{2,1}$)	**3.73**
Factor model	4.51
Model-based cluster clustering	3.97
Proposed method	4.53

Table 3.9 MSFEs in forecasting sexual transmitted disease rate in the 4 STD regions

	MSFE
Univariate AR	14.10
VAR	NA
Lasso	19.10
Lag-weighted lasso	19.27
HVAR-C	19.81
HVAR-OO	19.16
HVAR-E	18.75
STAR($2_{2,1}$)	10.91
Factor model	13.96
Model-based cluster clustering	12.56
Proposed method	**10.65**

A.1 Appendix

The data used in the presentation is a subset of that used in Stock and Watson (2009) and Koop (2013). Reader who is interested in this data can see further details in their papers. Variables that are originally at a monthly frequency are transformed to quarterly by taking average of 3 months in a quarter. Seasonally adjusts are taken if necessary. All variables are transformed to stationary by differencing. Table 3.10 contains brief description of each variable, and the aggregation group they are belong to, along with a transformation code, where 1 = first differencing of log of variables, 2 = second differencing of log of variables.

Table 3.10 Variables used in Sect. 3.5.1

Variable	Description	Code	Group
GDP252	Real personal consumption exp: quantity index	1	GDP
GDP253	Real personal consumption exp: durable goods	1	GDP
GDP254	Real personal consumption exp: nondurable goods	1	GDP
GDP255	Real personal consumption exp: services	1	GDP
GDP256	Real gross private domestic inv: quantity index	1	GDP
GDP257	Real gross private domestic inv: xed inv	1	GDP
GDP258	Real gross private domestic inv: nonresidential	1	GDP
GDP259	Real gross private domestic inv: nonres structure	1	GDP
GDP260	Real gross private domestic inv: nonres equipment	1	GDP
GDP261	Real gross private domestic inv: residential	1	GDP
GDP266	Real gov consumption exp, gross inv: federal	2	GDP
GDP267	Real gov consumption exp, gross inv: state and local	2	GDP
GDP268	Real final sales of domestic product	2	GDP
GDP269	Real gross domestic purchases	2	GDP
GDP271	Real gross national product	2	GDP
GDP272	Gross domestic product: price index	2	GDP
GDP274	Personal cons exp: durable goods, price index	2	GDP
GDP275	Personal cons exp: nondurable goods, price index	2	GDP
GDP276	Personal cons exp: services, price index	2	GDP
GDP277	Gross private domestic investment, price index	2	GDP
GDP278	Gross priv dom inv: fixed inv, price index	2	GDP
GDP279	Gross priv dom inv: nonresidential, price index	2	GDP
GDP280	Gross priv dom inv: nonres structures, price index	2	GDP
GDP281	Gross priv dom inv: nonres equipment, price index	2	GDP
GDP282	Gross priv dom inv: residential, price index	2	GDP
GDP284	Exports, price index	2	GDP
GDP285	Imports, price index	2	GDP
GDP286	Government cons exp and gross inv, price index	2	GDP
GDP287	Gov cons exp and gross inv: federal, price index	2	GDP
GDP288	Gov cons exp and gross inv: state and local, price index	2	GDP
GDP289	Final sales of domestic product, price index	2	GDP
GDP290	Gross domestic purchases, price index	2	GDP
GDP291	Final sales to domestic purchasers, price index	2	GDP
GDP292	Gross national products, price index	2	GDP
IPS11	Industrial production index: products total	1	IPS
IPS299	Industrial production index: final products	1	IPS
IPS12	Industrial production index: consumer goods	1	IPS
IPS13	Industrial production index: consumer durable	1	IPS
IPS18	Industrial production index: consumer nondurable	1	IPS
IPS25	Industrial production index: business equipment	1	IPS

(continued)

Table 3.10 (continued)

Variable	Description	Code	Group
IPS32	Industrial production index: materials	1	IPS
IPS34	Industrial production index: durable goods materials	1	IPS
IPS38	Industrial production index: nondurable goods material	1	IPS
IPS43	Industrial production index: manufacturing	1	IPS
IPS307	Industrial production index: residential utilities	1	IPS
IPS306	Industrial production index: consumer fuels	1	IPS
CES275	Avg hrly earnings, prod wrkrs, nonfarm-goods prod	2	CES
CES277	Avg hrly earnings, prod wrkrs, nonfarm-construction	2	CES
CES278	Avg hrly earnings, prod wrkrs, nonfarm-manufacturing	2	CES
CES003	Employees, nonfarm: goods-producing	1	CES
CES006	Employees, nonfarm: mining	1	CES
CES011	Employees, nonfarm: construction	1	CES
CES015	Employees, nonfarm: manufacturing	1	CES
CES017	Employees, nonfarm: durable goods	1	CES
CES033	Employees, nonfarm: nondurable goods	1	CES
CES046	Employees, nonfarm: service providing	1	CES
CES048	Employees, nonfarm: trade, transportation, and utilities	1	CES
CES049	Employees, nonfarm: wholesale trade	1	CES
CES053	Employees, nonfarm: retail trade	1	CES
CES088	Employees, nonfarm: financial activities	1	CES
CES140	Employees, nonfarm: government	1	CES

References

Bai, J., Ng, S.: Determining the number of factors in approximate factor models. Econometrica. **70**, 191–221 (2002)

Banfield, J., Raftery, A.: Model-based cluster Gaussian and non-Gaussian clustering. Biometrics. **49**, 803–821 (1993)

Box, G.E.P., Tiao, G.C.: A canonical analysis of multiple times series. Biometrika. **64**, 355–370 (1977)

Box, G.E.P., Jenkins, G.M., Reinsel, G.C., Ljung, G.M.: Time Series Analysis, Forecasting and Control, 5th edn. Wiley, Hoboken (2015)

Cattell, R.B.: The description of personality: basic traits resolved into clusters. J. Abnorm. Soc. Psychol. **38**, 476–506 (1943)

Cliff, A.D., Ord, J.: Model building and the analysis of spatial pattern in human geography. J. R. Stat. Soc. Ser. B. **37**, 297–348 (1975)

Fraley, C., Raftery, A.: Model-based cluster clustering, discriminant analysis, and density estimation. J. Am. Stat. Assoc. **97**, 458–470 (2002)

Gehman, A.: The effects of spatial aggregation on spatial time series modeling and forecasting, PhD dissertation, Temple University (2015)

Hamilton, J.D.: Time Series Analysis. Princeton University Press, Princeton (1994)

Hannan, E.J.: Multiple Time Series. Wiley, New York (1970)

Hsu, N., Hung, H., Chang, Y.: Subset selection for vector autoregressive processes using lasso. Comput. Stat. Data Anal. **52**, 3645–3657 (2008)

Kalman, R.E.: A new approach to linear filtering and prediction problems. J. Basic Eng. **82**, 35–45 (1960)
Kohn, R.: When is an aggregate of a time series efficiently forecast by its past? J. Econ. **18**, 337–349 (1982)
Koop, G.M.: Forecasting with medium and large Bayesian VARS. J. Appl. Econ. **28**, 177–203 (2013)
Lütkepohl, H.: Forecasting contemporaneously aggregated vector ARMA processes. J. Bus. Econ. Stat. **2**, 201–214 (1984)
Lütkepohl, H.: New Introduction to Multiple Time Series Analysis. Springer, Berlin (2007)
MacQueen, J.: Some methods for classification and analysis of multivariate observations. In: Cam, L.M., Neyman J. (eds.) Proceedings of the 5th Berkeley Symposium on Mathematical Statistics and Probability, vol. 1, pp. 281–297. University of California Press (1967)
Matteson, D.S., Tsay, R.S.: Dynamic orthogonal components for multivariate time series. J. Am. Stat. Assoc. **106**, 1450–1463 (2011)
McLachlan, G., Basford, K.E.: Mixture Models: Inference and Applications to Clustering. Marcel Dekker, New York (1988)
Nicholson, W.B., Bien, J., Matteson, D.S.: High dimensional forecasting via interpretable vector autoregression, arXiv:1412.5250v3 [stat.ME] (2018)
Pfeifer, P.E., Deutsch, S.J.: A three-stage iterative procedure for space-time modeling. Technometrics. **22**, 35–47 (1980a)
Pfeifer, P.E., Deutsch, S.J.: Identification and interpretation of the first order space-time ARMA models. Technometrics. **22**, 397–408 (1980b)
Pfeifer, P.E., Deutsch, S.J.: Stationary and inevitability regions for low order STARMA models. Commun. Stat. Simul. Comput. **9**, 551–562 (1980c)
Reinsel, G.C.: Elements of Multivariate Time Series Analysis, 2nd edn. Springer, New York (1997)
Rose, D.: Forecasting aggregates of independent ARIMA process. J. Econ. **5**, 323–345 (1977)
Scrucca, L.: Dimension reduction for model-based cluster clustering. Stat. Comput. **20**, 471–484 (2010)
Song, S., Bickel, P.: Large vector autoregressions, arXiv: 1106.3915v1 [stat.ML] (2011)
Stock, J.H., Watson, M.W.: Forecasting using principal components from a large number of predictors. J. Am. Stat. Assoc. **97**, 1167–1179 (2002a)
Stock, J.H., Watson, M.W.: Macroeconomic forecasting using diffusion index. J. Bus. Econ. Stat. **20**, 1147–1162 (2002b)
Stock, J.H., Watson, M.W.: Forecasting in dynamic factor models subject to structural instability. In: Shephard, N., Castle, J. (eds.) The Methodology and Practice of Econometrics: Festschrift in Honor of D.F. Hendry, chap. 7. Oxford University Press, Oxford (2009)
Tiao, G.C., Guttman, I.: Forecasting contemporal aggregate of multiple time Series. J. Econ. **12**, 219–230 (1980)
Tibshirani, R.: Regression shrinkage and selection via the lasso. J. R. Stat. Soc. Ser. B Methodol. **58**, 267–288 (1996)
Tryon, R.C.: Cluster analysis: correlation profile and orthometric (factor) analysis for the isolation of unities in mind and personality. Edwards Brothers, Ann Arbor (1939)
Tsay, R.S.: Multivariate Time Series Analysis with R and Financial Applications. Wiley, Hoboken (2013)
Wang, S., Zhou, J.: Variable selection for model-based high-dimensional clustering and its application to microarray data. Biometrics. **64**, 440–448 (2008)
Wang, Y., Tsay, R.S., Ledolter, J., Shrestha, K.M.: Forecasting simultaneously high-dimensional time series: a robust model-based clustering approach. J. Forecast. **32**, 673–684 (2013)
Ward, J.H.: Hierarchical grouping to optimize an objective function. J. Am. Stat. Assoc. **58**, 234–244 (1963)
Wei, W.W.S.: Time Series Analysis – Univariate and Multivariate Methods, 2nd edn. Pearson Addison-Wesley, Boston, MA (2006)

Wei, W.W.S., Abraham, B.: Forecasting contemporal time series aggregates. Commun. Stat. Theory Methods. **10**, 1335–1334 (1981)

Yuan, M., Lin, Y.: Model selection and estimation in regression with grouped variables. J. R. Stat. Soc. Ser. B. **68**, 49–67 (2006)

Chapter 4
A Powerful Retrospective Multiple Variant Association Test for Quantitative Traits by Borrowing Strength from Complex Genotypic Correlations

Xiaowei Wu

4.1 Introduction

The last decade has seen many successful applications of genome-wide association studies (GWASs) in identifying susceptibility loci for complex genetic diseases (Manolio 2010; McCarthy et al. 2008). Various methods for association testing have been developed to account for different study setups such as cross-sectional/longitudinal survey, family/population-based design, and binary/quantitative trait. Despite previous efforts, GWASs typically focus on the marginal association of trait phenotype with each individual single nucleotide polymorphism (SNP). Alternatively, assessing the joint effect of multiple variants in a predefined genomic region, i.e., SNP-set or gene-based association test, is believed to be more advantageous and will become the natural end point for association analysis in the post-GWAS era of dense genotyping and fine mapping (Neale and Sham 2004). This approach, formulated naturally from multiple regression model, has several appealing features. First, as genes are the functional unit of the human genome and remain highly consistent across diverse human populations, shifting from SNP-based to gene-based association analysis leads to more interpretable and replicable findings in gene function (Li et al. 2011) and gene-gene interaction (Ma et al. 2013). Second, by aggregating small signals from each single variant, especially for low-frequency minor alleles (Asimit and Zeggini 2010), gene-based association analysis may achieve improved power. In addition, the multiple testing problem is much simplified in gene-based analysis (Wu et al. 2010).

X. Wu (✉)
Department of Statistics, Virginia Tech, Blacksburg, VA, USA
e-mail: xwwu@vt.edu

L. Zhang et al. (eds.), *Contemporary Biostatistics with Biopharmaceutical Applications*, ICSA Book Series in Statistics,
https://doi.org/10.1007/978-3-030-15310-6_4

A number of multiple variant association tests have been developed by pooling univariate tests for individual variants. Depending on the form of the resulting test statistic, linear or quadratic (Derkach et al. 2013), these tests generally fall under two categories: *burden tests* and *kernel tests*. Burden tests group multiple variants into a single variable called genetic burden score, and perform univariate association testing based on the collapsed variable. Kernel tests, on the other hand, assumes random effects for individual variants, and test the regression coefficients of the variants by a variance-component score test. This formulation essentially leads to a weighted sum of score statistics for testing individual variant effects (Chen et al. 2013; Schaid et al. 2013; Schifano et al. 2012; Wang et al. 2014, 2013; Wu et al. 2011). Since these two approaches aggregate genetic signals at different levels, i.e., one applies linear combination on individual variants whereas the other on score statistics of individual variants, their performance depends strongly on the underlying assumptions of the genetic effects, such as proportions of the causal variants, directions of the associations (risk, protective or both), as well as variant frequencies (Ladouceur et al. 2012). In general, burden tests are not as robust as kernel tests. It has been shown that burden tests are more powerful when most variants to be tested are causal and have homogeneous effects in the magnitude and direction, whereas kernel tests are more powerful when the effects of causal variants are in different directions or a large proportion of neutral variants present (Chen et al. 2013; Lee et al. 2013; Schaid et al. 2013; Wang et al. 2012; Wu et al. 2011). To borrow strength from both approaches and avoid power loss in certain scenarios, methods have been developed to combine linear and quadratic statistics, such as SKAT-O (Lee et al. 2012, 2013) and MONSTER (Jiang and McPeek 2013). In particular, Jiang and McPeek (2013) generalized the SKAT-O method to allow relatedness among sampled individuals by using a mixed effects model that accounts for covariates and additive polygenic effects (Jiang and McPeek 2013). Under the assumption that random effects have mean zero and compound symmetric covariance, the resulted MONSTER fixed-ρ test statistic is shown to be a convex combination of the famSKAT (Chen et al. 2013; Schaid et al. 2013; Schifano et al. 2012) and famBT statistics (Chen et al. 2013), and the optimal weight for combination is sought through a grid search as exploited by SKAT-O.

In this paper, we consider multiple variant association testing for quantitative traits in a general study design where related individuals (e.g., family trios or pedigree samples) are allowed in dense genotyping GWASs. Under this setup, the genotype data exhibit complex correlations caused by both familial relation and linkage disequilibrium (LD). Borrowing strength from such genotypic correlations thus becomes the key to improve power for testing. Starting with burden tests, we note that several methods have been developed for both unrelated (Li and Leal 2008; Madsen and Browning 2009; Morgenthaler and Thilly 2007; Price et al. 2010) and related (Chen et al. 2013; Schaid et al. 2013) individuals. Most of these collapsing methods, however, are based on prospective regression models which treat genotypes as fixed explanatory variables. Though easy to implement, such methods cannot directly incorporate the LD correlations among variants. Recognizing this limitation, we propose PC-ABT, a novel principal-component-based adaptive-weight burden test. This method uses a retrospective score test to incorporate genotypic correlations, and employs "data-driven" weights to obtain maximized test statistic. In addition, PC-ABT is able to reduce the degree of freedom (df) of the null distribution to improve power by choosing major principal components of the genotype data. In what follows, we will start with development of three sequentially related multiple variant tests: the fixed-weight burden test, adaptive-weight burden test, and PC-ABT. These tests provide a step-by-step generalization of the single-variant MASTOR test (Jakobsdottir and McPeek 2013) for quantitative traits on related individuals. Extensive simulations are performed to assess the type I error rate of PC-ABT and compare its empirical power with other tests that allow related individuals. We then apply the proposed method to the systolic blood pressure data from the NHLBI "Grand Opportunity" Exome Sequencing Project (GO-ESP) for gene-based association analysis.

4.2 Methods

Suppose that in an association study, we sample a group of n individuals with known pedigree information for phenotype, covariate, and genotype data. The phenotype data consist of a quantitative trait, denoted by a vector $\boldsymbol{Y}$ of length n. The covariates contain several non-genetic variables such as age and sex. We include these covariates in an $n \times k$ design matrix $\boldsymbol{Z}$, with the first column being a vector of ones. To assess SNP-set or gene-based association, we consider a genomic region of m variants with allele frequencies $p_1, p_2, \cdots, p_m$. Assuming that each variant is biallelic and the alleles are arbitrarily labeled as "0" and "1", we write the genotype data as an $n \times m$ matrix $\boldsymbol{G} = [\boldsymbol{G}_1, \boldsymbol{G}_2, \cdots, \boldsymbol{G}_m]$ with the (i, j)th element coded as $G_{ij} = \frac{1}{2} \times$ (number of alleles of type 1 in individual i at variant site j), for $1 \leq i \leq n, 1 \leq j \leq m$.

Under this setting, two kinds of correlations in the genotype data $\boldsymbol{G}$ need to be taken into account. The correlation across rows describes the Mendelian inheritance (i.e., relatedness) of the sampled individuals, which can be characterized by a kinship matrix

$$\boldsymbol{\Phi} = \begin{pmatrix} 1+h_1 & 2\phi_{12} & \cdots & 2\phi_{1n} \\ 2\phi_{12} & 1+h_2 & \cdots & 2\phi_{2n} \\ \vdots & \vdots & \ddots & \vdots \\ 2\phi_{1n} & 2\phi_{2n} & \cdots & 1+h_n \end{pmatrix},$$

where h_i is the inbreeding coefficient of individual i, and ϕ_{ij} is the kinship coefficient between individuals i and $j, 1 \leq i, j \leq n$. The correlation across columns is caused by the non-random association of alleles at different loci, i.e., linkage disequilibrium. We denote the correlation matrix of LD by

$$\boldsymbol{R} = \begin{pmatrix} 1 & r_{12} & \cdots & r_{1m} \\ r_{12} & 1 & \cdots & r_{2m} \\ \vdots & \vdots & \ddots & \vdots \\ r_{1m} & r_{2m} & \cdots & 1 \end{pmatrix},$$

where $r_{ij} = (p_{11} - p_i p_j)/\sqrt{p_i(1-p_i)p_j(1-p_j)}$ is the correlation coefficient between variant i and variant j, $1 \le i \ne j \le m$. Here, p_i, p_j are the allele frequencies of variants i, j respectively, and p_{11} is the frequency of the haplotype having allele 1 at both variants.

4.2.1 *Retrospective, Fixed-Weight Burden Test*

In order to conveniently model genotypic correlations caused by both familial relation and LD, we treat genotypes as random and conduct a retrospective analysis based on $\boldsymbol{G}|(\boldsymbol{Y}, \boldsymbol{Z})$ (Jakobsdottir and McPeek 2013; Thornton and McPeek 2007) to derive the *fixed-weight burden test* (FBT). Following the burden test approach, we first construct a weighted sum burden score by

$$\boldsymbol{X} = \sum_{i=1}^{m} w_i \boldsymbol{G}_i = \boldsymbol{G}\boldsymbol{W}, \tag{4.1}$$

where $\boldsymbol{W} = [w_1, w_2, \cdots, w_m]^T$ is a prescribed weight vector of length m. After collapsing $\boldsymbol{G}$ into $\boldsymbol{X}$, we then apply MASTOR (Jakobsdottir and McPeek 2013), a single-variant, retrospective, quasi-likelihood score test that allows related individuals, to assess the genetic association between $\boldsymbol{Y}$ and $\boldsymbol{X}$ while adjusting for $\boldsymbol{Z}$. For the sake of understanding, a brief description of MASTOR is included in Appendix 1. Specifically, for the model $\boldsymbol{Y} = \boldsymbol{Z}\boldsymbol{\beta}_0 + \boldsymbol{\epsilon}, \boldsymbol{\epsilon} \sim N(\boldsymbol{0}, \boldsymbol{\Sigma}_0)$ under the null hypothesis that $\boldsymbol{Y}$ is not associated with $\boldsymbol{X}$, we let $\boldsymbol{V} = \widehat{\boldsymbol{\Sigma}}_0^{-1}(\boldsymbol{Y} - \boldsymbol{Z}\widehat{\boldsymbol{\beta}}_0)$ be the transformed phenotypic residual. Here, the trait covariance matrix $\boldsymbol{\Sigma}_0$ is assumed to take form $\sigma_e^2 \boldsymbol{I} + \sigma_a^2 \boldsymbol{\Phi}$, where σ_e^2 represents variance due to random measurement error, $\boldsymbol{I}$ is the identity matrix, and σ_a^2 stands for variance attributed to additive polygenic random effects. Using the known results in Jakobsdottir and McPeek (2013), we obtain the quasi-likelihood score test statistic

$$S = \frac{(\boldsymbol{V}^T \boldsymbol{X})^2}{\widehat{Var}_0(\boldsymbol{V}^T \boldsymbol{X}|\boldsymbol{Y}, \boldsymbol{Z})} = \frac{(\boldsymbol{V}^T \boldsymbol{X})^2}{\boldsymbol{V}^T \widehat{\boldsymbol{\Sigma}}_X \boldsymbol{V}}, \tag{4.2}$$

where $\widehat{\boldsymbol{\Sigma}}_X$ is an estimator of the covariance matrix of $\boldsymbol{X}$ under the null. For any fixed weight vector $\boldsymbol{W}$, this test statistic follows a null distribution of χ_1^2. It remains to derive $\boldsymbol{\Sigma}_X$ and then obtain an appropriate estimator $\widehat{\boldsymbol{\Sigma}}_X$ in (4.2).

We write the LD covariance matrix of $\boldsymbol{G}$ as $\boldsymbol{DRD}$, where $\boldsymbol{D} = diag\{\sigma_j\}, 1 \leq j \leq m$ is a diagonal matrix, and the diagonal element σ_j is the standard deviation of the jth variant $\boldsymbol{G}_j$. Let $\boldsymbol{G}_c$ be the vectorized genotype, i.e., $\boldsymbol{G}_c = [\boldsymbol{G}_1^T, \boldsymbol{G}_2^T, \cdots, \boldsymbol{G}_m^T]^T$, then $\boldsymbol{X} = (\boldsymbol{W}^T \otimes \boldsymbol{I})\boldsymbol{G}_c$, where $\otimes$ denotes the kronecker product. Assuming the covariance matrix of $\boldsymbol{G}_c$ is separable (Fuentes 2006), i.e., $\boldsymbol{\Sigma}_{\boldsymbol{G}_c} = (\boldsymbol{DRD}) \otimes \boldsymbol{\Phi}$ (Zhu and Xiong 2012), it follows that $\boldsymbol{\Sigma}_X = (\boldsymbol{W}^T \otimes \boldsymbol{I})[(\boldsymbol{DRD}) \otimes \boldsymbol{\Phi}](\boldsymbol{W}^T \otimes \boldsymbol{I})^T = (\boldsymbol{W}^T\boldsymbol{DRDW})\boldsymbol{\Phi}$. If we further assume that $\boldsymbol{R}$ and $\boldsymbol{\Phi}$ are known, an appropriate estimator of $\boldsymbol{\Sigma}_X$ would be

$$\widehat{\boldsymbol{\Sigma}}_X = (\boldsymbol{W}^T\widehat{\boldsymbol{D}}\boldsymbol{R}\widehat{\boldsymbol{D}}\boldsymbol{W})\boldsymbol{\Phi}. \tag{4.3}$$

When Hardy-Weinberg equilibrium (HWE) holds for the genomic region of interest, the jth diagonal element of $\widehat{\boldsymbol{D}}$ can be estimated by $\widehat{\sigma}_j = \sqrt{\widehat{p}_j(1-\widehat{p}_j)/2}, 1 \leq j \leq m$ where $\widehat{p}_j = (\boldsymbol{1}^T\boldsymbol{\Phi}^{-1}\boldsymbol{1})^{-1}\boldsymbol{1}^T\boldsymbol{\Phi}^{-1}\boldsymbol{G}_j$ is the best linear unbiased estimator (BLUE) (McPeek et al. 2004) of the allele frequency p_j. Here, $\boldsymbol{1}$ denotes a vector of ones. In practice, another more general and robust estimator $\widehat{\sigma}_j^2 = \boldsymbol{G}_j^T\boldsymbol{U}\boldsymbol{G}_j/(n-1)$ may be used instead (Thornton and McPeek 2010), where $\boldsymbol{U} = \boldsymbol{\Phi}^{-1} - \boldsymbol{\Phi}^{-1}\boldsymbol{1}(\boldsymbol{1}^T\boldsymbol{\Phi}^{-1}\boldsymbol{1})^{-1}\boldsymbol{1}^T\boldsymbol{\Phi}^{-1}$.

By Eqs. (4.1)–(4.3), we obtain the retrospective FBT statistic

$$S_{FBT} = \frac{[\boldsymbol{W}^T\boldsymbol{G}^T\boldsymbol{V}\boldsymbol{V}^T\boldsymbol{G}\boldsymbol{W}]}{[\boldsymbol{W}^T(\widehat{\boldsymbol{D}}\boldsymbol{R}\widehat{\boldsymbol{D}})\boldsymbol{W}][\boldsymbol{V}^T\boldsymbol{\Phi}\boldsymbol{V}]}, \tag{4.4}$$

which also has the null distribution of χ_1^2. Clearly, S_{FBT} is invariant to the scale of $\boldsymbol{W}$. We note that Eqs. (4.2)–(4.4) can be considered as generalizations of the MASTOR test to the multiple variant case by replacing the single variant with the weighted-sum burden score. As a special case, when $m = 1$, $\widehat{\boldsymbol{D}}\boldsymbol{R}\widehat{\boldsymbol{D}}$ in (4.3) reduces to the variance estimator of a single variant and hence S_{FBT} becomes the MASTOR statistic.

4.2.2 *Adaptive-Weight Burden Test to Maximize Test Statistic*

By collapsing multiple genetic variants into a burden score, the fixed-weight burden test is able to aggregate small signals from each single variant to gain increased power for association (especially for variants with low minor allele frequencies). However, because the combining weights are pre-specified and cannot adapt to data, this test, as well as other burden tests, may experience loss of power in the presence of both risk (i.e., positively associated) and protective (i.e., negatively associated) variants. Though several adaptive burden tests (Fang et al. 2014; Han and Pan 2010; Lin and Tang 2011; Liu and Leal 2010; Sha et al. 2012; Sha and Zhang 2014) have been proposed to overcome this deficiency, many of these weighting strategies cannot take into account the complex correlations caused by both sample relatedness and LD. Moreover, since the adaptive weights depend on the genotype data, derivation of the null distribution of the resulting test statistic becomes non-trivial. Some existing adaptive burden tests rely on permutations to evaluate the p-value. This severely restricts their applications as permutations are computationally expensive and not straightforward when related individuals are included in the sample. It is thus desirable to "let the data speak for themselves" by constructing data-adaptive weights that make full use of the genotypic information and lead to a statistic with explicit null distribution.

To fulfill this purpose, we consider maximizing the test statistic (4.4) which is of a generalized Rayleigh quotient form. Note that in general such an adaptive method of weight selection on linear statistics will lead to quadratic statistics, as pointed out by Derkach et al. (2013), Li and Lagakos (2006) among others. In our context, we will see that applying adaptive weights on the retrospective burden test ends up with a statistic of the *family-based kernel test* (famSKAT) (Chen et al. 2013), with the weight matrix determined by both the LD covariance estimate $\widehat{\boldsymbol{D}}\boldsymbol{R}\widehat{\boldsymbol{D}}$ and the kinship matrix $\boldsymbol{\Phi}$. Let $\boldsymbol{A} = \widehat{\boldsymbol{D}}\boldsymbol{R}\widehat{\boldsymbol{D}}$, $\boldsymbol{b} = \boldsymbol{G}^T\boldsymbol{V}$ and $\boldsymbol{B} = \boldsymbol{b}\boldsymbol{b}^T$. Assuming $\widehat{\boldsymbol{D}}\boldsymbol{R}\widehat{\boldsymbol{D}}$ is invertible, we can show that the optimal weight vector $\boldsymbol{W}^*$ that maximizes the FBT statistic satisfies

$$\boldsymbol{W}^* \propto \boldsymbol{A}^{-1}\boldsymbol{b} = (\widehat{\boldsymbol{D}}\boldsymbol{R}\widehat{\boldsymbol{D}})^{-1}\boldsymbol{G}^T\boldsymbol{V}. \tag{4.5}$$

Since $\boldsymbol{W}^*$ is determined by the data, i.e., genotypes $\boldsymbol{G}$, trait $\boldsymbol{Y}$, and covariates $\boldsymbol{Z}$, we call the burden test with such data-driven weights the *adaptive-weight burden test* (ABT). It follows by plugging (4.5) into (4.4) that the ABT statistic takes the form

$$S_{ABT} = \frac{\boldsymbol{V}^T\boldsymbol{G}(\widehat{\boldsymbol{D}}\boldsymbol{R}\widehat{\boldsymbol{D}})^{-1}\boldsymbol{G}^T\boldsymbol{V}}{\boldsymbol{V}^T\boldsymbol{\Phi}\boldsymbol{V}}. \tag{4.6}$$

Regarding the null distribution of S_{ABT}, we observe from (4.6) that in the numerator, $\boldsymbol{G}(\widehat{\boldsymbol{D}}\boldsymbol{R}\widehat{\boldsymbol{D}})^{-1/2}$ may be thought of as a decorrelated and standardized genotype matrix with cross-column covariance being transformed to identity. Therefore, S_{ABT} can be seen as the summation of m independent MASTOR statistics, hence follows χ^2_m distribution under the null hypothesis. More details on the theoretical justification of the null distribution can be found in Appendix 1. Using this explicit null distribution, the p-value calculation of ABT becomes straightforward. This property makes ABT more favorable to real, whole genome applications than other permutation-based approaches. It should be noted that this null distribution is achieved under the assumptions that the covariance matrix $\boldsymbol{\Sigma}_{G_c}$ of the vectorized genotype is separable, and $\boldsymbol{R}$ is known. In practice, when the separability condition is not satisfied, an alternative null distribution derived from the family-based kernel test will be used instead (see details in the following two paragraphs). When $\boldsymbol{R}$ is not known a priori, one may obtain its estimate from a reference population, e.g., the one provided by The 1000 Genomes Project (2010), or directly estimate the LD covariance matrix $\boldsymbol{DRD}$ from $\boldsymbol{G}$ in the absence of reference panel.

Two main features of ABT are revealed from the above derivations: (1) the weights of ABT are adaptive to risk and protective variants, which may be explained by the $\boldsymbol{W}^*$ expression (4.5), and (2) there is a connection between ABT and the family-based kernel test, which is implied by the ABT statistic (4.6). For elucidation purpose, let us switch the viewpoint from retrospective to prospective, and consider such a trait model

$$\boldsymbol{Y} = \boldsymbol{Z}\boldsymbol{\beta} + \boldsymbol{G}\boldsymbol{\gamma} + \boldsymbol{\epsilon}, \quad \boldsymbol{\epsilon} \sim N(\boldsymbol{0}, \sigma_e^2\boldsymbol{I} + \sigma_a^2\boldsymbol{\Phi}), \tag{4.7}$$

where $\boldsymbol{\beta}$ and $\boldsymbol{\gamma}$ represent the regression coefficients of non-genetic and genetic variables, respectively. On the one hand, from a fixed-effect perspective of multiple linear regression, it can be seen that the data-driven weights $\boldsymbol{W}^*$ in (4.5) have the same sign with the GLS estimation $\widehat{\boldsymbol{\gamma}}$ thus are adaptive to the direction of the true genetic effects. This result is further confirmed by simulations (see Appendix 2). This shows that, by maximizing the fixed-weight burden statistic, ABT is able to accommodate to the direction of individual variant effects through the adaptive weights. On the other hand, if we treat $\boldsymbol{\gamma}$ as random effect and assume that its jth component follows a distribution with mean 0 and variance $w_j^2\tau$, $1 \le j \le m$, the famSKAT (Chen et al. 2013) statistic for testing $\tau = 0$ can be obtained as

$$S_{famSKAT} = \boldsymbol{V}^T\boldsymbol{G}\boldsymbol{W}\boldsymbol{W}\boldsymbol{G}^T\boldsymbol{V}. \tag{4.8}$$

Here, $\boldsymbol{W} = diag\{w_j\}$ is a diagonal matrix playing the role of the square root of the $\boldsymbol{W}$ matrix defined in Chen et al. (2013). Comparing the ABT statistic (4.6) with the famSKAT statistic (4.8), we find that, although the FBT statistic is directly derived from burden test where the aggregation is on individual variants, after employing the data-driven weights $\boldsymbol{W}^*$ to S_{FBT}, the resulting ABT statistic is formally equivalent to the famSKAT statistic where the aggregation is on individual variant statistics. This finding extends the results of Derkach et al. (2013) and Li and Lagakos (2006) to pedigree structured data. More details about the relation between the ABT statistic and the famSKAT statistic are illustrated by simulations (see Appendix 3).

As an important note, famSKAT has been shown to have a null distribution of a mixture of independent χ_1^2's (Chen et al. 2013). Therefore by the similarity between (4.6) and (4.8), we can obtain another null distribution of ABT as $\sum_{i=1}^m \lambda_i \chi_{1,i}^2$, where λ_i's are the eigenvalues of the matrix $\boldsymbol{W}^{\#}\boldsymbol{G}^T\boldsymbol{P}\boldsymbol{G}\boldsymbol{W}^{\#}$ and $\chi_{1,i}^2$'s are independent χ_1^2 random variables. Here, $\boldsymbol{W}^{\#} = (\boldsymbol{V}^T\boldsymbol{\Phi}\boldsymbol{V})^{-1/2}(\widehat{\boldsymbol{D}}\boldsymbol{R}\widehat{\boldsymbol{D}})^{-1/2}$ is a weight matrix that recognizes ABT as a special famSKAT test, and $\boldsymbol{P} = \widehat{\boldsymbol{\Sigma}}_0^{-1} - \widehat{\boldsymbol{\Sigma}}_0^{-1}\boldsymbol{Z}(\boldsymbol{Z}^T\widehat{\boldsymbol{\Sigma}}_0^{-1}\boldsymbol{Z})^{-1}\boldsymbol{Z}^T\widehat{\boldsymbol{\Sigma}}_0^{-1}$. We provide simulations using unrelated individuals and common variants with strong LD (see Scenario S1 and Configuration C3 in Sects. 4.2.4 and 4.3.1) to validate the two asymptotic null distributions of S_{ABT}: χ_m^2 and χ_1^2 mixture, and compare them with the one obtained from permutation-based approach. The validation details are in Appendix 4. We found that the two null distributions are almost identical when condition (4.3) for deriving FBT and ABT is satisfied, that is, when the covariance structure in genotype data follows a kronecker product form, and the LD correlation matrix $\boldsymbol{R}$ is known. However, in practice if these assumptions are not satisfied (e.g., non-separable covariance, $\boldsymbol{R}$ unknown and hard to be estimated accurately), the null distribution of χ_1^2 mixture is more preferable in terms of controlling type-I error. We therefore adopt χ_1^2 mixture as the null distribution of ABT in the simulation studies and real data analysis later on, and use χ_m^2 for conceptual illustration purpose only.

As a burden test using adaptive weights $\boldsymbol{W}^*$ (also a special family-based kernel test with $\boldsymbol{W}^{\#}$), ABT is expected to outperform other burden tests with fixed weights in maintaining power when both risk and protective variants exist in the genomic region of interest. However, maximizing the FBT statistic may not always guarantee an optimal test since the power also depends on the null distribution. This motivates us to further enhance power by reducing the df of the null χ^2 distribution while maximizing the FBT statistic.

4.2.3 Principal-Component-Based Adaptive-Weight Burden Test to Enhance Power

The principal component analysis (PCA) is an orthogonal transformation that converts possibly correlated variables into linearly uncorrelated variables called principal components. Applying PCA reveals the internal structure of the data in a way that best explains the variance in the data. Under our model assumptions, the genotype data $\boldsymbol{G}$ exhibit correlations from both familial relation and LD, where the former (i.e., the kinship matrix $\boldsymbol{\Phi}$) is known in most pedigree-based studies but the latter (i.e., the LD correlation matrix $\boldsymbol{R}$) is usually assumed unknown. We notice that, in the numerator of the ABT statistic (4.6), the LD covariance estimator plays a key role in determining the df of the null χ^2 distribution. If we could find a more efficient representation of $\boldsymbol{G}$ through PCA, which can be used to approximate the LD covariance matrix such that the df of the null distribution can be reduced, then the power for association will be improved. It is worth noting that the principal component (PC) -based approach has been used to capture LD information within a candidate region and leads to fewer df, more powerful omnibus tests than genotype- and haplotype-based approaches (Gauderman et al. 2007). Here, by combining ABT (which maximizes the test statistic) with PCA (which reduces the df of the null distribution), we expect this *principal-component-based adaptive-weight burden test* (PC-ABT) to have enhanced power.

Previous efforts have been made to estimate parameters for a Kronecker product covariance structure based on Gaussian assumption (Srivastava et al. 2008). However, no work has been done with regards to categorical, non-Gaussian data such as the genotype data under our consideration. Since the kinship matrix $\boldsymbol{\Phi}$ is known, a natural way to estimate the LD covariance matrix is to use the sample covariance of the transformed genotype data $\boldsymbol{\Phi}^{-1/2}\boldsymbol{G}$. Suppose such an LD covariance matrix estimator can be factorized by spectral decomposition

$$\widehat{\boldsymbol{DRD}} = \boldsymbol{L}\boldsymbol{\Lambda}\boldsymbol{L}^T = \sum_{i=1}^{m} \lambda_i \boldsymbol{L}_i \boldsymbol{L}_i^T, \tag{4.9}$$

where $\boldsymbol{L}$ is an $m \times m$ orthogonal matrix whose columns $\boldsymbol{L}_i$, $1 \leq i \leq m$, are the eigenvectors of $\widehat{\boldsymbol{DRD}}$, and $\boldsymbol{\Lambda}$ is a diagonal matrix saving the corresponding eigenvalues $\lambda_1, \lambda_2, \cdots, \lambda_m$ in descending order. The PC decomposition (without centering) of $\boldsymbol{\Phi}^{-1/2}\boldsymbol{G}$ can be obtained by

$$\boldsymbol{T} = \boldsymbol{\Phi}^{-1/2}\boldsymbol{G}\boldsymbol{L}. \tag{4.10}$$

By (4.9) and (4.10), the adaptive weight vector (4.5) becomes $\boldsymbol{W}^* \propto \boldsymbol{L}\boldsymbol{\Lambda}^{-1}\boldsymbol{T}^T\boldsymbol{\Phi}^{1/2}\boldsymbol{V}$, and the ABT statistic (4.6) can be rewritten as

$$S_{ABT} = \frac{\boldsymbol{V}^T\boldsymbol{\Phi}^{1/2}\boldsymbol{T}\boldsymbol{\Lambda}^{-1}\boldsymbol{T}^T\boldsymbol{\Phi}^{1/2}\boldsymbol{V}}{\boldsymbol{V}^T\boldsymbol{\Phi}\boldsymbol{V}}. \tag{4.11}$$

Equation (4.11) provides the ABT statistic in the orthogonal, principal component space, with the LD covariance characterized by the diagonal eigenvalue matrix $\boldsymbol{\Lambda}$.

Now suppose the variances along some PCA loading axes are small, i.e., some diagonal components in the $\boldsymbol{\Lambda}$ matrix are small. By the dimension reduction property of PCA, we may approximate the LD covariance estimator $\widehat{\boldsymbol{DRD}}$ by keeping only the first q principal components. Let $\boldsymbol{L}_q$ be an $m \times q$ matrix containing the first q eigenvectors, $q < m$, and $\boldsymbol{\Lambda}_q$ be the corresponding $q \times q$ diagonal matrix, then the truncated PCA transformation

$$\boldsymbol{T}_q = \boldsymbol{\Phi}^{-1/2}\boldsymbol{G}\boldsymbol{L}_q \tag{4.12}$$

for dimension reduction leads to the PC-ABT statistic

$$S_{PC-ABT} = \frac{\boldsymbol{V}^T\boldsymbol{\Phi}^{1/2}\boldsymbol{T}_q\widehat{\boldsymbol{\Lambda}}_q^{-1}\boldsymbol{T}_q^T\boldsymbol{\Phi}^{1/2}\boldsymbol{V}}{\boldsymbol{V}^T\boldsymbol{\Phi}\boldsymbol{V}}. \tag{4.13}$$

With the first q "important" principal components retained and the rest "noise" components discarded, we conclude that S_{PC-ABT} follows a χ_q^2 distribution under the null hypothesis. In the presence of strong LD (hence some variants share commonalities in the direction and magnitude of their genetic effects), the genotype data can be represented more efficiently in the principal component space. By omitting the principal component axes along which the genotype data exhibit small variances, we lose only a commensurately small amount of information in the calculation of S_{PC-ABT}, but reduce the df of the null

χ^2 distribution from m to q. Therefore, the PC-ABT method is expected to achieve higher power than ABT. A commonly used strategy for selecting the number of PCs q is by checking the total percent of variance explained (PVE) in the genotype data: $\sum_{j=1}^{q} \lambda_j / \sum_{j=1}^{m} \lambda_j$ such that it exceeds some threshold.

We note that, in practice when using the mixture of independent χ_1^2's instead of χ_m^2 as the null distribution of S_{ABT}, the corresponding null distribution of S_{PC-ABT} will still be $\sum_{i=1}^{m} \lambda_i \chi_{1,i}^2$ however only the first q λ_i's become non-trivial because the $\boldsymbol{G}(\widehat{\boldsymbol{D}}\boldsymbol{R}\widehat{\boldsymbol{D}})^{-1}\boldsymbol{G}^T$ term in (4.6) has now been transformed into a more compact representation $\boldsymbol{G}\boldsymbol{L}_q\widehat{\boldsymbol{\Lambda}}_q^{-1}\boldsymbol{L}_q^T\boldsymbol{G}^T$ with rank q in (4.13).

4.2.4 Simulation Studies

We perform simulations to assess the type I error rate of PC-ABT and compare its power to that of FBT, ABT, famSKAT, and MONSTER. The simulation studies are conducted for the following four scenarios depending on the sample relatedness and the minor allele frequencies (MAFs) of the variants:

S1 Unrelated individuals and common variants: This scenario considers 1600 unrelated individuals and 50 variants with MAFs sampled independently from unif(0.1, 0.5).

S2 Unrelated individuals and rare variants: This scenario uses the same setting as Scenario S1 but the MAFs are sampled from unif(0.005, 0.05).

S3 Related individuals and common variants: Simulations in this scenario are based on an assumption that the samples are from 100 outbred, 3-generation families, each containing 16 individuals related as in Fig. 4.1. The MAFs use the same setting as Scenario S1.

S4 Related individuals and rare variants: This scenario is used to evaluate rare variant association testing on related individuals. The sample relatedness and the MAFs are set to be the same as in Scenarios S3 and S2, respectively.

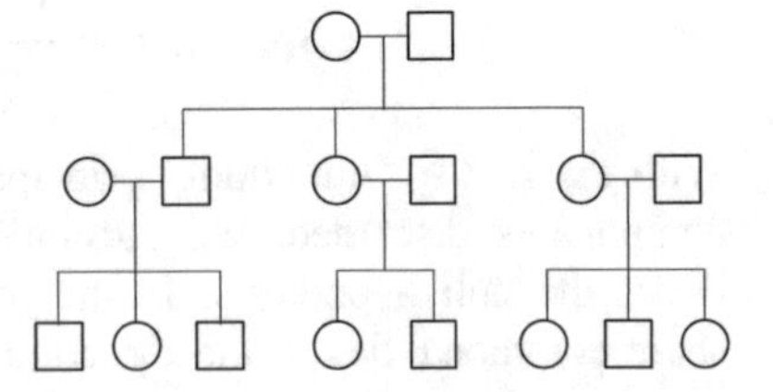

Fig. 4.1 Basic family structure of 16 members coming from three generations, used in simulation studies to generate genotype data for related individuals

In order to simulate LD correlations among variants, we consider a latent multivariate normal model MVN$(\boldsymbol{0}, \boldsymbol{\Omega})$ underlying the unrelated or founder individuals in the sample. By sampling independently from the multivariate normal distribution and dichotomizing the latent variables according to the MAFs of the variants, binary haplotypes of each unrelated or founder individual can be generated and then added up to form genotypes. When the sample contains individuals from families (Scenarios S3 and S4), the non-founders' genotype data can be generated by Mendelian "gene-dropping" the founders' haplotypes along generations, assuming no recombination within haplotypes. Clearly, the LD correlations among variants are controlled by the latent covariance $\boldsymbol{\Omega}$. Detailed settings of the LD correlations in type I error assessment and power comparison are described in Sects. 4.3.1 and 4.3.2.

After the genotype data are generated, we simulate the quantitative trait data according to model (4.7). In this model, we let the design matrix $\boldsymbol{Z}$ include the intercept and a non-genetic covariate sampled independently from standard normal distribution, and set the covariate coefficient vector $\boldsymbol{\beta} = (1, 0.6)^T$. For scenarios with unrelated individuals (S1 and S2), the variance components are set to be $\sigma_e^2 = 6, \sigma_a^2 = 0$, and for the other two scenarios, $\sigma_e^2 = 4, \sigma_a^2 = 2$. In the type I error simulations, we set the genetic effect vector $\boldsymbol{\gamma} = \boldsymbol{0}$, and in the power simulations, for each scenario, we consider three different $\boldsymbol{\gamma}$'s under the alternative hypothesis (see Table 4.2 in Sect. 4.3.2 for details).

4.2.5 *The NHLBI GO-ESP Data*

The NHLBI Framingham ESP Heart-GO is a sub-study of the Framingham Cohort (Splansky et al. 2007) for discovering novel genes in coding regions and mechanisms contributing to heart, lung, and blood disorders. This study contains exome sequence data and harmonized phenotype variables. Our use of this project data was approved by the Institutional Review Board of Virginia Tech. In the GO-ESP project (dbGaP Study Accession: phs000401.v12.p10), a total of 499 Framingham Heart Study (FHS) participants were selected for exome sequencing. Using Q/C metrics, 458 individuals are represented in the GO-ESP exome sequencing data in dbGaP, in which 198 are from 75 families and the rest are unrelated individuals. Repeated measurements were obtained for the sampled individuals at multiple time points (some individuals are from cohort 2 of the FHS with at most 8 measurements and some are from cohort 3 with at most 2 measurements). For adult individuals (with age $\geq$18 at their first measurement), we consider the average systolic blood pressure (SBP) across multiple time points as the quantitative trait. The log-transformed variable log(SBP) was adjusted for hypotenstive medication usage by adding a sensible constant (10 mmHg) if the corresponding patient was on treatment (Cui et al. 2003; Tobin et al. 2005). Six variables: age, sex, body mass index (BMI), smoking status, smoking history, and blood glucose were included as covariates, where smoking status is measured in number of cigarettes per day, smoking history indicates whether the patient is a current/former smoker or never smoked up to the first measurement. The raw exome sequencing data contain approximately 2,283,000 SNP variants on 24,484 genes. Since the primary purpose of this work is on gene-based association testing for rare variant analysis, we focus on genes containing rare variants with MAF<0.05. Given the limited sample size, we excluded genes with fewer than four rare variants from the analysis, as these genes provide little information about SBP association. Similar filtration procedure has been seen in Lee et al. (2012). After removing monomorphic and duplicated variants, a total of 380,772 SNP variants on 18,864 genes remained for analysis.

4.3 Results

The analysis results for both simulated and real data are reported in the following subsections.

4.3.1 Assessment of Type I Error

The type I error assessment is based on 20,000 simulated data replicates for each of the four genotypic scenarios S1–S4. In addition, we consider the following four configurations of the LD correlations among variants:

C1 $\boldsymbol{\Omega}$ is an identity matrix, indicating negligible LD.
C2 $\boldsymbol{\Omega}$ is a compound symmetric covariance matrix, i.e., $\boldsymbol{\Omega} = (1-\eta)\boldsymbol{I} + \eta\mathbf{1}\mathbf{1}^T$. Here we set $\eta = 0.4$ to represent moderate LD.
C3 $\boldsymbol{\Omega}$ is compound symmetric with $\eta = 0.7$, indicating strong LD.
C4 The LD correlations are set based on 50 neighboring SNPs (located on Chromosome 22, positions 16,990,110 to 17,567,009) in the FHS genotype data. This configuration is used to mimic LD in real data.

Table 4.1 reports the empirical type I error rates of PC-ABT for the number of principal components $q = 1, 25$, and 50. From this table, we observe that, in all scenarios and LD configurations, the empirical type I error rates of PC-ABT are not significantly different from the nominals when using different q's. This shows that PC-ABT is able to correctly control type I error, despite the number of principal components selected. Indeed, PC-ABT with $q = 1$ is similar to FBT (both collapse m variants to one however one adopts PCA whereas the other uses fixed weights), and PC-ABT with $q = m$ is equivalent to ABT. More complete type I error assessment results for five testing methods: FBT, famSKAT, ABT, MONSTER, and PC-ABT can be found in Appendix 5.

Table 4.1 Empirical type I error of PC-ABT for the number of principal components $q = 1, 25$, and 50

Scenario	# of PCs	$\alpha = 0.001$				$\alpha = 0.01$			
		C1	C2	C3	C4	C1	C2	C3	C4
S1	1	0.0007	0.0009	0.00075	0.0008	0.01085	0.00965	0.0088	0.0095
	25	0.0012	0.00075	0.0006	0.0009	0.0094	0.00945	0.00995	0.01015
	50	0.00075	0.0009	0.00105	0.0009	0.00975	0.00975	0.0094	0.0096
S2	1	0.0011	0.0009	0.00105	0.0007	0.0102	0.01015	0.0096	0.0096
	25	0.0011	0.0007	0.00110	0.0007	0.00995	0.00875	0.00925	0.01
	50	0.0009	0.0007	0.001	0.0006	0.01025	0.00910	0.0096	0.0091
S3	1	0.0013	0.0007	0.00125	0.001	0.01035	0.00985	0.0107	0.01135
	25	0.0011	0.0009	0.00095	0.0009	0.0091	0.00865	0.0092	0.0102
	50	0.00095	0.00075	0.0009	0.0008	0.0104	0.00905	0.00875	0.0093
S4	1	0.001	0.0014	0.0013	0.0013	0.01065	0.01105	0.01055	0.01025
	25	0.001	0.00115	0.0013	0.0011	0.00875	0.00985	0.01025	0.0109
	50	0.0008	0.0007	0.00115	0.001	0.00955	0.00920	0.0088	0.0097

The type I error rate estimates are calculated as the proportion of p-values smaller than nominal under the null hypothesis based on 20,000 simulated data replicates

4.3.2 Power Comparison to Other Burden Tests

We calculate the empirical power based on 1000 data replicates generated under the alternative $\boldsymbol{\gamma} \neq 0$ with three different compositions of risk/protective/neutral variants: (I) 10%/10%/80%, (II) 20%/20%/60%, and (III) 30%/30%/40%. In these simulations, we set the LD correlations among causal variants according to configuration C1 in Sect. 4.3.1, whereas for non-causal variants, the LD correlations are set using different configurations C1–C4. To demonstrate the power of PC-ABT in an appropriate range, the signal to noise ratio (SNR) has been controlled by adjusting the magnitude of $\boldsymbol{\gamma}$ with respect to the prespecified variance component parameters σ_e^2 and σ_a^2. These magnitude settings are listed in Table 4.2.

Table 4.2 Genetic effect settings in power simulation

Scenario	Composition of risk/protective/neutral variants		
	I: 10%/10%/80%	II: 20%/20%/60%	III: 30%/30%/40%
S1	unif(0.1, 0.6)	unif(0.1, 0.5)	unif(0.1, 0.4)
S2	unif(0.1, 1.5)	unif(0.1, 1.3)	unif(0.1, 1.2)
S3	unif(0.1, 0.6)	unif(0.1, 0.5)	unif(0.1, 0.4)
S4	unif(0.1, 1.5)	unif(0.1, 1.3)	unif(0.1, 1.2)

For each scenario, different compositions of risk/protective/neutral variants, I, II, and III are considered. The magnitude of the genetic effect for causal variants is sampled from uniform distributions

When calculating the power of PC-ABT, the appropriate number of principal components is determined according to the LD configurations in the simulated neutral variants. Clearly, higher LD among neutral variants leads to more redundant information in genetic association hence a smaller number of principal components should be retained. In particular, for LD configurations C1, C2, C3, and C4, we set the number of principal components to guarantee that the total PVE in the genotype data >95%, 85%, 80%, and 80%, respectively. The empirical power for all five methods: FBT, famSKAT, MONSTER, ABT, and PC-ABT at nominals 0.001 and 0.01 are reported in Figs. 4.2 and 4.3.

We observe that, for the simulated data including risk/protective/neutral variants, FBT always loses power because the genetic effects from causal variants are cancelled or diluted. In most of the cases, MONSTER performs slightly better than famSKAT. When the LD correlation among variants is negligible (Configuration C1), the four kernel-based methods, famSKAT, MONSTER, ABT, and PC-ABT perform similarly, whereas in the presence of moderate (Configuration C2) and strong LD (Configuration C3), ABT outperforms famSKAT and MONSTER because the latter two use prespecified weights (for MONSTER, referring to $\boldsymbol{W}$ not ρ) which cannot incorporate the LD correlation information. We also see that, by retaining appropriate number of principal components, PC-ABT is able to achieve higher power than ABT in most of the cases. The power difference is especially highlighted for Configurations C2 and C3 where considerable correlation exists among the variants. This is because the genotype data in these configurations can be efficiently represented in the principal component space and hence truncated PCA helps improve power by reducing the df of the null χ^2 distribution.

4.3.3 Analysis of Rare Variant Association in GO-ESP Data

We focus on rare variant (defined as MAF < 0.05) association in the GO-ESP Data. After excluding genes with fewer than four rare variants, we applied five methods: FBT, famSKAT, ABT, MONSTER, and PC-ABT to identify genes associated with the average systolic blood pressure. A total of 18,864 genes were analyzed (see Sect. 4.2). Figure 4.4 presents the quantile-quantile (Q-Q) plot of the p-values calculated by using the five methods. For FBT, famSKAT, and MONSTER, the Madsen-Browning weights (Madsen and Browning 2009) were used. For PC-ABT, we chose the number of principal components to guarantee that the total PVE in the genotype data >95%. Given the sample size is relatively small, no p-value achieved the adjusted genome-wide significance of 2.65×10^{-6} based on Bonferroni correction. We observe that the Q-Q plots of FBT, famSKAT, and PC-ABT were close to the 45° line, suggesting that the these methods worked well and properly controlled type I error rates. The Q-Q plot of ABT was skewed downward, indicating the conservativeness of this method, and the Q-Q plot of MONSTER had a slightly anticonservative pattern.

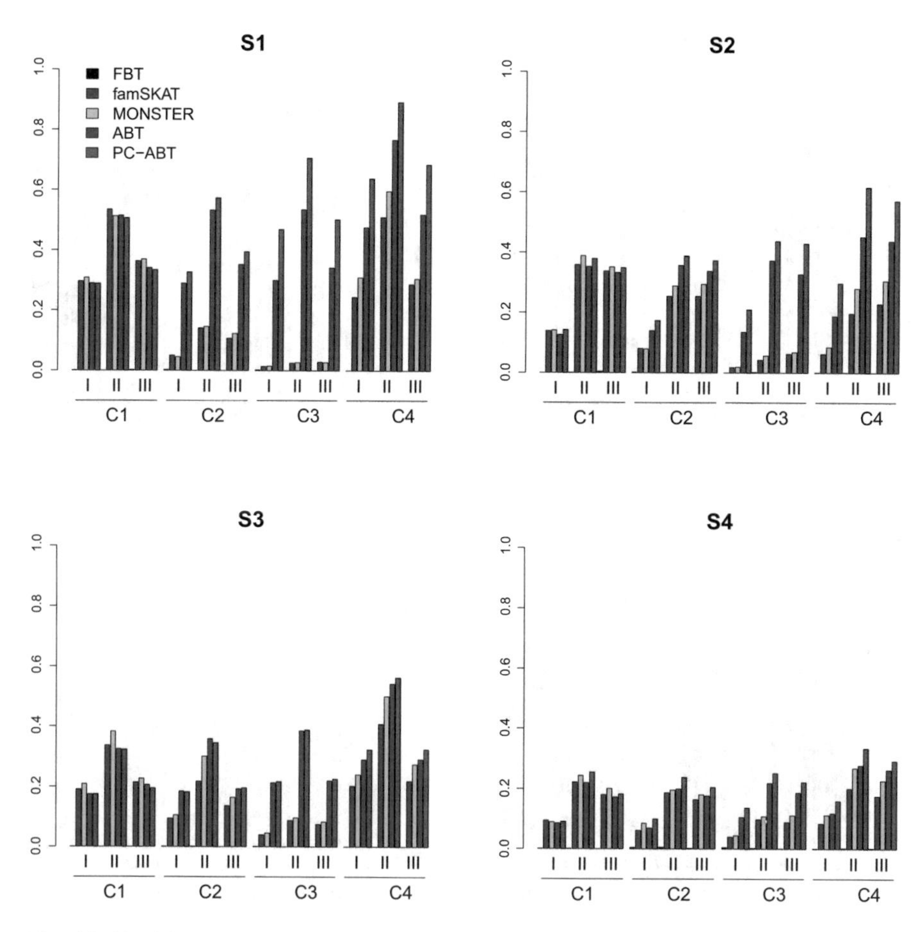

Fig. 4.2 Empirical power of FBT, famSKAT, MONSTER, ABT, and PC-ABT, at $\alpha = 0.001$, based on 1000 simulated data replicates. S1: unrelated individuals and common variants; S2: unrelated individuals and rare variants; S3: related individuals and common variants; S4: related individuals and rare variants; C1: negligible LD; C2: moderate LD; C3: high LD; C4: LD based on real data; I: 10%/10%/80%; II: 20%/20%/60%; III: 30%/30%/40%

Table 4.3 reports ten top-ranked PC-ABT p-values for testing the association between systolic blood pressure and gene regions on all 22 chromosomes in the GO-ESP data. Comparing the p-values obtained from five methods, we see that the FBT p-values are quite different from those obtained from kernel-based methods: famSKAT, MONSTER, ABT, and PC-ABT. This indicates that FBT and the other four methods evaluate different aspects of association patterns. The famSKAT p-values show remarkable correlations with those obtained from MONSTER (empirical correlation between $-\log_{10} pval_{\text{famSKAT}}$ and $-\log_{10} pval_{\text{MONSTER}}$ calculated from all 18,864 genes is 0.976), indicating that MONSTER, as a

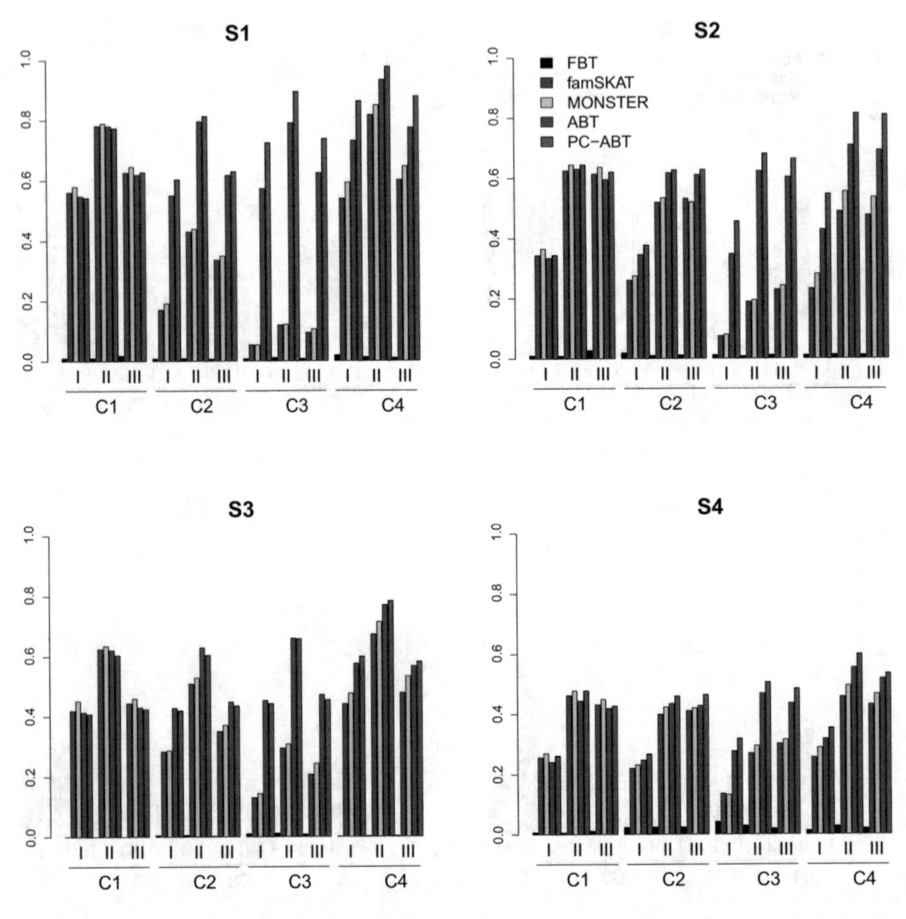

Fig. 4.3 Empirical power of FBT, famSKAT, MONSTER, ABT, and PC-ABT, at $\alpha = 0.01$, based on 1000 simulated data replicates. S1: unrelated individuals and common variants; S2: unrelated individuals and rare variants; S3: related individuals and common variants; S4: related individuals and rare variants; C1: negligible LD; C2: moderate LD; C3: high LD; C4: LD based on real data; I: 10%/10%/80%; II: 20%/20%/60%; III: 30%/30%/40%

unified method formed by combining burden and kernel tests, tends to weight more on its kernel component rather than burden component. With the number of principal components chosen according to PVE >95%, there are 17 genes with p-values $<10^{-3}$ by PC-ABT but only 3 by ABT. Among the top-ranked genes, *ALOX12* has been found involved in the regulation of key oxylipin metabolic genes in circulating peripheral blood mononuclear cells (Berthelot et al. 2015) and in angiotensin-II induced signaling in vascular smooth muscle cells (VSMCs) (Weisinger et al. 2007). *THBS1* (*TSP1*) is an important regulator of VSMC physiology. This gene has been found to significantly alter the expression of microRNAs in VSMCs, which indicates possible mechanism by which THBS1 contributes to atherosclerosis and intimal hyperplasia (Maier et al. 2016).

Table 4.3 Analysis of the systolic blood pressure data from GO-ESP using different tests: FBT, famSKAT, ABT, MONSTER, and PC-ABT

Gene	Chr	# of SNPs	*P*-value calculated by using				
			FBT	famSKAT	MONSTER	ABT	PC-ABT
ALOX12	17	30	8.72×10^{-2}	3.77×10^{-2}	3.61×10^{-2}	6.99×10^{-4}	1.94×10^{-5}
RIPK4	21	37	1.15×10^{-1}	1.15×10^{-1}	1.25×10^{-1}	2.30×10^{-2}	1.80×10^{-4}
XKR5	8	24	9.61×10^{-4}	6.30×10^{-3}	6.54×10^{-3}	2.26×10^{-2}	2.30×10^{-4}
JAK3	19	38	2.26×10^{-4}	6.03×10^{-6}	4.37×10^{-6}	6.80×10^{-3}	3.56×10^{-4}
TSGA13	7	11	7.32×10^{-1}	1.75×10^{-3}	1.02×10^{-3}	2.32×10^{-3}	3.58×10^{-4}
CARNS1	11	18	1.62×10^{-1}	2.56×10^{-2}	2.24×10^{-2}	1.31×10^{-2}	5.18×10^{-4}
THBS1	15	36	4.55×10^{-1}	7.75×10^{-2}	7.62×10^{-2}	2.55×10^{-2}	5.45×10^{-4}
CHODL	21	9	3.26×10^{-1}	4.79×10^{-3}	4.88×10^{-3}	2.04×10^{-2}	6.54×10^{-4}
SMAP2	1	14	2.88×10^{-1}	7.24×10^{-2}	7.53×10^{-2}	3.57×10^{-3}	6.74×10^{-4}
OR51E2	11	8	9.74×10^{-1}	9.69×10^{-3}	8.43×10^{-3}	3.43×10^{-3}	7.42×10^{-4}

Systolic blood pressure was averaged across multiple time points, log-transformed, and adjusted for age, sex, BMI, smoking status, smoking history, and glucose. For FBT, famSKAT, and MONSTER, the Madsen-Browning weights were used. For PC-ABT, the number of principal components was chosen to guarantee that the total percent of variance explained in the genotype data >95%

4.4 Discussion

Gene-based association testing can be constructed by pooling a set of univariate tests on individual variants within gene region. A straightforward approach is by using burden tests which collapse multiple variants into a single genetic burden score. In order to achieve a powerful gene-based test, two important issues during collapsing need to be taken into account: (1) The weights should be data-driven instead of prescribed so as to accommodate the presence of both risk and protective variants; (2) Since the effective number of tests per gene depends highly on the LD correlation among variants, it is crucial to appropriately adjust the degree of freedom of the null distribution in the test to avoid possible power loss. To address these two issues, we propose PC-ABT, a novel principal-component-based adaptive-weight burden test, which incorporates the complex genotypic correlations to improve power. Compared with other multiple variant tests, PC-ABT is advantageous in the following aspects: First, it uses a retrospective model to directly characterize genotypic correlations caused by both familial relation and LD, thereby overcoming the deficiency of existing prospective-model-based tests. Moreover,

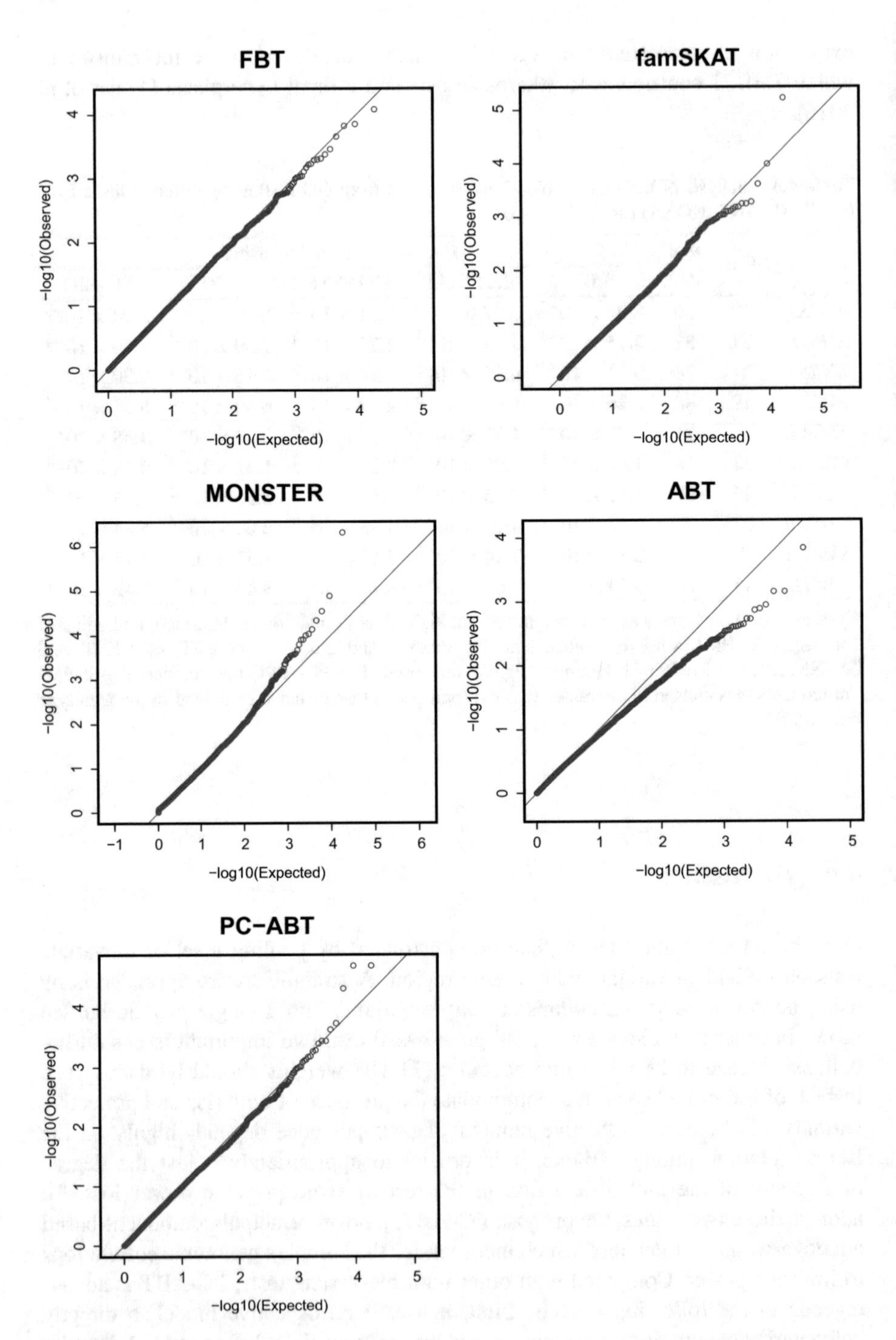

Fig. 4.4 Q-Q plots of the observed versus expected *p*-values for the GO-ESP exome sequence data. The x axis represents $-\log_{10}$(expected *p*-values), and the y axis represents $-\log_{10}$(observed *p*-values). A total of 18,864 genes with at least four rare variants were tested for associations with systolic blood pressure using the GO-ESP data, based on five methods: FBT, famSKAT, MONSTER, ABT, and PC-ABT

modeling phenotypes as fixed is also theoretically appealing because it makes fewer assumptions about phenotypic covariance structure (Price et al. 2011) and is more convenient to incorporate partially missing data in genotypes (Jakobsdottir and McPeek 2013; McPeek 2012; Thornton and McPeek 2007). Second, by maximizing the fixed-weight burden statistic, PC-ABT is able to assign weights that are adaptive to the direction of individual variant effects. Third, PC-ABT represents the LD covariance matrix by an efficient way through principal component decomposition, which reduces the df of the null distribution and eventually helps achieve increased power.

We demonstrate the performance of PC-ABT in terms of type I error and empirical power through extensive simulations based on different genotypic scenarios and LD configurations. Our simulation results show that, PC-ABT correctly controls the type I error, and is generally more powerful than other multiple variant tests under various simulation settings. We note that since PC-ABT and ABT are based on burden tests using adaptive weights, they essentially belong to the kernel test category. This has been shown in Sect. 4.2.2 as a generalization of the result by Derkach et al. (2013) and Li and Lagakos (2006). Special attention needs to be paid on the null distributions of ABT and PC-ABT statistics. As clarified in Sect. 4.2.2, when condition (4.3) for deriving FBT (hence ABT) is satisfied, i.e., when the covariance structure in genotype data follows a kronecker product form, and the LD correlation matrix $\boldsymbol{R}$ is known, the ABT statistic has a null distribution of χ_m^2 (see Appendices 1 and 4 for justification), and hence the PC-ABT null distribution is χ_q^2. However, there are cases in real data applications or simulations where condition (4.3) may not be satisfied (e.g., the correlations caused by LD and sample relatedness are not separable). Therefore in practice, we suggest to use the mixture of χ_1^2 as the null distribution, which follows by treating ABT as a special famSKAT test with weights $\boldsymbol{W}^{\#}$. Correspondingly, the null distribution of PC-ABT statistic is also mixture of χ_1^2 with the first q λ_i's being non-trivial, as noted in Sect. 4.2.3.

One critical problem remains in choosing the appropriate number of PCs when applying PC-ABT. Clearly if we keep all PCs in analysis, then PC-ABT is equivalent to ABT. In practice the number of PCs can be chosen such that the total percent of variance explained exceeds some threshold. This threshold may be determined depending on the (estimated) LD levels in the genotype data: the higher LD, the smaller threshold would be used.

Appendix 1: Description of MASTOR and Theoretical Justification of the Null Distribution of S_{ABT}

MASTOR (Jakobsdottir and McPeek 2013) is a retrospective, quasi-likelihood score test for testing single-variant association with a quantitative trait in samples with related individuals. Considering a biallelic genetic variant $\boldsymbol{X}$ of interest (an example in the general setting described in Sect. 4.2.1 is to let $\boldsymbol{X} = \boldsymbol{G}_j, 1 \leq j \leq m$), the MASTOR statistic (for complete data) takes the form

$$S_{MAS} = \frac{(\boldsymbol{V}^T \boldsymbol{X})^2}{(\boldsymbol{V}^T \boldsymbol{\Phi} \boldsymbol{V})\widehat{\sigma}_X^2}.$$

In this expression, $\boldsymbol{V} = \widehat{\boldsymbol{\Sigma}}_0^{-1}(\boldsymbol{Y} - \boldsymbol{Z}\widehat{\boldsymbol{\beta}}_0)$ is the transformed phenotypic residual obtained from the null model $\boldsymbol{Y} = \boldsymbol{Z}\boldsymbol{\beta}_0 + \boldsymbol{\epsilon}, \boldsymbol{\epsilon} \sim N(\boldsymbol{0}, \boldsymbol{\Sigma}_0)$, where $\boldsymbol{\beta}_0$ represents the coefficient of regressing quantitative trait $\boldsymbol{Y}$ on non-genetic covariates $\boldsymbol{Z}$, and $\boldsymbol{\Sigma}_0$ is the trait covariance matrix under the null, usually with a variance component form $\sigma_e^2 \boldsymbol{I} + \sigma_a^2 \boldsymbol{\Phi}$. The variance of variant $\boldsymbol{X}$ is denoted by σ_X^2. When Hardy-Weinberg equilibrium is assumed for this variant, σ_X^2 can be estimated by $\widehat{\sigma}_X^2 = \widehat{p}(1 - \widehat{p})/2$, where $\widehat{p} = (\boldsymbol{1}^T \boldsymbol{\Phi}^{-1} \boldsymbol{1})^{-1} \boldsymbol{1}^T \boldsymbol{\Phi}^{-1} \boldsymbol{X}$ is the best linear unbiased estimator (McPeek et al. 2004) of the allele frequency p of $\boldsymbol{X}$, and $\boldsymbol{1}$ denotes a vector with every element equal to 1.

Now in Sect. 4.2.2, we have obtained the ABT statistic

$$S_{ABT} = \frac{\boldsymbol{V}^T \boldsymbol{G}(\widehat{\boldsymbol{D}}\boldsymbol{R}\widehat{\boldsymbol{D}})^{-1} \boldsymbol{G}^T \boldsymbol{V}}{\boldsymbol{V}^T \boldsymbol{\Phi} \boldsymbol{V}}.$$

Let $\widetilde{\boldsymbol{G}} = \boldsymbol{G}(\widehat{\boldsymbol{D}}\boldsymbol{R}\widehat{\boldsymbol{D}})^{-1/2}$ be a decorrelated version of the genotype matrix in which the across-column covariance has been transformed to identity, and let $\widetilde{\boldsymbol{G}}_j$ be the jth column of $\widetilde{\boldsymbol{G}}$. By linear algebra,

$$S_{ABT} = \sum_{j=1}^{m} \frac{(\boldsymbol{V}^T \widetilde{\boldsymbol{G}}_j)^2}{\boldsymbol{V}^T \boldsymbol{\Phi} \boldsymbol{V}}.$$

This is essentially the summation of m independent MASTOR statistics (in observing the uncorrelatedness and joint normality of $\boldsymbol{V}^T \widetilde{\boldsymbol{G}}_j$), each formulated from a transformed variant $\widetilde{\boldsymbol{G}}_j$ (note the variance estimate is 1 after transformation). Hence S_{ABT} follows χ_m^2 distribution under the null hypothesis.

Appendix 2: Additional Simulation Results Show That the Data-Driven Weights W^* Is Adaptive to the Direction of True Genetic Effects

In order to understand how the data-driven weights $\boldsymbol{W}^*$ (defined in Eq. (4.5) of the main text) help gain power in association testing, we compare the signs of $\boldsymbol{W}^*$ to those of the genetic effects $\boldsymbol{\gamma}$ using the simulated data sets in the power analysis.

Figure 4.5, Panels a–d, present boxplots of the weights $\boldsymbol{W}^*$ based on 5000 simulated data replicates in Scenario S2 with genetic effect Setting III, for LD Configurations C1–C4, respectively. We note that, in this setting, the first 30% components of $\boldsymbol{\gamma}$ are set to be positive, the next 30% are negative, and the remaining 40% are zeros. The boxplots clearly demonstrates that on average, the weights $\boldsymbol{W}^*$ is able to track the direction of true genetic effects, thus result in stronger association on the weighted sum genetic score.

Appendix 3: Additional Simulation Results Show the Relation Between the ABT Statistic and the famSKAT Statistic

We show in Fig. 4.6, Panels a–d, the scatter plots of the numerator of the ABT statistic vs. the famSKAT statistic based on 5000 simulated data replicates in Scenario S3 with genetic effect Setting II, for LD Configurations C1–C4, respectively. We observe that, when the LD correlation is negligible (Panel a), the numerator of the ABT statistic behaves similarly as the famSKAT statistic because in Eq. (4.6) of the main text, $(\widehat{\boldsymbol{D}}\boldsymbol{R}\widehat{\boldsymbol{D}})^{-1}$ is equivalent to the Madsen-Browning weights used in calculating the famSKAT statistic. As the LD correlation increases (Panels b, c, and d), the two statistics become less and less consistent because in calculating the famSKAT statistic, the Madsen-Browning weights only depend on individual variants, whereas in calculating the ABT statistic, the weight of an individual variant statistic is also affected by other variants on linked sites, as seen from the weight matrix $(\widehat{\boldsymbol{D}}\boldsymbol{R}\widehat{\boldsymbol{D}})^{-1}$ in Eq. (4.6) of the main text.

Appendix 4: Additional Simulation Results to Validate the Asymptotic Null Distribution of S_{PC-ABT} via Permutation Based Approach

We perform 1000 permutations to the simulated data under Scenario S1 (unrelated individuals and common variants) and configuration C3 (strong LD with $\eta = 0.7$). Figure 4.7 shows the asymptotic null distributions of S_{PC-ABT} for the number of principal components $q = 1, 25$, and 50, together with the corresponding empirical CDFs obtained via permutation. Note that two different asymptotic distributions are shown in this figure, one is χ_q^2, the other is a mixture of χ_1^2 distribution, obtained by applying adaptive weights $W^{\#}$ in the famSKAT method. In Fig. 4.8, panels a, b, and

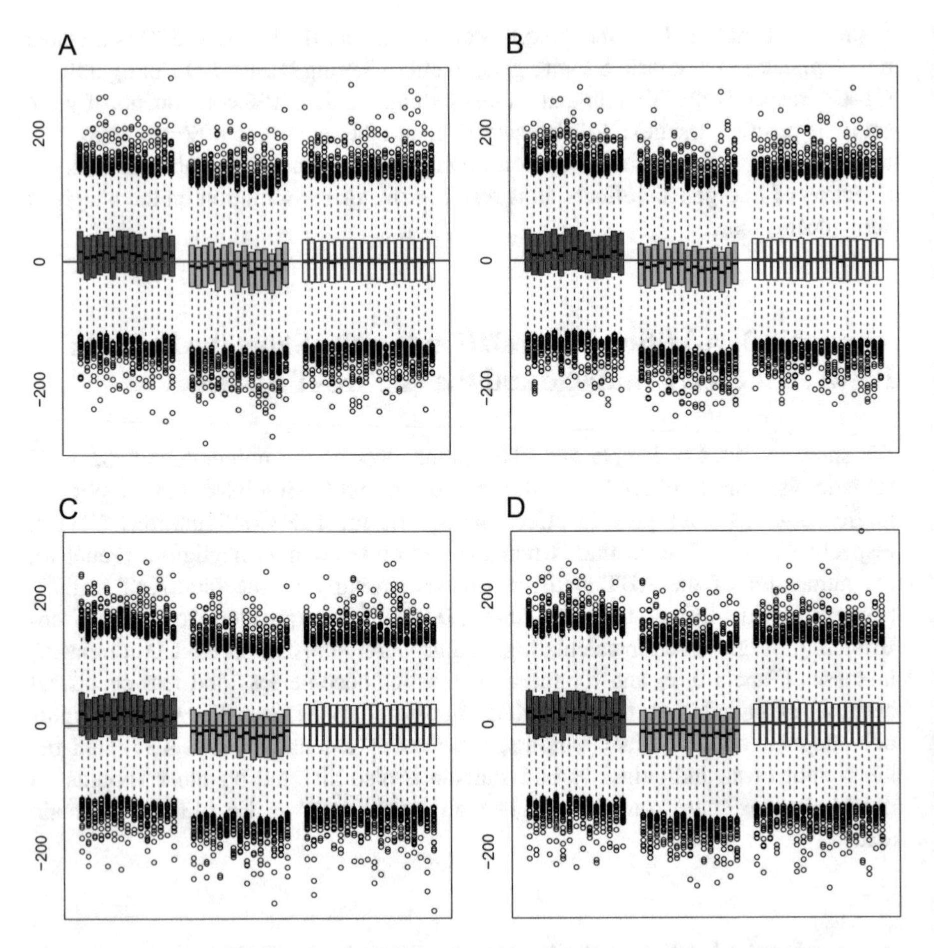

Fig. 4.5 Boxplot of W^* based on 5000 simulated data replicates in Scenario S2 with genetic effect Setting III. The adaptive weights of risk, protective, and neutral variants are marked with red, green, and white color, respectively. Panel **a**: Configuration C1; Panel **b**: Configuration C2; Panel **c**: Configuration C3; Panel **d**: Configuration C4

c, we compare in log scale the empirical p-values via permutation based approach against the p-values from the asymptotic distribution (mixture of χ_1^2) for the number of principal components $q = 1, 25$, and 50, respectively. Panel d of Fig. 4.8 further reports the correlation between $-\log_{10}$(empirical p-values via permutation) and $-\log_{10}$(p-values based on the asymptotic distribution) for the number of principal components $q = 1, 2, \cdots, 50$.

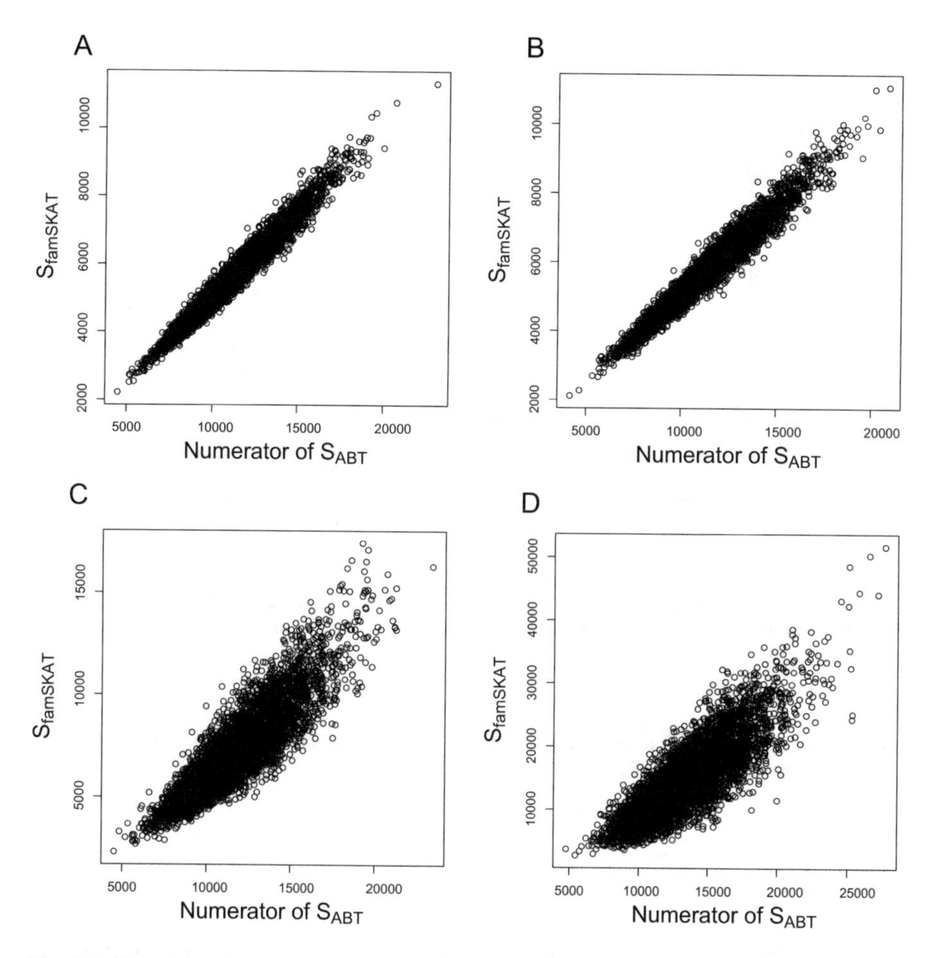

Fig. 4.6 Comparison between $S_{famSKAT}$ and the numerator of S_{ABT} based on 5000 simulated data replicates in Scenario S3 with genetic effect Setting II. Panel **a**: Configuration C1; Panel **b**: Configuration C2; Panel **c**: Configuration C3; Panel **d**: Configuration C4

Appendix 5: Additional Simulation Results for Type I Error Evaluation

We provide additional simulation results for type I error evaluation. Table 4.4 lists the empirical type I error rates of five testing methods: FBT, famSKAT, ABT, MONSTER, and PC-ABT for the combinations of four scenarios (S1, S2, S3, and S4) and four LD configurations (C1, C2, C3, and C4), based on 20,000 simulated data replicates. Figures 4.9, 4.10, 4.11, and 4.12 show the Q-Q plots of the PC-ABT p-values under the null hypothesis for Scenarios S1, S2, S3, and S4, respectively. The number of principal components is chosen to guarantee that the total percent variance explained (PVE) >90%.

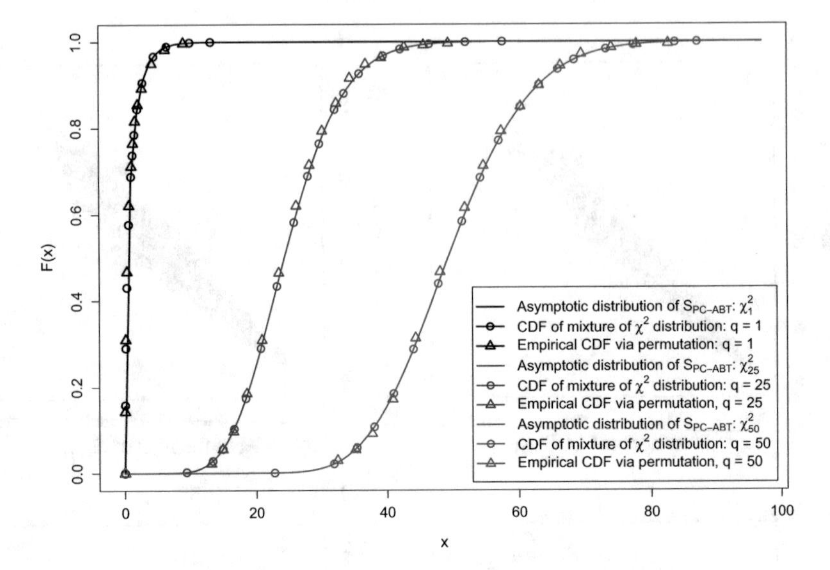

Fig. 4.7 Asymptotic null distribution and permutation based empirical distribution of S_{PC-ABT} (for the number of principal components $q = 1, 25$, and 50) under Scenario S1 and LD configuration C3

Acknowledgements This research was funded by 4-VA, a collaborative partnership for advancing the Commonwealth of Virginia.

References

Asimit, J., Zeggini, E.: Rare variant association analysis methods for complex traits. Ann. Rev. Genet. **44**, 293–308 (2010)

Berthelot, C.C., et al.: Changes in PTGS1 and ALOX12 gene expression in peripheral blood mononuclear cells are associated with changes in arachidonic acid, oxylipins, and oxylipin/fatty acid ratios in response to Omega-3 fatty acid supplementation. PLoS One **10**(12), e0144,996 (2015)

Chen, H., Meigs, J.B., Dupuis, J.: Sequence kernel association test for quantitative traits in family samples. Genet. Epidemiol. **37**(2), 196–204 (2013)

Cui, J.S., Hopper, J.L., Harrap, S.B.: Antihypertensive treatments obscure familial contributions to blood pressure variation. Hypertension **41**(2), 207–210 (2003)

Derkach, A., Lawless, J.F., Sun, L.: Assessment of pooled association tests for rare variants within a unified framework. Stat. Sci. **29**(2), 302–321 (2013)

Fang, S., Zhang, S., Sha, Q.: Detecting association of rare variants by testing an optimally weighted combination of variants for quantitative traits in general families. Ann. Hum. Genet. **77**(6), 524–534 (2014)

Fuentes, M.: Testing for separability of spatial-temporal covariance functions. J. Stat. Plan. Inference. **136**, 447–466 (2006)

Gauderman, W.J., Murcray, C., Gilliland, F., Conti, D.V.: Testing association between disease and multiple SNPs in a candidate gene. Genet. Epidemiol. **31**(5), 383–395 (2007)

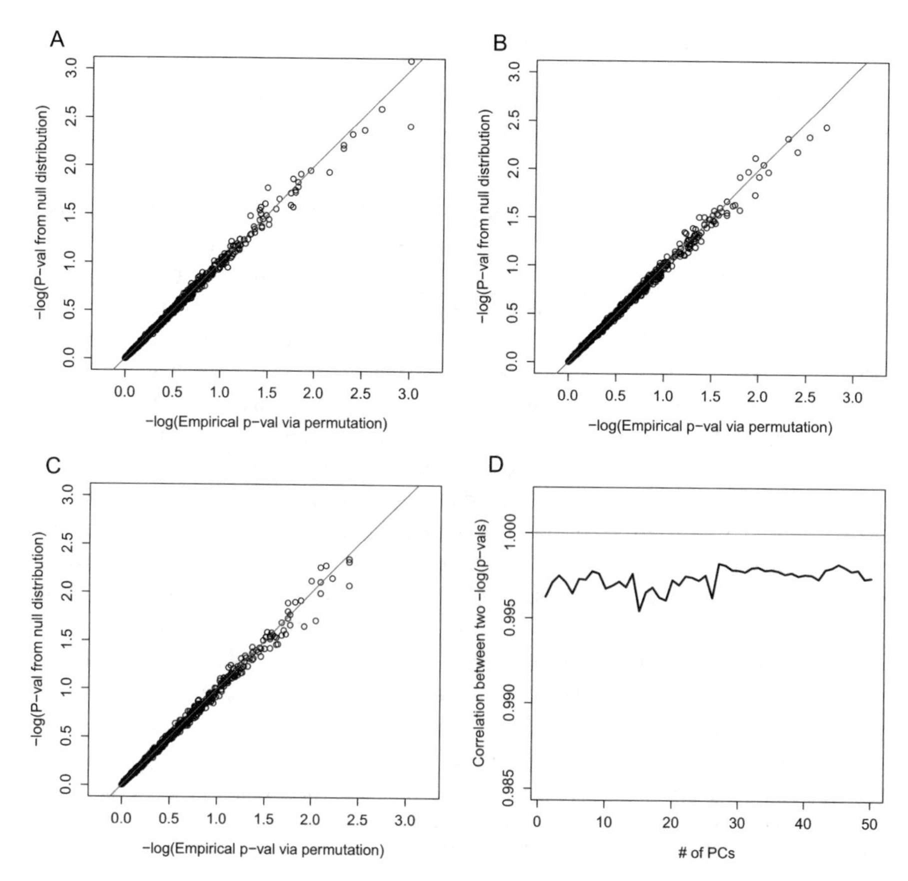

Fig. 4.8 Comparing empirical p-values via permutation and p-values based on the asymptotic distribution. Panel **a**: scatter plot in log scale for the number of principal components $q = 1$; Panel **b**: for $q = 25$; Panel **c**: for $q = 50$; Panel **d**: correlation between $-\log_{10}$(empirical p-values via permutation) and $-\log_{10}(p$-values based on the asymptotic distribution) for $q = 1, 2, \cdots, 50$

Han, F., Pan, W.: A data-adaptive sum test for disease association with multiple common or rare variants. Hum. Hered. **70**(1), 42–54 (2010)

Jakobsdottir, J., McPeek, M.S.: Mastor: Mixed-model association mapping of quantitative traits in samples with related individuals. Am. J. Hum. Genet. **92**, 652–666 (2013)

Jiang, D., McPeek, M.S.: Robust rare variant association testing for quantitative traits in samples with related individuals. Genet. Epidemiol. **38**(1), 1–20 (2013)

Ladouceur, M., Dastani, Z., Aulchenko, Y.S., Greenwood, C.M., Richards, J.B.: The empirical power of rare variant association methods: Results from Sanger sequencing in 1998 individuals. PLoS Genet. **8**(2), e1002,496 (2012)

Lee, S., Emond, M.J., Bamshad, M.J., Barnes, K.C., Rieder, M.J., Nickerson, D.A., NHLBI GO Exome Sequencing Project-ESP Lung Project Team, Christiani, D.C., Wurfel, M.M., Lin, X.: Optimal unified approach for rare-variant association testing with application to small-sample case-control whole-exome sequencing studies. Am. J. Hum. Genet. **91**, 224–237 (2012)

Lee, S., Wu, M.C., Lin, X.: Optimal tests for rare variant effects in sequencing association studies. Biostatistics **13**(4), 762–775 (2013)

Table 4.4 Empirical type I error of five testing methods

Scenario	# of PCs	$\alpha = 0.001$				$\alpha = 0.01$			
		C1	C2	C3	C4	C1	C2	C3	C4
S1	FBT	0.001	0.00095	0.001	0.00095	0.0099	0.00995	0.01005	0.01
	famSKAT	0.001	0.001	0.0009	0.0009	0.01	0.01005	0.00985	0.0099
	ABT	0.0008	0.0011	0.00085	0.00105	0.01005	0.01	0.0098	0.0099
	MONSTER	0.0009	0.0009	0.0011	0.0009	0.00975	0.01005	0.01015	0.01005
	PC-ABT*	0.001	0.0011	0.00105	0.0009	0.01035	0.00985	0.0101	0.0105
S2	FBT	0.00085	0.00095	0.0009	0.00085	0.0099	0.00985	0.01	0.0098
	famSKAT	0.00105	0.0009	0.009	0.00095	0.00995	0.0099	0.01005	0.00995
	ABT	0.001	0.001	0.0011	0.0008	0.01005	0.0099	0.01	0.01015
	MONSTER	0.00105	0.00105	0.0008	0.001	0.0099	0.01	0.0098	0.0097
	PC-ABT*	0.00115	0.00095	0.00095	0.0008	0.0099	0.01	0.0099	0.01025
S3	FBT	0.00105	0.00105	0.001	0.001	0.01	0.0099	0.0099	0.0101
	famSKAT	0.00085	0.0009	0.00095	0.0009	0.00995	0.01	0.01	0.0103
	ABT	0.001	0.0009	0.0009	0.00095	0.01	0.01005	0.0101	0.0102
	MONSTER	0.00095	0.0009	0.00085	0.00085	0.01	0.0102	0.01	0.0101
	PC-ABT*	0.00095	0.00085	0.00095	0.00095	0.0102	0.0102	0.0103	0.0098
S4	FBT	0.0008	0.001	0.0009	0.00110	0.0099	0.00995	0.0098	0.01005
	famSKAT	0.001	0.00105	0.00105	0.001	0.0098	0.00995	0.01	0.0099
	ABT	0.0009	0.00085	0.00095	0.00095	0.0098	0.0099	0.00975	0.01015
	MONSTER	0.0009	0.00105	0.00085	0.001	0.01005	0.0099	0.0098	0.01005
	PC-ABT*	0.0009	0.00085	0.00125	0.0009	0.0102	0.0101	0.0098	0.0097

The type I error rate estimates are calculated as the proportion of p-values smaller than nominal under the null hypothesis based on 20,000 simulated data replicates. PC-ABT*: the number of principal components is chosen to guarantee that the total percent variance explained (PVE) >90%

Li, Q.H., Lagakos, S.W.: On the relationship between directional and omnibus statistical tests. Scand. J. Stat. **33**, 239–246 (2006)

Li, B., Leal, S.M.: Methods for detecting associations with rare variants for common diseases: application to analysis of sequence data. Am. J. Hum. Genet. **83**, 311–321 (2008)

Li, M.X., Gui, H.S., Kwan, J.S., Sham, P.C.: GATES: a rapid and powerful gene-based association test using extended Simes procedure. Am. J. Hum. Genet. **88**, 283–293 (2011)

Lin, D.Y., Tang, Z.Z.: A general framework for detecting disease associations with rare variants in sequencing studies. Am. J. Hum. Genet. **89**, 354–367 (2011)

Liu, D.J., Leal, S.M.: A novel adaptive method for the analysis of next-generation sequencing data to detect complex trait associations with rare variants due to gene main effects and interactions. PLoS Genet. **6**, e1001,156 (2010)

Ma, L., Clark, A.G., Keinan, A.: Gene-based testing of interactions in association studies of quantitative traits. PLoS Genet. **9**, e1003,321 (2013)

Madsen, B.E., Browning, S.R.: A groupwise association test for rare mutations using a weighted sum statistic. PLoS Genet. **5**, e1000,384 (2009)

Maier, K.G., Ruhle, B., Stein, J.J., Gentile, K.L., Middleton, F.A., Gahtan, V.: Thrombospondin-1 differentially regulates microRNAs in vascular smooth muscle cells. Mol. Cell. Biochem. **412**(1–2), 111–117 (2016)

Manolio, T.A.: Genomewide association studies and assessment of the risk of disease. N. Engl. J. Med. **363**(2), 166–176 (2010)

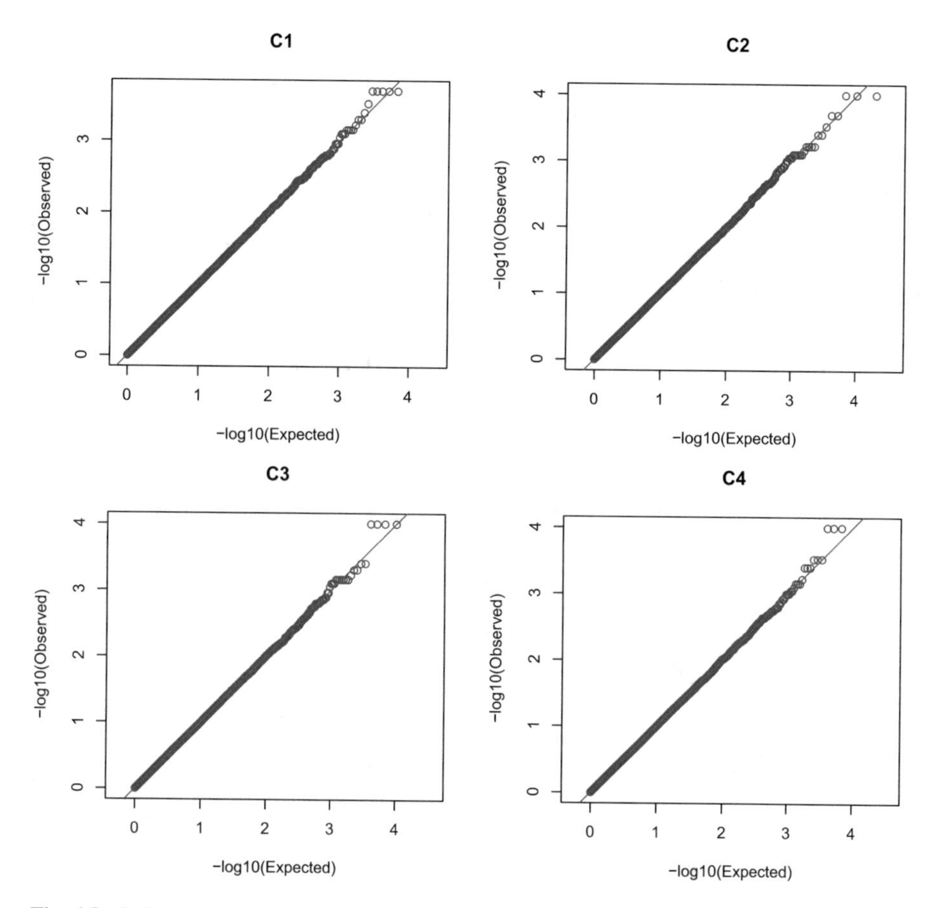

Fig. 4.9 Q-Q plots of the PC-ABT p-values under the null hypothesis for Scenario S1 based on 20,000 simulated data replicates. In each simulation, the number of principal components is chosen such that PVE $>90\%$

McCarthy, M.I., Abecasis, G.R., Cardon, L.R., Goldstein, D.B., Little, J., Ioannidis, J.P., Hirschhorn, J.N.: Genome-wide association studies for complex traits: consensus, uncertainty and challenges. Nat. Rev. Genet. **9**(5), 356–369 (2008)

McPeek, M.S.: BLUP genotype imputation for case control association testing with related individuals and missing data. J. Comp. Biol. **19**(6), 756–765 (2012)

McPeek, M.S., Wu, X., Ober, C.: Best linear unbiased allele-frequency estimation in complex pedigrees. Biometrics **60**, 359–367 (2004)

Morgenthaler, S., Thilly, W.G.: A strategy to discover genes that carry multi-allelic or mono-allelic risk for common diseases: a cohort allelic sums test (CAST). Mutat. Res. **615**, 28–56 (2007)

Neale, B.M., Sham, P.C.: The future of association studies: Gene-based analysis and replication. Am. J. Hum. Genet. **75**, 353–362 (2004)

Price, A.L., Kryukov, G.V., de Bakker, P.I., Purcell, S.M., Staples, J., Wei, L.J., Sunyaev, S.R.: Pooled association tests for rare variants in exon-resequencing studies. Am. J. Hum. Genet. **86**, 832–838 (2010)

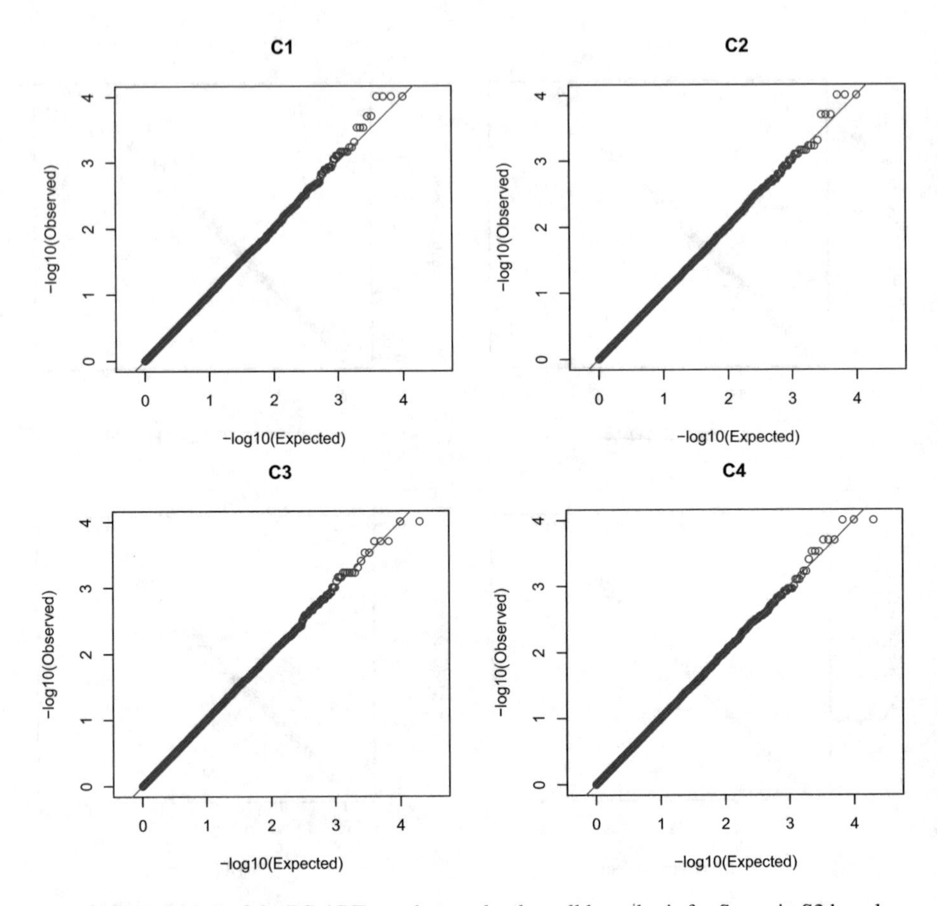

Fig. 4.10 Q-Q plots of the PC-ABT p-values under the null hypothesis for Scenario S2 based on 20,000 simulated data replicates. In each simulation, the number of principal components is chosen such that PVE >90%

Price, A.L., Zaitlen, N.A., Reich, D., Patterson, N.: New approaches to population stratification in genome-wide association studies. Nat. Rev. Genet. **11**(7), 459–463 (2011)

Schaid, D.J., McDonnell, S.K., Sinnwell, J.P., Thibodeau, S.M.: Multiple genetic variant association testing by collapsing and kernel methods with pedigree or population structured data. Genet. Epidemiol. **37**(5), 409–418 (2013)

Schifano, E.D., Epstein, M.P., Bielak, L.F., Jhun, M.A., Kardia, S.L., Peyser, P.A., Lin, X.: SNP set association analysis for familial data. Genet. Epidemiol. **36**(8), 797–810 (2012)

Sha, Q., Wang, X., Wang, X., Zhang, S.: Detecting association of rare and common variants by testing an optimally weighted combination of variants. Genet. Epidemiol. **36**(6), 561–571 (2012)

Sha, Q., Zhang, S.: A novel test for testing the optimally weighted combination of rare and common variants based on data of parents and affected children. Genet. Epidemiol. **38**(2), 135–143 (2014)

Splansky, G.L., et al.: The third generation cohort of the National Heart, Lung, and Blood Institute's Framingham Heart Study: design, recruitment, and initial examination. Am. J. Epidemiol. **165**(11), 1328–1335 (2007)

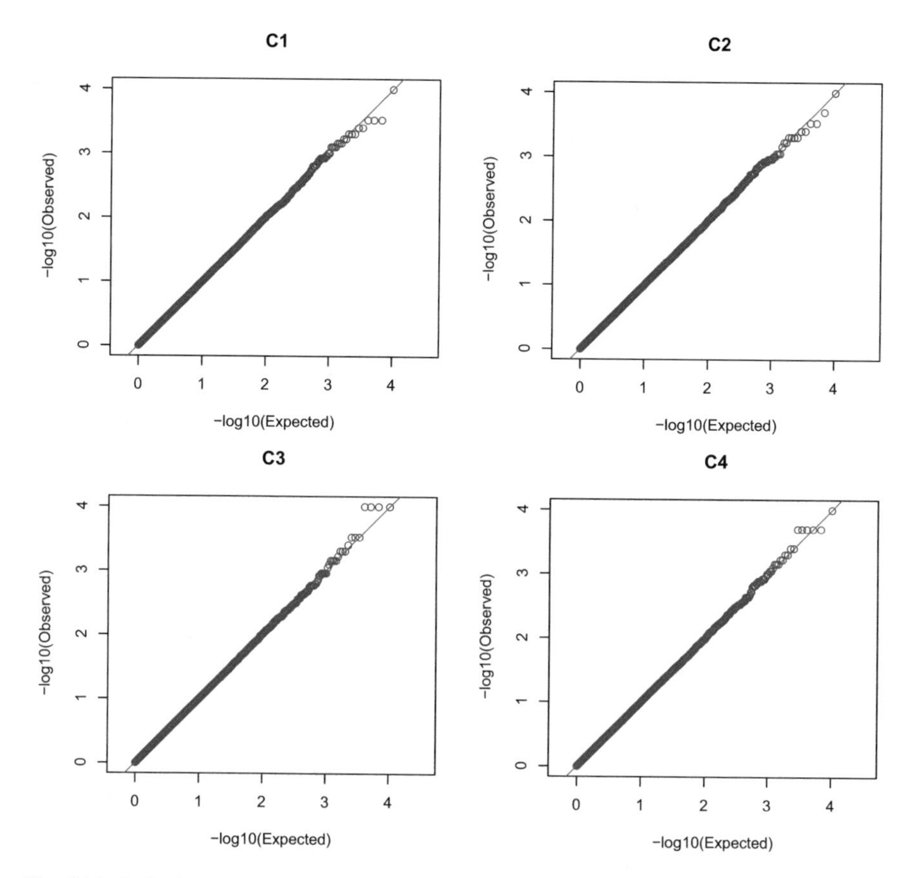

Fig. 4.11 Q-Q plots of the PC-ABT p-values under the null hypothesis for Scenario S3 based on 20,000 simulated data replicates. In each simulation, the number of principal components is chosen such that PVE $>90\%$

Srivastava, M.S., von Rosen, T., von Rosen, D.: Models with a Kronecker product covariance structure: estimation and testing. Math. Methods Stat. **17**(4), 357–370 (2008)

The 1000 Genomes Project Consortium: A map of human genome variation from population-scale sequencing. Nature **467**, 1061–1073 (2010)

Thornton, T., McPeek, M.S.: Case-control association testing with related individuals: a more powerful quasi-likelihood score test. Am. J. Hum. Genet. **81**, 321–337 (2007)

Thornton, T., McPeek, M.S.: ROADTRIPS: Case-control association testing with partially or completely unknown population and pedigree structure. Am. J. Hum. Genet. **86**, 172–184 (2010)

Tobin, M.D., Sheehan, N.A., Scurrah, K.J., Burton, P.R.: Adjusting for treatment effects in studies of quantitative traits: antihypertensive therapy and systolic blood pressure. Stat. Med. **24**, 2911–2935 (2005)

Wang, Y., Chen, Y.H., Yang, Q.: Joint rare variant association test of the average and individual effects for sequencing studies. PLoS One **7**, e32,485 (2012)

Wang, X., Morris, N.J., Zhu, X., Elston, R.C.: A variance component based multi-marker association test using family and unrelated data. BMC Genet. **14**, 17 (2013)

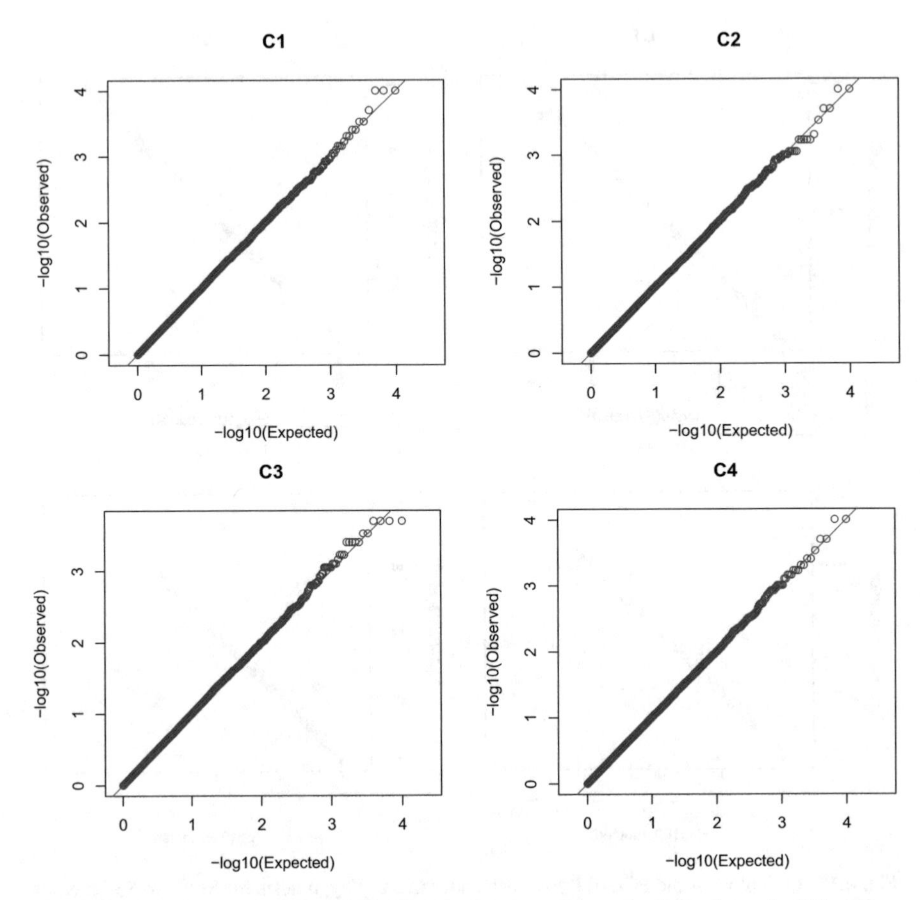

Fig. 4.12 Q-Q plots of the PC-ABT p-values under the null hypothesis for Scenario S4 based on 20,000 simulated data replicates. In each simulation, the number of principal components is chosen such that PVE >90%

Wang, X., Lee, S., Zhu, X., Redline, S., Lin, X.: GEE-based SNP set association test for continuous and discrete traits in family based association studies. Genet. Epidemiol. **37**(8), 778–786 (2014)

Weisinger, G., Limor, R., Marcus-Perlman, Y., Knoll, E., Kohen, F., Schinder, V., Firer, M., Stern, N.: 12S-lipoxygenase protein associates with alpha-actin fibers in human umbilical artery vascular smooth muscle cells. Biochem. Biophys. Res. Commun. **356**(3), 554–560 (2007)

Wu, M.C., Kraft, P., Epstein, M.P., Taylor, D.M., Chanock, S.J., Hunter, D.J., Lin, X.: Powerful SNP-set analysis for case-control genome-wide association studies. Am. J. Hum. Genet. **86**, 929–942 (2010)

Wu, M.C., Lee, S., Cai, T., Li, Y., Boehnke, M., Lin, X.: Rare-variant association testing for sequencing data with the sequence kernel association test. Am. J. Hum. Genet. **89**, 82–93 (2011)

Zhu, Y., Xiong, M.: Family-based association studies for next-generation sequencing. Am. J. Hum. Genet. **90**, 1028–1045 (2012)

Chapter 5
Inference of Gene Regulatory Network Through Adaptive Dynamic Bayesian Network Modeling

Yaqun Wang, Scott A. Berceli, Marc Garbey, and Rongling Wu

5.1 Background

Thousands of genes on the genome encode the products essential for cell division and differentiation toward the phenotypic formation of organisms. How the properties of these products, including abundance, mutual interactions, and temporal pattern, determine the process of life is governed by regulatory networks of genes. A gene regulatory network (GRN) is formed by a set of genes in a cell which interact with each other through their RNA and protein products and regulated by the transcription factors that activate the expression of particular genes (Brazhnik et al. 2002). Knowledge about the structure and organization of GRN can help us identify the causal regulations involved in metabolic and physiological processes within cells. With the availability of high-throughput data, increasing efforts have been made to reconstruct GRN by developing either model based or machine

Y. Wang
Department of Statistics, The Pennsylvania State University, University Park, PA, USA

Department of Biostatistics, Rutgers, The State University of New Jersey, New Brunswick, NJ, USA
e-mail: yw505@sph.rutgers.edu

S. A. Berceli
Department of Surgery, University of Florida, Gainesville, FL, USA
e-mail: bercesa@surgery.ufl.edu

M. Garbey
Department of Computer Science, University of Houston, Houston, TX, USA
e-mail: garbey@cs.uh.edu

R. Wu (✉)
Department of Public Health Sciences, The Pennsylvania State University, Hershey, PA, USA
e-mail: rwu@phs.psu.edu

L. Zhang et al. (eds.), *Contemporary Biostatistics with Biopharmaceutical Applications*, ICSA Book Series in Statistics,
https://doi.org/10.1007/978-3-030-15310-6_5

learning based approaches (Barabasi et al. 2011; Zhu et al. 2012; Zhang et al. 2013; Wang et al. 2013). These approaches have played an important role in referring the complex regulatory mechanisms that underlie biological functions and phenotypic characteristics (Gerstein et al. 2012; Hurley et al. 2012; Ortiz-Gutiérrez et al. 2015). To better separate direct regulations from indirect ones among genes within a GRN, Zhang et al. (2015) proposed a concept of conditional mutual inclusive information and implemented it into a computer algorithm for quantifying the mutual information between two genes given a third one.

Given that life is a dynamic process (de Lichtenberg et al. 2005), a considerable body of modeling studies has begun to reconstruct dynamic GRN from expression data measured across a time and space scale (Li et al. 2011). The formation of any biological characteristics activated by developmental signals is contingent on dynamic changes of gene expression. For example, in flowering plants, embryogenesis undergoes three distinct phases, asymmetric cell divisions to establish apical–basal polarity (early phase), the initiation of major organs and primordia (intermediate phase) and the mature embryo (late phase) (De Smet et al. 2010). By genome-wide profiling of gene expression during a complete developmental process from the zygote to the mature embryo in *Arabidopsis thaliana*, Xiang et al. (2011) constructed stage-specific regulatory networks, which provide an important foundation for understanding the dynamic pattern of pathway interactions during embryogenesis. The application of stage-specific regulatory networks to study the genetic underpinnings of trait development has now become a routine approach in a wide range of biological areas from plant biology to cancer biology (Zhang et al. 2015; Yosef et al. 2013; Kourou et al. 2015).

Approaches for reconstructing dynamic GRN from time course gene expression data have been well developed, including dynamic Boolean networks and probabilistic Boolean networks (Akutsu et al. 2000; Martin et al. 2007) and dynamic Bayesian networks (Murphy and Mian 1999; Friedman et al. 2000; Zou and Conzen 2005; Ogami et al. 2012; Godsey 2013; Kim et al. 2003) among others. By integrating expression data measured at multiple time points, these approaches have been used to infer the temporal change of the structure and topological features of multiple interactions within genomic networks during a period of biological process. However, they may suffer the limitation of being unable to manipulate sparse, unevenly-spaced expression data which are quite popular in practice. On the other hand, there has been increasing recognition of using multiple different experiments to reconstruct a comprehensive GRN, in which data were rarely measured at the same schedule (Hecker et al. 2009; Greenfield et al. 2010). As a consequence, the statistical issue of simultaneous use and modeling of irregular data from different experiments should be addressed.

In this article, we present and validate a computational procedure for dynamic GRN reconstruction from sparse, irregular gene expression data by interpolating those missing points in time course measurements. The idea of interpolation used to model GRN is not new. Wessels et al. (2001) and Bansal et al. (2006) proposed cubic interpolation for GRN modeling. Yu et al. (2004) devised a linear interpolation method for dynamic Bayesian network construction. By implementing a parametric (such as Fourier series approximation) or nonparametric (such as Legendre orthogonal polynomials) function whose optimal order is determined by information criteria, our interpolation approach is adaptive, assuring the best function to fit a given expression dataset and, thus, capturing dynamic features of genes precisely. Different from the previous work, we integrate functional clustering (Wang et al. 2011) into the DBN modeling framework by which to infer GRN based on functional clusters of genes. Functional clustering classifies gene profiles into distinct categories according to their similarity, and estimates a functional nonlinear curve for the mean dynamic expression of genes within the same cluster. By interpolating missing data based on the functional curve, an evenly-spaced, regular time course data can be obtained from which DBN is used to infer GRN among gene clusters. Our approach can handle any dynamic gene expression data, regardless of its sparsity and irregularity, thereby providing a broader application in computational biology. In addition, the clustering method jointly model gene expression from multiple environments and make it possible to compare regulation effects between genes in distinct conditions. Our model focuses on microarray gene expression data and has been implemented in R combining with Matlab. The codes are available on website https://sites.google.com/site/yxw179/software.

5.2 Methods

5.2.1 Dynamic Bayesian Network Modeling

Consider a hypothetical gene network (Fig. 5.1; Brazhnik et al. 2002), in which three different levels of regulation exist: genes, proteins and metabolites. Here we assume that genes do not directly affect each other but interact through the action of their specific products, proteins, metabolites, or protein-metabolite complexes. Gene 2 is regulated by the protein product of the gene 1 and by the complex 3–4 formed by the products of gene 3 and gene 4. The regulation of gene 4 is made by the metabolite 2 which in turn is produced by protein 2. Based on these webs of regulation, we can construct a gene network which describes how one gene interacts with others (indicated by dashed lines in Fig. 5.1).

A Bayesian network (BN) approach derived from the combination of graph theory and probability theory can be used to yield topologies or qualitative networks of interactions between the genes. A BN is considered as a directed acyclic graph $G(X,E)$, where X is a set of nodes, x_i's, which are random variables representing genes' expression, and E is a set of edges which indicate the dependencies between nodes (Aluru 2005). The nodes follow conditional probability mass function $P(x_i|Pa(x_i))$, where $Pa(x_i)$ is the set of parents of node x_i. The Markov assumption is encoded implicitly in a BN; i.e., each node is independent of its non-descendants given its parents. Therefore, the joint distribution of all nodes can be decomposed down into the conditional distributions of the nodes as

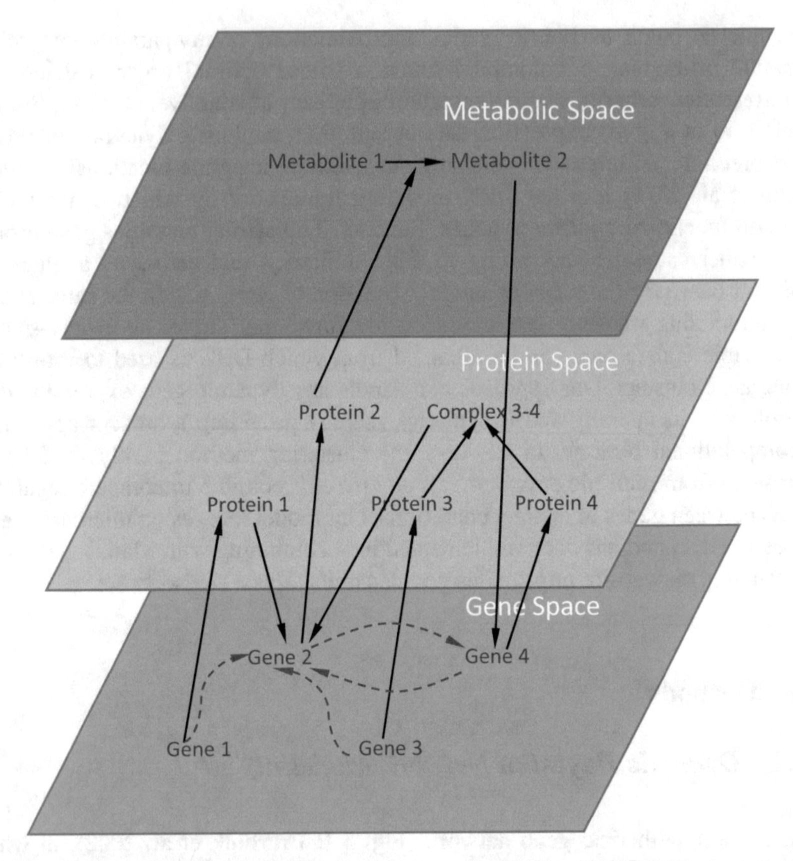

Fig. 5.1 A hypothetical gene network, modified from Brazhnik et al. (2002). It shows multiple levels of regulation by genes, proteins and metabolites. Direct interaction is indicated by solid lines. The dashed lines in the gene level indicate regulation between genes which is an abstraction of interactions over the three levels

$$P\left(x_1, x_2, \cdots, x_n\right) = \prod_{i=1}^{n} P\left(x_i \middle| Pa\left(x_i\right)\right). \tag{5.1}$$

Figure 5.2a shows a sample of BN, under the Markov assumption, we have

$$P\left(A, B, C, D, E\right) = P(A)P(B)P\left(C|A, B\right) P\left(D|B\right) P\left(E|C\right) \tag{5.2}$$

To handle dynamic gene expression, dynamic Bayesian network (DBN) (Murphy and Mian 1999; Friedman et al. 1998) is developed by taking into account the time components, i.e., two copies of the same BN are used to model a state transition of gene network from time t to time t+1. In a DBN as shown in Fig. 5.2b, the state of A is affected by B and itself but the state at a previous time.

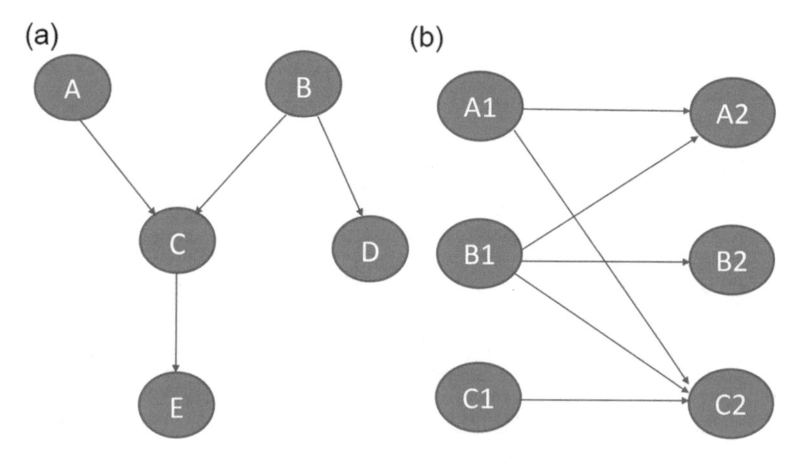

Fig. 5.2 A diagrammatical construction of gene regulatory networks. (**a**) In a Bayesian network, the Markov assumption is encoded implicitly. For example, Given its parents A and B, C is independent of D. (**b**) In a dynamic Bayesian network, the time component is taken into account such that the state of A is affected by B and itself but the state at a previous time

The DBN approach for reconstruction of GRN based on gene expression data includes the following steps (Zou and Conzen 2005): (1) *Discretizing the expression levels.* The expression levels for all genes are discretized as 1 (down-regulation) or 2 (up-regulation) by using a cut-off value of fold-change comparing with a baseline expression level. (2) *Realigning expression levels for potential regulator and target genes.* Expression levels for a potential regulator and target genes will be realigned according to the transcriptional time lag which is defined as the difference between the time when the regulator gene to encode its protein product and the time when the transcription of the target gene to be affected by this regulator protein. Suppose we have two hypothetical genes, gene A and its potential target gene B, and their expression levels are measured at six evenly spaced time points t_1–t_6, expressed as $A_{t_1}, \ldots, A_{t_6}$ and $B_{t_1}, \ldots, B_{t_6}$, respectively (Fig. 5.3a). If we decide the time lag is one time unit, then A_{t_1} will be aligned with B_{t_2}, A_{t_2} will be aligned with B_{t_3} and so on. (3) *Determining regulators by calculating conditional probabilities and marginal likelihood scores.* In this step, conditional probabilities (target gene give potential regulators) and marginal likelihood scores will be calculated using the realigned expression levels. The potential regulators which have the highest marginal likelihood score will be selected as regulators.

We follow the three steps of DBN as described above on these two genes to discretize the expression levels (using onefold as the cut-off), realign them (using one time unit as the time lag) and calculate the conditional probabilities of gene B with respect to its potential regulator gene A (Table 5.1). Intuitively, since $P(B = 1|A = 1) = 1$ and $P(B = 2|A = 2) = 0.67$, we would consider gene A as a regulator of gene B. The basic condition of using DBN is that it requires the expression levels measured at evenly spaced time points because the time points are

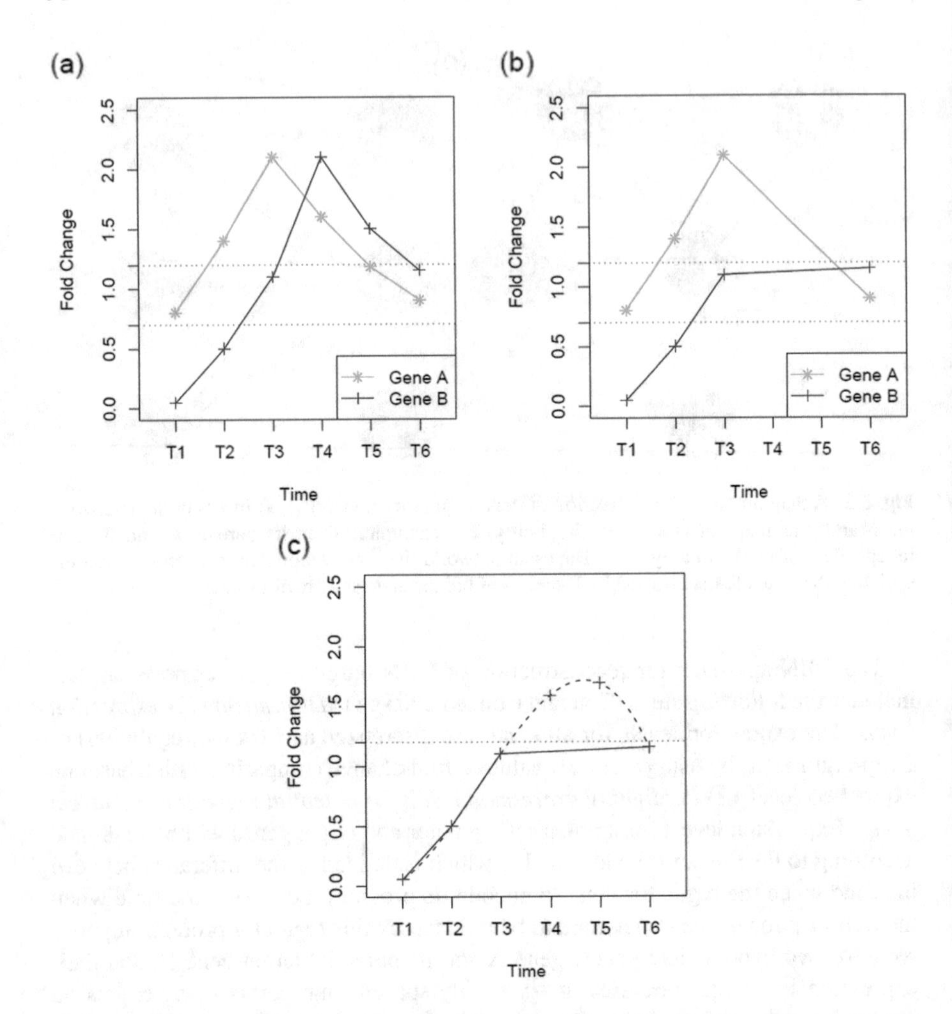

Fig. 5.3 Time courses of gene expression for two hypothetical genes A and B. (**a**) Original expression levels measured at six time points evenly. (**b**) Expression levels of both genes without measurements at time t_4 and t_5. For gene B, the information is misleading since it seems that it has no up-regulation at all. (**c**) Estimated expression levels for gene B at time t_4 and t_5 according to the trend of time t_1–t_3

realigned one by one in step 2. If this condition was not satisfied, two issues would arise.

Suppose we do not measure the expression levels of gene A and B at time t_4 and t_5 (Fig. 5.3b). This will lead to two problems as following: (1) *Mismatching*. In this situation, we could still align A_{t_1} with B_{t_2} and A_{t_2} with B_{t_3}, but we cannot align A_{t_3} with B_{t_4} because it is missing. We cannot align A_{t_3} with B_{t_6} either since the time period between them is much different from that between A_{t_1} and B_{t_2}. (2) *Losing information*. Since expression levels at time t_4 and t_5 are missing, the information

Table 5.1 Three steps for DBN modeling

(1) Discretized expression levels for Gene A and B						
	t_1	t_2	t_3	t_4	t_5	t_6
Gene A	1	2	2	2	1	1
Gene B	1	1	1	2	2	1

(2) Realigned expression levels for Gene A and B					
	$t_{1(2)}$	$t_{2(3)}$	$t_{3(4)}$	$t_{4(5)}$	$t_{5(6)}$
Gene A	1	2	2	2	1
Gene B	1	1	2	2	1

(3) Conditional probabilities of gene B given gene A		
	$A = 1$	$A = 2$
$B = 1$	1	0.33
$B = 2$	0	0.67

of gene B is misleading since it seems that it has no up-regulation at all. If we still apply DBN, we would only have information from two pairs: A_{t_1} with B_{t_2} and A_{t_2} with B_{t_3}. With conditional probabilities of $P(B = 1|A = 1) = 1$, $P(B = 1|A = 2) = 1$ and $P(B = 2|A = 2) = 0$, it is difficult to decide whether gene A is a regulator of gene B.

5.2.2 *Interpolation by a Parametric or Nonparametric Function*

A natural way to solve the above problems is to restore the information missed at time t_4 and t_5. Indeed, this can be done if sufficient data is observed. For example, for gene B, we would expect that up-regulation should take place between time t_3 and t_6, as shown in Fig. 5.3c by a dashed line, if the expression level would develop following the trend of time t_1–t_3. If this dashed line could be estimated as a function of time, it is straightforward to interpolate the values of expression levels at time t_4 and t_5 (as marked in red in Fig. 5.3c) based on such a function. However, if this function is estimated from time-dependent observations of a single gene, we may not eliminate the effect of measurement noises. Since many genes may share a similar biological function, they could be classified into the same group with an indistinguishable time course expression pattern. These genes can be put together to provide a more precise estimation of functional curve.

Functional clustering, aimed to group those genes of similar function, can serve as a tool to estimate functional curves. Kim et al. (2008, 2010) implemented a Fourier series approximation to model periodic patterns of gene expression, whereas nonparametric approaches based on Functional Data Analysis (FNDA) (Song et al. 2007), B-splines (Luan and Li 2003) and Legendre Orthogonal Polynomials (LOP) (Wang et al. 2011) were developed to characterize time-varying expression levels when no explicit parametric function can be used. These approaches consider the mean of a cluster as a representative gene, thereby providing a more stable and accurate interpolation of missing points. In this section, we present a procedure for identifying gene regulatory network based on time course gene expression data in the four following steps:

- Step 1: Clustering genes into different groups by parametric or nonparametric functional clustering and estimating the mean function for each cluster;
- Step 2: Interpolating missing values in uneven intervals to obtain evenly spaced measurements;
- Step 3: Constructing the GRN using the DBN model to identify the effects of regulation due to interactions between clusters;
- Step 4: Analyzing gene functions by Gene Ontology to explore the biological relevance of gene clusters in the reconstructed regulatory network.

Step 1. Clustering Genes into Different Groups Consider a high-dimensional set of genes (say n) measured at multiple time points from different experiments. Thus, it is possible that different genes are measured with different time schedules. For a particular process, such as embryogenesis, gene expression levels may be measured more densely in an early stage than late stage, making the time intervals of measurement unevenly-spaced. Overall, we have a sparse, irregular time course gene expression data for GRN reconstruction.

Let $\mathbf{y}_i = (y_i(t_1), \ldots, y_i(t_{T_i}))$ denote a vector of expression levels for gene i ($i = 1, \ldots, n$) measured at time points $(t_1, \ldots, t_{T_i})$ in a experiment. Note that time points are gene-specific. We assume that these n genes can be classified into m clusters because of their similarity and differences. This can be expressed by a mixture model in which there are m components. Each gene arises from one and only one of the m possible components. We further assume that $\mathbf{y}_i$ is a realization of a mixture of m multivariate normal distributions with the density function specified as

$$\mathbf{y}_i \sim f_i\left(\mathbf{y}_i; \boldsymbol{\omega}_i, \boldsymbol{\mu}_i, \boldsymbol{\Sigma}_i\right) = \omega_{1|i} f_{1|i}\left(\mathbf{y}_i; \boldsymbol{\mu}_{1|i}, \boldsymbol{\Sigma}_i\right) + \cdots + \omega_{m|i} f_{m|i}\left(\mathbf{y}_i; \boldsymbol{\mu}_{m|i}, \boldsymbol{\Sigma}_i\right) \tag{5.3}$$

where $\boldsymbol{\omega}_i = (\omega_{1|i}, \ldots, \omega_{m|i})$ is a vector of non-negative proportions for the m possible clusters that sum to unity and $f_{j|i}(\mathbf{y}_i; \boldsymbol{\mu}_{j|i}, \boldsymbol{\Sigma}_i)$ denotes the density function for gene cluster j ($j = 1, \ldots, m$), a multivariate normal with mean vector $\boldsymbol{\mu}_{j|i} = (\mu_{j|i}(t_1), \ldots, \mu_{j|i}(t_{T_i}))$ and the common $T_i \times T_i$ covariance matrix $\boldsymbol{\Sigma}_i$. Let $\boldsymbol{\mu}_i = (\boldsymbol{\mu}_{1|i}, \ldots, \boldsymbol{\mu}_{J|i})$ contain the cluster-specific mean vectors for gene i.

Parametric functional clustering implements an explicit mathematical equation to approximate time-dependent expression. If the genes are periodically regulated (Rustici et al. 2004), Kim et al. (2008, 2010) used the Fourier series function, showing adequate power to capture the temporal expression pattern of oscillating genes. In the case where no explicit mathematical equation is available, Wang et al. (2011) deployed a flexible approach based on LOP to model gene-specific function curves for each cluster. Both parametric and nonparametric approaches allow handling the sparsity of time points in gene expression data. Also, by determining the best order of Fourier series or LOP by information criteria, both approaches can provide an optimal function for modeling time course expression levels for each cluster from a given dataset.

Increasing power of functional clustering also results from the parsimonious modeling of the covariance structure by a few parameters. Parametric, nonparametric or semiparametric approaches have been used to model the covariance matrix $\boldsymbol{\Sigma}_i$, each with specific strengths and weakness. Li et al. (2010) proposed a general parametric approach for covariance modeling through a general autoregressive moving-average process of order (p, q), the so-called ARMA(p, q). These authors derived the EM algorithm to estimate the ARMA parameters that model the covariance structure within a mixture model framework. The orders p and q of the ARMA process that provide the best fit are identified by model selection criteria.

To integrate data from multiple expression experiments, the model proposed by Wang et al. (2011) can be applied. The authors consider gene expression patterns of multiple experiments jointly and group genes into clusters based on the joint patterns. Consequently, after clustering, a gene will have the same cluster label for multiple experiments and the same number of clusters will be obtained for each experiment. The optimal number of clusters is determined by Bayesian information criterion (BIC). In Step 3, network inference is conducted on the clustering results from this model and enables one to compare gene regulation effects in distinct environments.

Step 2. Interpolating Missing Values in Uneven Intervals For DBN modeling, we interpolate missing data adaptively to satisfy the requirement of evenly spaced intervals. The mean vectors for each cluster which can be expressed as a function

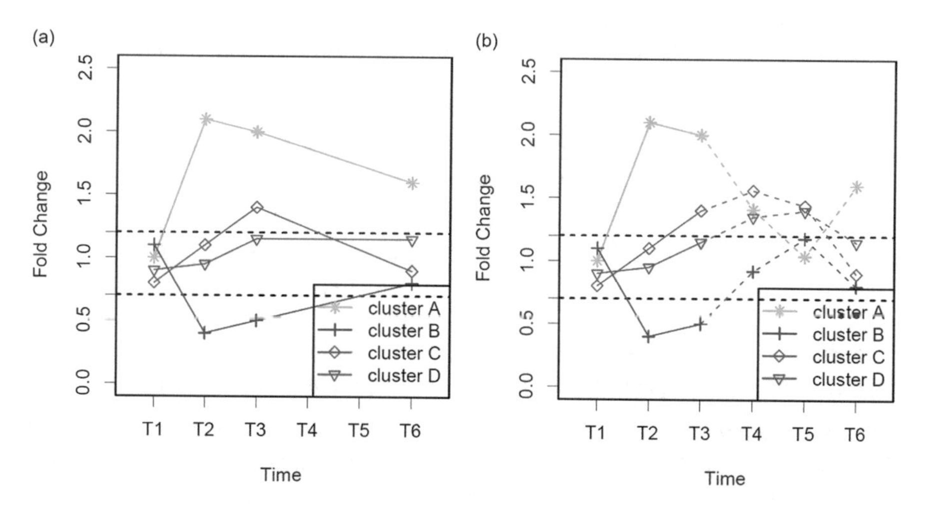

Fig. 5.4 Time courses of gene expression for four hypothetical clusters. (**a**) Originally, Their expression levels measured at unevenly spaced time points t_1, t_2, t_3 and t_6, the information at t_4 and t_5 is missing. (**b**) After interpolation based on LOP, all of these gene clusters have evenly spaced time course measurements of expression levels

of time have been obtained from step 1. Here, as an example, we describe step 2 by using LOP-based nonparametric functional clustering.

Suppose we have four hypothetical gene clusters A, B, C and D whose expression levels are measured at unevenly spaced time points t_1, t_2, t_3 and t_6 (Fig. 5.4a), rather than six evenly spaced time points t_1–t_6 as required in Zou and Conzen (2005). Let $\mathbf{u}_A$, ..., $\mathbf{u}_D$ denote the base means of these clusters, respectively. According to Wang et al. (2011), their mean expression vectors $\boldsymbol{\mu}_A$, ..., $\boldsymbol{\mu}_D$ are determined by $M\mathbf{u}_A$, ..., $M\mathbf{u}_D$, where M is a $(4 \times r)$ matrix constructed by the LOP (with r being the optimal order of LOP). For example, the mean vector of cluster A is expressed as

$$\boldsymbol{\mu}_A = \begin{pmatrix} \mu_{A_1} \\ \mu_{A_2} \\ \mu_{A_3} \\ \mu_{A_6} \end{pmatrix} = M\mathbf{u}_A = \begin{pmatrix} P_0\left(t_1^*\right) & P_1\left(t_1^*\right) & \cdots & P_r\left(t_1^*\right) \\ P_0\left(t_2^*\right) & P_1\left(t_2^*\right) & \cdots & P_r\left(t_2^*\right) \\ P_0\left(t_3^*\right) & P_1\left(t_3^*\right) & \cdots & P_r\left(t_3^*\right) \\ P_0\left(t_6^*\right) & P_1\left(t_6^*\right) & \cdots & P_r\left(t_6^*\right) \end{pmatrix} \begin{pmatrix} u_{A_1} \\ u_{A_2} \\ \vdots \\ u_{A_r} \end{pmatrix} \tag{5.4}$$

where μ_{A_1}, μ_{A_2}, μ_{A_3} and $\mu_{A_6}$are the expression values of cluster A at time points t_1, t_2, t_3 and t_6, respectively; and t^*_1, t^*_2, t^*_3 and t^*_6 are the normalized time points using the formula:

$$t^* = -1 + \frac{2\left(t - t_1\right)}{t_6 - t_1} \tag{5.5}$$

To interpolate expression values at t_4 and t_5, we first calculate the rescaled time values t^*_4 and t^*_5 and then insert two rows corresponding to t_4 and t_5 into matrix M, obtaining the interpolated $\boldsymbol{\mu}_A$, denoted as $\hat{\boldsymbol{\mu}}_A$, by the following equation:

$$\hat{\boldsymbol{\mu}}_A = \begin{pmatrix} \mu_{A_1} \\ \mu_{A_2} \\ \mu_{A_3} \\ \hat{\mu}_{A_4} \\ \hat{\mu}_{A_5} \\ \mu_{A_6} \end{pmatrix} = \begin{pmatrix} P_0\left(t_1^*\right) & P_1\left(t_1^*\right) & \cdots & P_r\left(t_1^*\right) \\ P_0\left(t_2^*\right) & P_1\left(t_2^*\right) & \cdots & P_r\left(t_2^*\right) \\ P_0\left(t_3^*\right) & P_1\left(t_3^*\right) & \cdots & P_r\left(t_3^*\right) \\ P_0\left(t_4^*\right) & P_1\left(t_4^*\right) & \cdots & P_r\left(t_4^*\right) \\ P_0\left(t_5^*\right) & P_1\left(t_5^*\right) & \cdots & P_r\left(t_5^*\right) \\ P_0\left(t_6^*\right) & P_1\left(t_6^*\right) & \cdots & P_r\left(t_6^*\right) \end{pmatrix} \begin{pmatrix} u_{A_1} \\ u_{A_2} \\ \vdots \\ u_{A_r} \end{pmatrix} \tag{5.6}$$

Similarly, we can have $\hat{\boldsymbol{\mu}}_B$, $\hat{\boldsymbol{\mu}}_C$ and $\hat{\boldsymbol{\mu}}_D$. As shown in Fig. 5.4b, all of these gene clusters have evenly spaced time course measurement of gene expression.

Step 3. Constructing the GRN Using the DBN Model We follow the improved DBN approach by Zou and Conzen (2005) to take advantage of its high efficiency and accuracy. According to these authors, only those genes are considered as potential regulators when they have either earlier or simultaneous expression changes (up- or down-regulation) compared to the targets. The up- and down-regulation are defined as $\geq$1.2-fold and $\leq$0.7-fold, relative to the baseline gene expression. These relatively modest cutoffs are used to avoid missing any genes with potentially important changes in gene expression although these changes could be small. Per these cutoffs, we determine the initial regulation time points of gene clusters A, ..., D, after which the regulators of genes that change later in expression are viewed as those genes that change earlier or simultaneously. As shown in Fig. 5.4b, cluster A has an initial up-regulation at t_2 while cluster B is also initially regulated at t_2 but with down-regulation. Cluster C and D initially change expression at time t_3 and t_4, respectively. Since cluster A, B and C each have an earlier change in expression than cluster D, the former is selected as potential regulators of the latter. Similarly, we can decide potential regulators, cluster A for cluster B and cluster A and B for cluster C.

Based on the determined initial regulation time points, we could also decide the transcriptional time lag between regulator and target genes. We calculate the time difference between the initial regulation time points for a potential regulator and its target gene, which is considered as a more accurate estimation of the corresponding transcriptional time lag (Zou and Conzen 2005). In this way the time lag between cluster D and its potential regulators cluster C is estimated as one unit. Similarly, the time lags between cluster D and B, D and A are estimated as two time units. According to the time lags between potential regulators and its target clusters, potential regulators are grouped into different categories with regulators in a category of the same time lag in terms of the target clusters. The reason for this grouping is that different regulators may have different time frames when interacting with targets. By grouping we analyze regulators separately, with a possibility to identify co-regulators. As an example, cluster D has two groups of potential regulators; one group, including cluster A and B, has the time lag of two time units, and the other group, including cluster C, has the time lag of one time unit. It is here possible that cluster A and B are the co-regulators of cluster D.

After determining potential regulators for each cluster and calculating the corresponding time lags, the DBN framework developed by Murphy and Mian (1999), is applied for network inference. Though continuous data can be directly analyzed by DBN, the assumptions of continuous DBN may not be satisfied in certain domain. In particular, continuous models assume additive influence of multiple regulators on a target, but it may not be a case in gene regulation. Discrete network is chosen for our data set by discretizing continuous gene expression data. Two categories are used for discretization with onefold as a cutoff, instead of $\geq$1.2-fold and $\leq$0.7-fold since the relative increase or decrease in expression levels is more important than the absolute expression value during the inference of relationships between potential regulators and targets. Specifically, "2" is assigned if the expression level is equal to or higher than onefold; otherwise "1" is assigned. The discretized expression levels for cluster A, B, C and D are shown in Table 5.2.

Another import step for the DBN algorithm is to align the expression levels for potential regulators and targets according to the relevant transcriptional time lags between them. Suppose the time lag between a regulator and its target is Δt. Then the expression level of the regulators at time t_1 will be aligned with the expression level of the target at $t_1 + \Delta t$. In this way, an $(R \times K)$ matrix will be constructed for the regulators and targets, where R is the number of potential regulators with

Table 5.2 Discretized expression levels for gene cluster A, B, C and D

	t_1	t_2	t_3	t_4	t_5	t_6
Cluster A	1	2	2	2	2	2
Cluster B	2	1	1	1	2	1
Cluster C	1	2	2	2	2	1
Custer D	1	1	2	2	2	2

the same time lag plus one which represents the target, and K is the number of time points between $t_1 + \Delta t$ and t_6. As an example, for cluster D, we have calculated its time lag with its potential regulator, cluster A, which is two time units. Therefore, we align the expression level of cluster A at t_1 with the expression level of cluster D at t_3, obtaining a (2×4) matrix expressed as

$$\begin{pmatrix} 1\,2\,2\,2 \\ 2\,2\,2\,2 \end{pmatrix} \tag{5.7}$$

where the first row is for cluster A and second row for cluster D.

As seen from above, the potential regulators of cluster D have been classified into two groups according to their time lags: one group of cluster A and B with two time units as the time lag and another group of cluster C with one time unit as the time lag. To identify all possible co-regulators of cluster D, all subsets of each group are generated and the relationships between all subsets of co-regulators are examined. For the first group, the possible subsets are [cluster A], [cluster B] and [cluster A, cluster B] and the subset of second group is [cluster C]. For each subset, a matrix like (5.7) is constructed on the basis of the corresponding time lags and number of regulators. We then calculate the conditional probabilities of target clusters with respect to their regulators based on the matrix containing aligned expression levels. For each target cluster, we calculated marginal likelihood scores for every subset of potential regulators using their conditional probabilities. The one of the highest score is selected as the final regulator for this target. The conditional probabilities of cluster D given cluster A are shown in Table 5.1.

We use the algorithm proposed by Murphy and Mian (1999) to make DBN inference. The idea of this algorithm is to select the optimal model that maximizes the following conditional probability,

$$P\,(G|D) = \frac{P\,(D|G)\,P(G)}{P(D)}, \tag{5.8}$$

where G denotes the network structure and D denotes the observed data.

Step 4. Analyzing Gene Functions Genes with a similar profile pattern in each cluster usually share the common biological functions. Gene ontology (GO) analysis enable us to figure out what function is shared by genes in a cluster. Therefore, based on the clustering results in step 1 and regulation network established in step 2, we perform function analysis in this step. Different from traditional clustering approaches, Wang et al.'s (2011) functional clustering allows environment-dependent expression plasticity to be clustered, producing results directly related to the mechanistic machineries of gene expression induced by environmental signals. Through GO analysis, we can shed more light on the regulation mechanisms underlying cellular and physiological processes.

5.3 Results and Discussion

5.3.1 Real Data Analyses

We demonstrated the application of the proposed procedure for GRN reconstruction by analyzing a real data set of time course gene expression from the surgery study of a rabbit bilateral vein graft construct (Fernandez et al. 2004; Jiang et al. 2004). The study involved two different environments, created by two distinct blood flows (differing by sixfold) in vein grafts for New Zealand White rabbits (weighing 3.0–3.5 kg) resulting from the treatment of bilateral jugular vein interposition grafting and unilateral distal carotid artery branch ligation, respectively. With a segment of the vein retained at the time of implantation for baseline morphometric measurements, vein grafts were harvested at 2 h, 28, 90 and 180 days after implantation. Expression of 14,958 genes was recorded for each of these time points under both of treatments, high flow and low flow. By combining the dynamic expression data from the two treatments, Wang et al. (2011) used the LOP-based functional clustering model to identify eight gene clusters jointly for both treatments, denoted as A (0.0116), B (0.1023), C (0.3354), D (0.3831), E (0.1134), F (0.0359), G (0.0100) and H (0.0083), where the numbers in parentheses are the proportions of genes belonging to a particular cluster. These clusters each display different patterns of environment-induced changes in gene expression trajectories. We treated the mean expression curve for each cluster as a representative profile. Since expression values were not measured at evenly spaced time intervals, our adaptive DBN model was used to reconstruct GRN, respectively, for high and low flows.

Figure 5.5 illustrates three different networks of gene expression under high and low flows and the difference of gene expression between the two flows. It is interesting to see that the structure of GRN is different dramatically between the two flows, although with some extent of similarity. Under high flow, cluster A is regulated jointly by cluster F and G. Meanwhile, cluster F and G are regulated by cluster H and B respectively (Fig. 5.5a). Under low flow, cluster A is regulated only by cluster F, whereas the latter is regulated by two clusters, H and G (Fig. 5.5b). Thus, cluster A is regulated directly by cluster G under high flow, but such a regulation operates through an indirect way under low flow. Under high flow, cluster B plays a role in regulating cluster G, but this regulation role disappears under low flow. For cluster C, D and E, since their expression patterns are relatively flat over

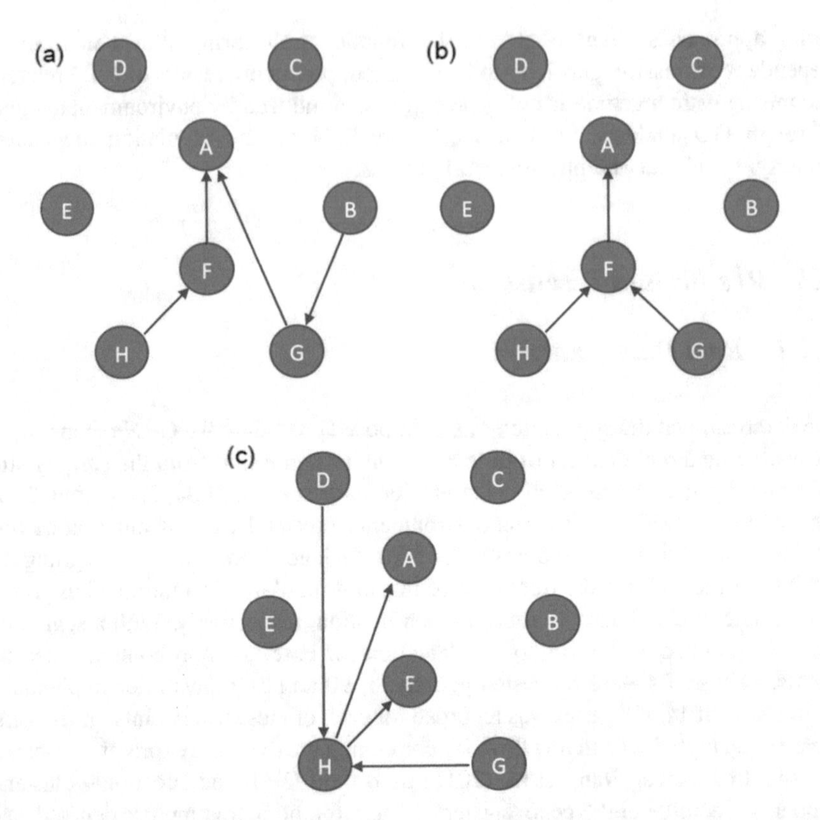

Fig. 5.5 The reconstruction of gene regulatory networks from the first surgical study of rabbits under high blood flow (**a**), under low blood flow (**b**), and for expression plasticity between high flow and low flow (**c**)

time in both environments, with no up- or down-regulation (Wang et al. 2011), they are not regulated by any clusters and also do not regulate other clusters.

We further made an inference of GRN based on the gene expression plasticity between high and low flows. Gene expression plasticity is defined as the environmentally induced alteration of gene expression, which is a capacity for the organism to respond to its environment. Let $\mu_j^H(t^*)$ and $\mu_j^L(t^*)$ denote the mean expression of cluster j at time t^* under high flow and low flow, respectively. The expression plasticity of this cluster is defined as

$$\Delta\mu_j(t^*) = \mu_j^H(t^*) - \mu_j^L(t^*) = P_r(t^*)\left(u_{jr}^H - u_{jr}^L\right). \tag{5.9}$$

The regulatory network based on expression plasticity data emphasizes the similarities of gene clusters in terms of their pattern of differential expression over two different flows (Fig. 5.5c). It was observed that cluster B, C and E which are not expressed differentially between two flows have no regulation effects. This is consistent with Wang et al.'s (2011) finding that their expression difference is close to zero. On the other hand, the other clusters are heavily involved in the regulation (Fig. 5.5c) since they have significant differential expression according to hypothesis tests performed in Wang et al. (2011). It appears that cluster H plays a multiple role in affecting the structure of GRN by regulating cluster A and F and by being regulated by cluster D and G. Given this, cluster H links the mutual relationships between discrete clusters D, A, F and G.

We applied our adaptive DBN model to analyze a different data set for another surgery study of rabbits. The same grafting procedure was used to obtain two treatments, high blood flow and low blood flow. Under each treatment, expressions levels were recorded for each of 9272 genes at time points 2h, 1, 3, 7, 14 and 28 days after implantation. Wang et al.'s (2011) clustering model was used to classify these genes into 29 different clusters. Our adaptive DBN model infers three gene networks for each flow and for expression plasticity between the two flows (Fig. 5.6). There is much similarity in the structure of GRN between the two flows, although specific differences exist. For example, cluster 1 links many other clusters through its active regulation (i.e., it regulates other clusters) or passive regulation (i.e., it is regulated by other clusters). Following cluster 1, cluster 23 is an important link for overall network under high flow, but it is replaced by cluster 20 under low flow. It is observed that the network has a much sparse structure for the expression plasticity between the two flows, compared to those under each flow. Obviously, by comparing these differences, one can better understand the regulatory mechanisms underlying the cellular processes toward environment-induced adaption.

5.4 Computer Simulation

Yu et al. (2004) used simulation studies to investigate the influence of interpolation on DBN modeling. Their results showed that DBN can benefit from moderate data interpolation by reducing false positives. Here, we performed computer simulation to evaluate the performance of our adaptive model by answering the following questions: Is LOP-based interpolation better than non-interpolation in the case of missing data? Is LOP interpolation is better than linear-interpolation? What is the difference between even interpolation and uneven interpolation?

5.4.1 Simulation Process

We generated simulation data of mean vectors for every cluster at every time point with the process as follows:

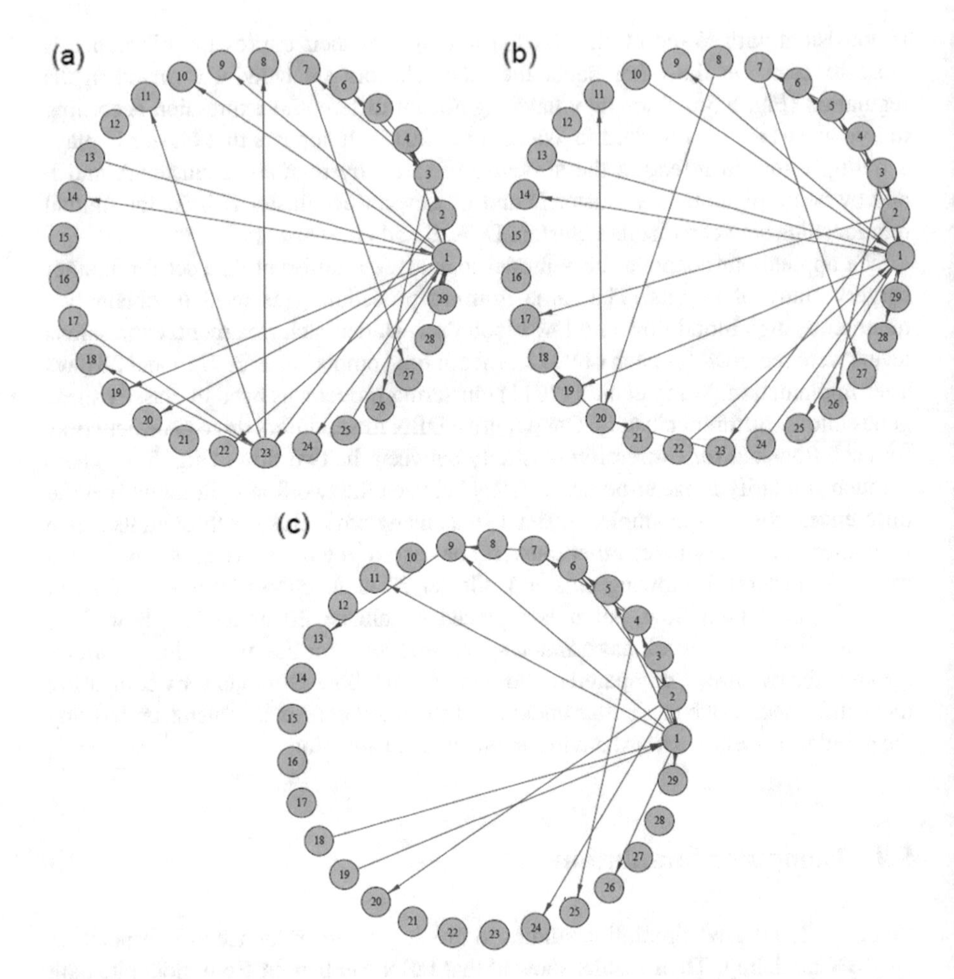

Fig. 5.6 The reconstruction of gene regulatory networks from the second surgical study of rabbits under high blood flow (**a**), under low blood flow (**b**), and for expression plasticity between high flow and low flow (**c**)

$$\mathbf{Y_{t+1}} = \mathbf{Y_t} + \mathbf{R}\,(\mathbf{Y_t} - \mathbf{C}) + \varepsilon \tag{5.10}$$

where $\mathbf{Y}_t$ is a vector of mean expression levels for all clusters in a regulatory network at time t; $\mathbf{R}$ is a design matrix used to define the regulatory relationships between clusters; $\mathbf{C}$ is the vector of constitutive expression values for each cluster (Yu et al. 2004); and ε is the noise drawn from a normal distribution. Let $r_{j_1 j_2}$ denote the entry of $\mathbf{R}$ at row j_1 and column j_2. Then the regulatory relationship between cluster j_1 and cluster j_2 could be interpreted completely by $r_{j_1 j_2}$. If $r_{j_1 j_2} = 0$, then cluster j_2 has no regulation upon cluster j_1. If $r_{j_1 j_2} > 0$, cluster j_2 activates cluster j_1. Otherwise, cluster j_2 represses cluster j_1. Moreover, the strength of regulation is defined by the magnitude of $r_{j_1 j_2}$. Similar to Yu et al. (2004), 0 and 100 are set as the minimum and maximum expression values and 50 set as constitutive value for all simulated clusters. Therefore, a cluster with expression value larger than 50 triggers an effect on the direction specified in **R** while a cluster with expression value less than 50 has an effect on the opposite direction specified in **R**.

The above procedure was used to generate mean expression values for each cluster. The expression values for clusters with no regulators (entries of the corresponding row in **R** are all zero) could be generated in a different way by moving these clusters in a random walk according noise term ε. However, we let the trajectory of expression level for a cluster move along a curve specified by the LOP. It is assumed that expression values of genes within each cluster follow a multivariate normal distribution with cluster-specific mean vectors and covariance matrix. We assumed that the covariance follows an autoregressive structure described by a correlation and variance (see Wang et al. 2011).

As shown in Wang et al. (2011), LOP-based functional clustering performs very well in classifying genes into distinct clusters. Here, we focused on the inference of GRN from our procedure. The interpolation of expression levels was first conducted for missing data, followed by GRN reconstruction. To evaluate the performance of our interpolation method, we defined two measurements, positive predictive value (PPV) and false negative rate (FNR) as follows:

$$PPV = \frac{TP}{TP + FP} \tag{5.11}$$

$$FNR = \frac{FN}{TN + FN} \tag{5.12}$$

where TP denotes the true positive (regulatory relationships exist in both inference network and true network); FN denotes the false negative (regulatory relationships exist only in true network); TN denotes the true negative (regulatory relationships do not exist in either network); and FP denotes the false positive (regulatory relationships exist only in inference network).

In each of the six randomly simulated networks, we generated 20 genes and about 10 regulatory relationships (Fig. 5.7). For each relationship, we randomly assigned a possible regulation strength value, 0.05, 0.1, 0.15 or 0.2. For each network, we generated 100 sets of expressions so that there are a total of 600 simulated networks. By comparing the inference networks with true ones, PPV and FRN were obtained.

5.4.2 *Results*

Whether Is LOP Interpolation Helpful for DBN Inference? We first picked up the simulation expression values at time point 10, 50, 90, . . . , etc. in each

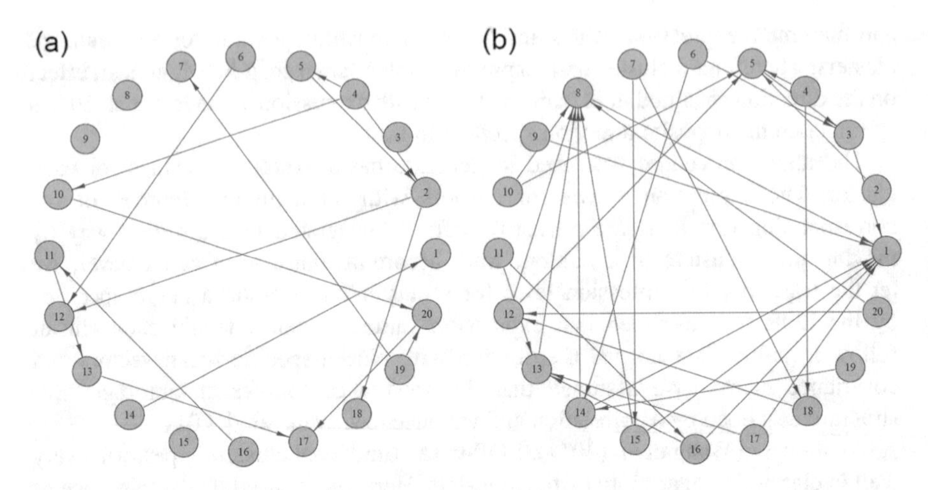

Fig. 5.7 Networks in simulation studies. (**a**) A true network is generated with 20 genes and 15 regulation effects. (**b**) The reconstructed network by our proposed model shows that most edges in the true network can be identified while several false edges are produced

simulation run. We then interpolated one or three points in each interval with LOP to generate two more data sets; thus, we had 3 data sets: $\{\mathbf{Y_{10}}, \mathbf{Y_{50}}, \mathbf{Y_{90}}, \cdots\}$, $\left\{\mathbf{Y_{10}}, \hat{\mathbf{Y}}_{\mathbf{30}}, \mathbf{Y_{50}}, \cdots\right\}$ and $\left\{\mathbf{Y_{10}}, \hat{\mathbf{Y}}_{\mathbf{20}}, \hat{\mathbf{Y}}_{\mathbf{30}}, \hat{\mathbf{Y}}_{\mathbf{40}}, \mathbf{Y_{50}}, \cdots\right\}$, where $\hat{\mathbf{Y}}$ denotes the interpolated values. From the comparison of a recovered network (Fig. 5.7b) with true one (Fig. 5.7a), it was observed that most edges in the true network could be identified while several false edges had also been produced.

We evaluated the overall quality of all recovered networks from those simulated data sets. The results show that interpolation does help to reduce the FNR; the FNR for the non-interpolated data is 0.59 (0.117), where the number in parentheses is standard error, while it is 0.41 (0.110) for the data with three interpolation points for each interval, where the numbers in parentheses are corresponding standard deviations. Moreover, interpolation also improves the PPV from 0.20 (0.056) to 0.24 (0.048). Therefore, GRN reconstruction can benefit from interpolation with the LOP.

Whether Is LOP Interpolation Better Than Linear Interpolation? Following the same scenario above, we interpolated one or three points in each interval with linear method, which generated two more data sets. The results in Fig. 5.8a, b show that LOP interpolation has better performance than linear method. The FNR is reduced by linear interpolation but only from 0.59 (0.117) to 0.50 (0.124), when there are three points interpolated. Liner interpolation also has less improvement of PPV, from 0.20 (0.056) to 0.22 (0.045).

What is the Difference Between Even Interpolation and Uneven Interpolation? We picked up the simulated expression values at time point 10, 20, 30, ..., etc. in each simulation run. We randomly dropped 1–3 consecutive

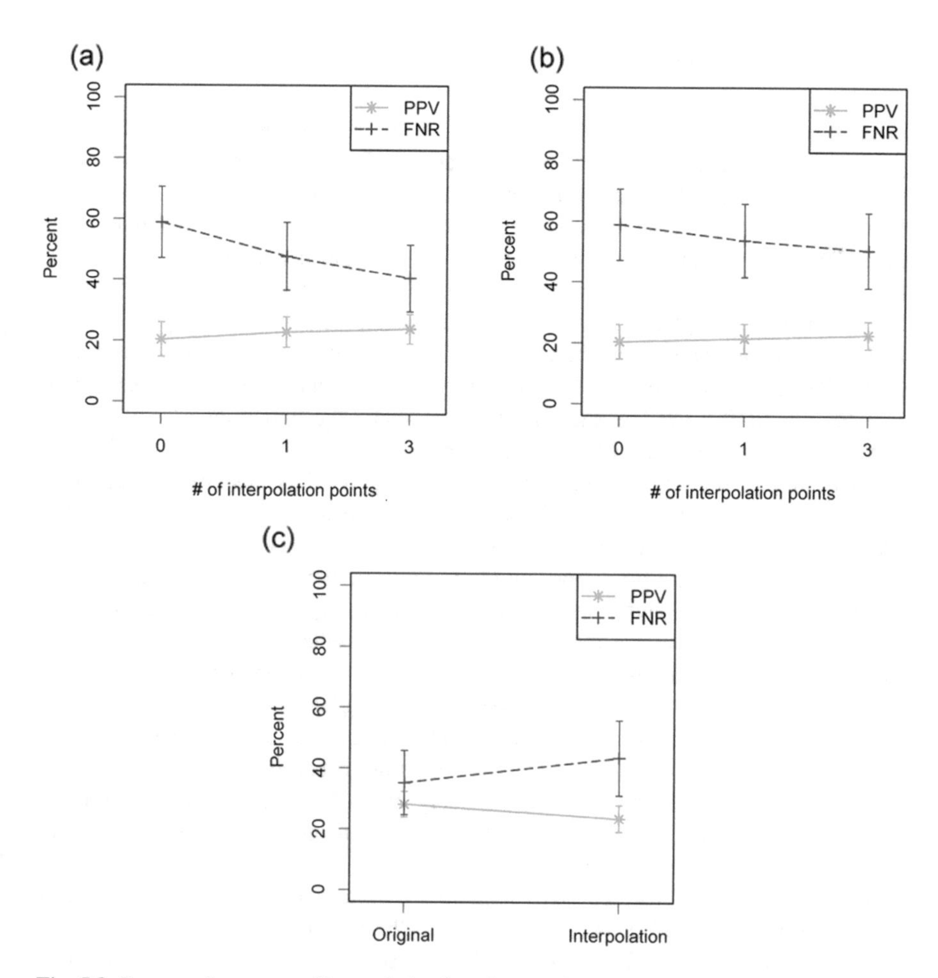

Fig. 5.8 Power and accuracy of interpolation-based network reconstruction. (**a**) With interpolation or 1 or 3 time points by LOP, PPV has been improved and FNR has been reduced. (**b**) Linear interpolation is also helpful for network reconstruction but not as good as LOP interpolation. (**c**) Unevenly interpolation by LOP is acceptable comparing to originally complete data with slightly decreased PPV and increased FNR

points, leading to unevenly spaced expression data. Suppose there are $\{\mathbf{Y_{10}}, \mathbf{Y_{20}}, \mathbf{Y_{40}}, \mathbf{Y_{80}}, \mathbf{Y_{110}}, \cdots\}$ after dropping. By interpolating the missing points by LOP, we got $\left\{\mathbf{Y_{10}}, \mathbf{Y_{20}}, \hat{\mathbf{Y}}_{\mathbf{30}}, \mathbf{Y_{40}}, \hat{\mathbf{Y}}_{\mathbf{50}}, \hat{\mathbf{Y}}_{\mathbf{60}}, \cdots\right\}$. The quality of the recovered network from the LOP-interpolated data was compared with the one from the original data set of $\{\mathbf{Y_{10}}, \mathbf{Y_{20}}, \mathbf{Y_{30}}, \mathbf{Y_{40}}, \mathbf{Y_{50}}, \cdots\}$ (Fig. 5.8c). It was observed that the quality of reconstructed network form LOP-interpolated data is not too much worse than that from true data. The FNR rises from 0.35 (0.105) to 0.44 (0.123), while the PPV drops from 0.28 (0.042) to 0.24 (0.044).

5.5 Conclusions

Many biological processes, including plant and animal development, disease pathogenesis and surgical recovery, are coordinated by cell-to-cell communications under the regulation of genes. High-throughput measurement techniques have now made it feasible to identify tens of thousands of genes at a time involved in sensing external cues. The understanding of the relationships between genes and biological functions has become one of the hottest and most promising aspects in contemporary biology (Barabasi et al. 2011; Gerstein et al. 2012). However, the dynamic interplay of genes is highly complex and cannot be understood by a simple approach (Brazhnik et al. 2002). The reconstruction of gene regulatory networks has proven to be a valuable tool for identifying the key mechanisms that shape the dynamics of cellular and transcriptional processes (Zhu et al. 2012; Yosef et al. 2013; Hecker et al. 2009).

Modelling of biological regulatory networks regulated by gene expression using dynamic Bayesian networks has been popular since Murphy and Mian's (1999) pioneering work. However, the requirement of evenly spaced measurements limits its widespread application. Time course records of gene expression are usually based on the distinct phases of biological processes (Quint et al. 2012), some of which receives more dense measurements than the others. Furthermore, increasing computational studies tend to integrate gene expression data from different experiments, in order to gain a comprehensive regulatory network underlying a biological phenomenon (Hecker et al. 2009). Because of these, the time course data of gene expression for GRN reconstruction are generally sparse and irregular. Despite tremendous efforts to model sparsely measured gene networks (Yu et al. 2004), a systematical procedure for DBN modeling using such imperfect data has still not been available in the literature.

In this article, we reformed DBN modeling by interpolating missing data points based on functional clustering (Kim et al. 2008, 2010; Song et al. 2007; Luan and Li 2003; Wang et al. 2011). The new model can handle any dynamic gene expression data, no matter they are evenly spaced or not, thereby providing a broader tool in computational biology. The model was used to analyze two time course data sets of gene expression measured for vein bypass grafts in rabbits that receive two distinct treatments, high and low blood flow. The similarity and difference in the structure and organization of genetic networks can be identified under high and low flow, providing new insights into the mechanisms of how genes regulate each other to determine final phenotypic formation. We have performed extensive simulation studies to demonstrate the practical usefulness and utility of the new model. It should be noted that the functional clustering model we used is under the assumption of independence among different clusters. A general model that does not rely on this assumption has been developed by Zhang (2013). The implementation of Zhang's (2013) epistatic clustering may glean additional insight into the results of clustering dynamically differentiated genes and GRN reconstruction.

The past decade has witnessed tremendous milestones in high-throughput sequencing and large-scale data generation because of improvement in the accuracy of these techniques and their cost reduction for the required sample size. These developments enable researchers to not only dissect genomes but also unravel the regulatory interactions that allow genomes to regulate cellular structure, function, and behavior. The new model modified from a commonly used network modelling approach will find its widespread application given the popularity of collecting and using high-throughput expression data in human and other model or non-model systems. The model emphasizes on transcriptional data, but can be refined and extended to integrate multiple data types, such as mRNA and microRNA (miRNA) expression data, TF DNA-binding data, and protein interaction data (Bolouri 2014). Also, the model should be linked to complex phenotypes or diseases within a causal-effect network framework toward identifying phenotype- or disease-causing perturbations. The model can be further perfected to readily determine the time of onset and duration of transcriptional activity and the magnitude of expression of particular genes. Finally, the nature and topological features of regulatory networks may vary among different individuals, thus the identification and mapping of network-controlling quantitative trait loci (*n*QTLs) would be important for the prediction of network behavior.

Declaration

Ethics approval and consent to participate

Not applicable.

Consent for publication

Not applicable.

Competing interests

The authors declare that they have no competing interests.

Authors' contributions

RW and MG conceived and designed the experiments. SAB performed the experiments and cleaned the data set. YW analyzed the data. SAB and YW contributed materials/analysis tools. YW and RW wrote the paper. MG and SAB revised the paper. All authors read and approved the final version of manuscript.

Availability of data and material

The data that support the findings of this study are available from publications as specified in the main text. The codes are available from website https://sites.google.com/site/yxw179/software.

Abbreviations

GRN, gene regulatory network; BN, Bayesian network; DBN, dynamic Bayesian network; FNDA, functional data analysis; LOP, Legendre orthogonal polynomials; ARMA, autoregressive moving-average; BIC, Bayesian information criterion; GO, gene ontology; PPV, positive predictive value; FNR, false negative rate; TP, true positive; FN, false negative; TN, true negative; FP, false positive; *n*QTLs, network-controlling quantitative trait loci.

Acknowledgements This research was supported by National Institute of Health grants 1U10HL098115, 5U01HL119178 and 5UL1TR000127. Any opinions, findings, and conclusions or recommendations expressed in this publication are those of the authors and do not necessarily reflect the views of the National Institutes of Health.

References

Akutsu, T., Miyano, S., Kuhara, S.: Inferring qualitative relations in genetic networks and metabolic pathways. Bioinformatics. **16**, 727–734 (2000)

Aluru, S.: Handbook of Computational Molecular Biology. CRC Press, Boca Raton (2005)

Bansal, M., Della Gatta, G., Di Bernardo, D.: Inference of gene regulatory networks and compound mode of action from time course gene expression profiles. Bioinformatics. **22**, 815–822 (2006)

Barabasi, A.-L., Gulbahce, N., Loscalzo, J.: Network medicine: a network-based approach to human disease. Nat. Rev. Genet. **12**, 56–68 (2011)

Bolouri, H.: Modeling genomic regulatory networks with big data. Trends Genet. **30**, 182–191 (2014)

Brazhnik, P., de la Fuente, A., Mendes, P.: Gene networks: how to put the function in genomics. Trends Biotechnol. **20**, 467–472 (2002)

de Lichtenberg, U., Jensen, L.J., Brunak, S., et al.: Dynamic complex formation during the yeast cell cycle. Science. **307**, 724–727 (2005)

De Smet, I., Lau, S., Mayer, U., et al.: Embryogenesis - the humble beginnings of plant life. Plant J. **61**, 959–970 (2010)

Fernandez, C.M., Goldman, D.R., Jiang, Z., et al.: Impact of shear stress on early vein graft remodeling: a biomechanical analysis. Ann. Biomed. Eng. **32**, 1484–1493 (2004)

Friedman, N., Murphy, K., Russell, S.: Learning the structure of dynamic probabilistic networks. In: Proceedings of the Fourteenth Conference on Uncertainty in Artificial Intelligence, pp. 139–147. Morgan Kaufmann Publishers, San Francisco (1998)

Friedman, N., Linial, M., Nachman, I., et al.: Using Bayesian networks to analyze expression data. J. Comput. Biol. **7**, 601–620 (2000)

Gerstein, M.B., Kundaje, A., Hariharan, M., et al.: Architecture of the human regulatory network derived from ENCODE data. Nature. **489**, 91–100 (2012)

Godsey, B.: Improved inference of gene regulatory networks through integrated Bayesian clustering and dynamic modeling of time-course expression data. PLoS One. **8**, e68358 (2013)

Greenfield, A., Madar, A., Ostrer, H., et al.: DREAM4: combining genetic and dynamic information to identify biological networks and dynamical models. PLoS One. **5**, e13397 (2010)

Hecker, M., Lambeck, S., Toepfer, S., et al.: Gene regulatory network inference: data integration in dynamic models–a review. Biosystems. **96**, 86–103 (2009)

Hurley, D., Araki, H., Tamada, Y., et al.: Gene network inference and visualization tools for biologists: application to new human transcriptome datasets. Nucleic Acids Res. **40**, 2377–2398 (2012)

Jiang, Z., Wu, L., Miller, B.L., et al.: A novel vein graft model: adaptation to differential flow environments. Am. J. Phys. Heart Circ. Phys. **286**, H240–H245 (2004)

Kim, S.Y., Imoto, S., Miyano, S.: Inferring gene networks from time series microarray data using dynamic Bayesian networks. Brief. Bioinform. **4**(3), 228–235 (2003)

Kim, B.-R., Zhang, L., Berg, A., et al.: A computational approach to the functional clustering of periodic gene-expression profiles. Genetics. **180**, 821–834 (2008)

Kim, B.-R., McMurry, T., Zhao, W., et al.: Wavelet-based functional clustering for patterns of high-dimensional dynamic gene expression. J. Comput. Biol. **17**, 1067–1080 (2010)

Kourou, K., Exarchos, K.P., Papaloukas, C., Fotiadis, D.I.: A Bayesian network-based approach for discovering oral cancer candidate biomarkers. In: Engineering in Medicine and Biology Society (EMBC), 2015 37th Annual International Conference of the IEEE, pp. 7663–7666. IEEE, New York (2015)

Li, N., McMurry, T., Berg, A., et al.: Functional clustering of periodic transcriptional profiles through ARMA(p,q). PLoS One. **5**(4), e9894 (2010)
Li, Z., Li, P., Krishnan, A., Liu, J.: Large-scale dynamic gene regulatory network inference combining differential equation models with local dynamic Bayesian network analysis. Bioinformatics. **27**, 2686–2691 (2011)
Luan, Y., Li, H.: Clustering of time-course gene expression data using a mixed-effects model with B-splines. Bioinformatics. **19**(4), 474–482 (2003)
Martin, S., Zhang, Z., Martino, A., et al.: Boolean dynamics of genetic regulatory networks inferred from microarray time series data. Bioinformatics. **23**, 866–874 (2007)
Murphy, K., Mian, S.: Modelling gene expression data using dynamic Bayesian networks. Technical Report, Computer Science Division, University of California, Berkeley (1999)
Ogami, K., Yamaguchi, R., Imoto, S., et al.: Computational gene network analysis reveals TNF-induced angiogenesis. BMC Syst. Biol. **6**(Suppl 2), S12 (2012)
Ortiz-Gutiérrez, E., García-Cruz, K., Azpeitia, E., Castillo, A., de la Paz Sánchez, M., Álvarez-Buylla, E.R.: A dynamic gene regulatory network model that recovers the cyclic behavior of Arabidopsis thaliana cell cycle. PLoS Comput. Biol. **11**(9), e1004486 (2015)
Quint, M., Drost, H.G., Gabel, A., et al.: A transcriptomic hourglass in plant embryogenesis. Nature. **490**, 98–101 (2012)
Rustici, G., Mata, J., Kivinen, K., et al.: Periodic gene expression program of the fission yeast cell cycle. Nat. Genet. **36**, 809–817 (2004)
Song, J.J., Lee, H.J., Morris, J.S., Kang, S.: Clustering of time-course gene expression data using functional data analysis. Comput. Biol. Chem. **31**(4), 265–274 (2007)
Wang, Y., Xu, M., Wang, Z., et al.: How to cluster gene expression dynamics in response to environmental signals. Brief. Bioinform. **13**, 162–174 (2011)
Wang, J., Chen, B., Wang, Y., et al.: Reconstructing regulatory networks from the dynamic plasticity of gene expression by mutual information. Nucleic Acids Res. **41**, e97 (2013)
Wessels, L.F., van Someren, E.P., Reinders, M.J., et al.: A comparison of genetic network models. Pac. Symp. Biocomput. **6**, 508–519 (2001)
Xiang, D., Venglat, P., Tibiche, C., et al.: Genome-wide analysis reveals gene expression and metabolic network dynamics during embryo development in Arabidopsis. Plant Physiol. **156**, 346–356 (2011)
Yosef, N., Shalek, A.K., Gaublomme, J.T., et al.: Dynamic regulatory network controlling TH17 cell differentiation. Nature. **496**, 461–468 (2013)
Yu, J., Smith, V.A., Wang, P.P., et al.: Advances to Bayesian network inference for generating causal networks from observational biological data. Bioinformatics. **20**, 3594–3603 (2004)
Zhang, J.: Epistatic clustering: a model-based approach for identifying links between clusters. J. Am. Stat. Assoc. **108**, 1366–1384 (2013)
Zhang, X., Liu, K., Liu, Z.-P., et al.: NARROMI: a noise and redundancy reduction technique improves accuracy of gene regulatory network inference. Bioinformatics. **29**, 106–113 (2013)
Zhang, X., Zhao, J., Hao, J.K., et al.: Conditional mutual inclusive information enables accurate quantification of associations in gene regulatory networks. Nucleic Acids Res. **43**(5), e31 (2015)
Zhu, H., Rao, R.S.P., Zeng, T., et al.: Reconstructing dynamic gene regulatory networks from sample-based transcriptional data. Nucleic Acids Res. **40**, 10657–10667 (2012)
Zou, M., Conzen, S.D.: A new dynamic bayesian network (dbn) approach for identifying gene regulatory networks from time course microarray data. Bioinformatics. **21**, 71–79 (2005)

Chapter 6
Maximin Designs for Ultra-Fast Functional Brain Imaging

R. Alghamdi, A. Alrumayh, and M.-H. Kao

6.1 Introduction

Functional brain imaging is widely used in various fields such as cognitive neuroscience, economics, education, medical science, and psychology in studying functions of the human brain. In many functional neuroimaging studies, a sequence of mental stimuli (e.g., images or sounds) is presented to the experimental subject while time series data are collected from the subject's brain via a brain mapping technique such as functional magnetic resonance imaging (fMRI) or functional near-infrared spectroscopy (fNIRS). The collected data are analyzed to make inferences about the underlying brain activity evoked by the stimuli. For such experiments, a key first step is to judiciously select a good stimulus sequence to allow the collection of informative data for making valid statistical inferences. This is an important, yet challenging experimental design issue. In this study, we are concerned with this design issue, and present an efficient approach for tackling it. In contrast to existing works on this research line, we consider experiments where an ultra-fast brain mapping technique with a high temporal resolution is utilized. We propose an efficient approach to obtain high-quality designs (stimulus sequences) for such experiments.

For clarity, we put our focus on the general linear model of the following form:

$$\mathbf{y} = \mathbf{X}\mathbf{h} + \mathbf{S}\boldsymbol{\gamma} + \boldsymbol{\varepsilon}, \tag{6.1}$$

where $\mathbf{y} = (y_1, y_2, \ldots, y_T)^T$ is the response vector, the superscript $\mathtt{T}$ denotes the transpose, $\mathbf{h}$ is the parameter vector of interest, $\boldsymbol{\gamma}$ represents some nuisance

R. Alghamdi · A. Alrumayh · M.-H. Kao (✉)
School of Mathematical and Statistical Sciences, Arizona State University, Tempe, AZ, USA
e-mail: ralghamd@asu.edu; aalrumay@asu.edu; mkao3@asu.edu

L. Zhang et al. (eds.), *Contemporary Biostatistics with Biopharmaceutical Applications*, ICSA Book Series in Statistics,
https://doi.org/10.1007/978-3-030-15310-6_6

parameter vector, $\mathbf{X}$ is the design matrix determined by the selected design, $\mathbf{S}$ is a pre-specified matrix, and $\boldsymbol{\varepsilon}$ is error. This model is widely used in functional brain imaging studies, and several design approaches have been proposed for obtaining high-quality designs for a precise estimate of $\mathbf{h}$. For example, Buračas and Boynton (2002) advocated the use of m-sequence-based designs, Aguirre et al. (2011) proposed to consider De Bruijn sequences, and Kao (2014) and Cheng and Kao (2015) showed that designs generated from some Hadamard matrices are optimal for estimating $\mathbf{h}$ under certain conditions. Computer algorithms such as the genetic algorithms of Wager and Nichols (2003) and Kao et al. (2009), and the exchange algorithm of Saleh et al. (2017) are also available for obtaining good designs.

A major restriction of the previous works is that they focused only on brain mapping techniques having relatively low temporal resolutions; the time to repetition, τ_{TR}, which is the time between the collections of y_t and y_{t+1}, is often at least a couple of seconds (e.g., $\tau_{TR} = 2\,\text{s}$). With new advances in neuroscience, many recent studies involve the use of pioneering technology that allows a relatively high temporal resolution with τ_{TR} being only tens or hundreds of milliseconds. As an example, Proulx et al. (2014) considered ultra-fast fMRI, called MR-Encephalography, with $\tau_{TR} = 100\,\text{ms}$. Several other brain mapping techniques such as fNIRS can also attain a similar or an even higher temporal resolution; see also Table 1 of Scholkmann et al. (2014). For these studies, the previously mentioned design approaches can become clumsy in identifying good designs. This is mainly due to the greatly enlarged dimension of the design matrix $\mathbf{X}$, which makes it very difficult, if not infeasible, to compute and compare the statistical efficiencies of the many competing designs. In addition, the efficiency of the design will depend on the values of the error correlation parameters, which are almost always uncertain at the design stage. We thus would like to obtain a design that performs relatively well across the possible values of these correlation parameters. This, unfortunately, is notoriously difficult (see also, Kao and Mittelmann 2014), and it makes the present design issue very challenging.

Here, we propose an efficient approach for tackling this challenging design issue. After introducing the relevant background information in the next section, we describe our proposed approach in Sect. 6.3. Case studies for demonstrating the usefulness of our approach are presented in Sect. 6.4. The conclusion and a discussion can be found in Sect. 6.5.

6.2 Background

For clarity, we describe our experimental setting by considering an fMRI experiment where a sequence of mental stimuli of Q different types (e.g., Q different images) is presented to an experimental subject. Starting at Time 0, each stimulus can possibly occur every τ_{ISI} seconds (e.g., $\tau_{ISI} = 4\,\text{s}$). The duration of the experiment is $(N-1)\tau_{ISI}$ for some integer N. For a typical 5 to 10 min study, N can be tens or hundreds. We use $\mathbf{d} = (d_1, \ldots, d_N)$ to represent a design (i.e. a stimulus sequence), where $d_n \in \{0, 1, \ldots, Q\}$ determines the type of stimulus to be presented at Time $(n-1)\tau_{ISI}$. Specifically, $d_n = 0$ indicates no stimulus presentation, whereas $d_n = q > 0$ means that a qth-type stimulus will be presented. Throughout the experiment, an fMRI scanner repeatedly scans each voxel (3D imaging unit) of the subject's brain every τ_{TR} seconds to collect the blood oxygenation level dependent (BOLD) signals, $\mathbf{y}$. Note that τ_{TR} can be much smaller than τ_{ISI}, and for simplicity, we further assume that τ_{TR} divides τ_{ISI}. The collected $\mathbf{y}$ is analyzed for studying the hemodynamic response function (HRF) evoked by each stimulus. Specificially, the HRF is a function of time describing the (error-free) change over time of the BOLD signals following a stimulus onset. Its duration τ_{dur} is relatively long, and counting from the onset of the stimulus, it may take $\tau_{dur} = 32$ s for the HRF to completely return to the baseline (Friston et al. 2007). Studying the HRF helps to understand the effect of each stimulus type to the brain. With Q stimulus types in the study, we have Q HRFs to estimate.

Model (6.1) is commonly used for analyzing the response $\mathbf{y}$. In particular, we set $\mathbf{h} = (\mathbf{h}_1^T, \ldots, \mathbf{h}_Q^T)^T$, where $\mathbf{h}_q = (h_{q1}, \ldots, h_{qK})^T$ is the HRF parameter vector for the qth-type stimulus; h_{qk} represents the kth height of the HRF evaluated $(k-1)\tau_{TR}$ seconds after the stimulus onset, $K = \lfloor \tau_{dur}/\tau_{TR} \rfloor + 1$, and $\lfloor a \rfloor$ is the integer part of a. By estimating $\mathbf{h}_q$, the properties of the HRF of the qth stimulus type can be investigated. The design matrix $\mathbf{X}$ in Model (6.1) is a 0–1 matrix that can be partitioned as $\mathbf{X} = [\mathbf{X}_1, \ldots, \mathbf{X}_Q]$, where $\mathbf{X}_q$ is the design matrix for $\mathbf{h}_q$. The elements of the design matrix $\mathbf{X}_q$ are determined by the selected design $\mathbf{d}$ and the pre-specified τ_{TR} and τ_{ISI}; see Saleh et al. (2017), and Kao et al. (2012). Specifically, the (t, k)th element of $\mathbf{X}_q$ is 1 when h_{qk} contributes to y_t, and is 0 otherwise; $t = 1, \ldots, T,\ k = 1, \ldots, K$. The nuisance term $\mathbf{S}\boldsymbol{\gamma}$ in (6.1) allows for a drift/trend of $\mathbf{y}$, and it may, for example, be set to represent a polynomial drift; e.g., Liu (2004). The error term $\boldsymbol{\varepsilon}$ in Model (6.1) is commonly assumed to follow a stationary $AR(p)$ process; see also Worsley et al. (2002).

Following Kiefer (1959), the quality of a design $\mathbf{d}$ in estimating $\mathbf{h}$ is evaluated by some function (i.e. optimality criterion) of the information matrix $\mathbf{M}(\mathbf{d}; \boldsymbol{\phi})$ of $\mathbf{h}$:

$$\mathbf{M}(\mathbf{d}; \boldsymbol{\phi}) = \mathbf{X}^T\{\Sigma(\boldsymbol{\phi})^{-1} - \Sigma(\boldsymbol{\phi})^{-1}\mathbf{S}[\mathbf{S}^T\Sigma(\boldsymbol{\phi})^{-1}\mathbf{S}]^{-1}\mathbf{S}^T\Sigma(\boldsymbol{\phi})^{-1}\}\mathbf{X}. \quad (6.2)$$

Here, $\Sigma(\boldsymbol{\phi}) = cov(\boldsymbol{\varepsilon})$ is the variance-covariance matrix of $\boldsymbol{\varepsilon}$; and the vector $\boldsymbol{\phi}$ represents the parameters in $\Sigma(\boldsymbol{\phi})$. $\mathbf{M}(\mathbf{d}; \boldsymbol{\phi})$ depends on the design $\mathbf{d}$ through the design matrix $\mathbf{X}$. For a $\mathbf{d}$ that makes $\mathbf{h}$ estimable, $\mathbf{M}(\mathbf{d}; \boldsymbol{\phi})$ is the inverse of the variance-covariance matrix $cov(\hat{\mathbf{h}})$ of $\hat{\mathbf{h}}$, the generalized least-squares estimate of $\mathbf{h}$. We aim at selecting a $\mathbf{d}$ that, in some sense, maximizes $\mathbf{M}(\mathbf{d}; \boldsymbol{\phi})$ so that the estimate of $\mathbf{h}$ is precise. The value of the error variance σ^2 (> 0) is assumed fixed across designs, and it does not affect the selection of an optimal design. Without loss of generality, we set $\sigma^2 = 1$. Popularly used optimality criteria for design selection include the A-optimality criterion, $F(\mathbf{d}; \boldsymbol{\phi}) = m/trace[\mathbf{M}^{-1}(\mathbf{d}; \boldsymbol{\phi})]$, and the D-optimality criterion, $F(\mathbf{d}; \boldsymbol{\phi}) = det[\mathbf{M}(\mathbf{d}; \boldsymbol{\phi})]^{1/m}$, where m is the size of $\mathbf{M}(\mathbf{d}; \boldsymbol{\phi})$. Here, we formulate these criteria as the larger-the-better criteria, and set the criterion value to 0 for those designs making the information matrix $\mathbf{M}(\mathbf{d}; \boldsymbol{\phi})$ singular.

Maximizing the A-criterion allows us to identify a design that minimizes the average variance of the parameter estimates, and the D-criterion helps to minimize the volume of a confidence ellipsoid for $\mathbf{h}$. Several authors consider the A-criterion for selecting fMRI designs; e.g., Friston et al. (1999), Dale (1999), Buračas and Boynton (2002), and Liu and Frank (2004). The D-criterion is also not uncommon (Maus et al. 2010). In our case study, we use the A-optimality criterion, but D- or other criteria can also be considered.

A design, $\mathbf{d}_{\phi}^{*}$, maximizing the optimality criterion $F(\mathbf{d}; \boldsymbol{\phi})$ with a given $\boldsymbol{\phi}$-value is considered as a locally optimal design (Chernoff 1953). Under the selected optimality criterion, the design $\mathbf{d}_{\phi}^{*}$ is optimal for the given value of $\boldsymbol{\phi}$, but its statistical efficiency for another $\boldsymbol{\phi}$-value is not guaranteed. When the value of $\boldsymbol{\phi}$ is uncertain, we normally would like to find a design that is rather robust to a misspecification of the $\boldsymbol{\phi}$-value. A possible approach for obtaining such a design is the maximin approach considered in Maus et al. (2010), and Kao et al. (2013). In particular, this is to obtain a maximin design $\mathbf{d}_{Mm}^{*}$ that maximizes

$$\min_{\phi \in \Omega} RE(\mathbf{d}; \mathbf{d}_{\phi}^{*}) = \min_{\phi \in \Omega} \frac{F(\mathbf{d}; \boldsymbol{\phi})}{F(\mathbf{d}_{\phi}^{*}; \boldsymbol{\phi})}, \tag{6.3}$$

where Ω is the parameter space of $\boldsymbol{\phi}$. Three layers of optimization are involved in solving this problem. In the innermost layer, a locally optimal design $\mathbf{d}_{\phi}^{*}$ that maximizes $F(\mathbf{d}; \boldsymbol{\phi})$ will need to be identified for each of the (many) possible values of $\boldsymbol{\phi}$. With these $\mathbf{d}_{\phi}^{*}$s, we then find $\min_{\phi \in \Omega} RE(\mathbf{d}; \mathbf{d}_{\phi}^{*})$ for each candidate design $\mathbf{d}$ in the second layer of optimization. The outermost layer is to achieve a design yielding the maximal value of $\min_{\phi \in \Omega} RE(\mathbf{d}; \mathbf{d}_{\phi}^{*})$. Solving such a maximin design problem normally requires much computational effort. This is especially true for our current setting. The main difficulty is that the dimension of the information matrix $\mathbf{M}(\mathbf{d}; \boldsymbol{\phi})$ in (6.2) is typically large (e.g., hundreds by hundreds), making it very time consuming (or even infeasible) to evaluate the $F(\mathbf{d}; \boldsymbol{\phi})$-values for each candidate design $\mathbf{d}$ at each $\boldsymbol{\phi}$-value in both the innermost and outermost layers of optimization. In the next section, we propose an efficient shortcut method that allows us to achieve a design yielding a high value of min RE.

6.3 Our Proposed Approach

Our proposed approach for obtaining $\mathbf{d}_{Mm}^{*}$ that maximizes min RE of (6.3) consists of two components. The first component involves the formulation of an easy-to-compute surrogate criterion, $F_s(\mathbf{d}; \boldsymbol{\phi})$, of $F(\mathbf{d}; \boldsymbol{\phi})$. To achieve such a surrogate, we first judiciously select a subsample of the columns of the design matrix $\mathbf{X}$ to allow for a reduction in the dimension of the information matrix $\mathbf{M}(\mathbf{d}; \boldsymbol{\phi})$. The selected columns form a submatrix $\mathbf{X}_s$ of $\mathbf{X}$, which in turn gives the following information matrix of a reduced size.

$$\mathbf{M}_s(\mathbf{d};\boldsymbol{\phi}) = \mathbf{X}_s^T\{\Sigma(\boldsymbol{\phi})^{-1} - \Sigma(\boldsymbol{\phi})^{-1}\mathbf{S}[\mathbf{S}^T\Sigma(\boldsymbol{\phi})^{-1}\mathbf{S}]^{-1}\mathbf{S}^T\Sigma(\boldsymbol{\phi})^{-1}\}\mathbf{X}_s; \quad (6.4)$$

the notation used here is the same as those in (6.2). The surrogate criterion $F_s(\mathbf{d};\boldsymbol{\phi})$ is then formulated based on $\mathbf{M}_s(\mathbf{d};\boldsymbol{\phi})$. Specifically, $F_s(\mathbf{d};\boldsymbol{\phi}) = m_s/trace[\mathbf{M}_s^{-1}(\mathbf{d};\boldsymbol{\phi})]$ for the A-criterion, and $F_s(\mathbf{d};\boldsymbol{\phi}) = det[\mathbf{M}_s(\mathbf{d};\boldsymbol{\phi})]^{1/m_s}$ for the D-criterion; m_s is the size of $\mathbf{M}_s(\mathbf{d};\boldsymbol{\phi})$. It is noteworthy that $\mathbf{M}_s(\mathbf{d};\boldsymbol{\phi})$ is the information matrix of $\mathbf{h}_s$ in the following model:

$$\mathbf{y} = \mathbf{X}_s\mathbf{h}_s + \mathbf{S}\boldsymbol{\gamma} + \boldsymbol{\varepsilon},$$

where $\mathbf{h}_s$ is a subvector of $\mathbf{h}$ of (6.1), and $\mathbf{X}_s$ is the corresponding design matrix; this model is a submodel of (6.1).

The previously mentioned simple idea of subsampling the columns of $\mathbf{X}$ turns out to be effective for solving our problem. We observe that the designs optimizing $F_s(\mathbf{d};\boldsymbol{\phi})$ tend also to yield very high efficiency under $F(\mathbf{d};\boldsymbol{\phi})$. The key issue in the implementation of this idea is the selection of $\mathbf{X}_s$. Unfortunately, an imprudently selected subsampling plan such as a random sampling can give a poor result. With τ_{TR} being a divisor of τ_{ISI}, we find it useful to consider the following three steps for obtaining $\mathbf{X}_s$:

Step X-1 Partition $\mathbf{X}$ as $\mathbf{X} = [\mathbf{X}_1, \ldots, \mathbf{X}_Q]$ where $\mathbf{X}_q$ is the 0–1 design matrix for stimuli of the qth type.
Step X-2 Set $\tau_s = \tau_{ISI}/\tau_{TR}$. Keep Columns 1, $(1+\tau_s)$, $(1+2\tau_s)$, ... of $\mathbf{X}_q$ and leave out the other columns to form $\mathbf{X}_{sq}$ for $q = 1, \ldots, Q$.
Step X-3. Set $\mathbf{X}_s = [\mathbf{X}_{s1}, \mathbf{X}_{s2}, \ldots, \mathbf{X}_{sQ}]$.

Our experience suggests that some other subsampling plans such as those by replacing τ_s with its divisors can also lead to a good surrogate for $F(\mathbf{d};\boldsymbol{\phi})$. Nevertheless, the previously mentioned procedure gives a greater reduction in the dimension of the information matrix and is thus recommended.

In addition, we borrow the Kriging method that is widely considered in computer experiments (Santner et al. 2003) to approximate (the surface of) the RE-values of each design $\mathbf{d}$ over the parameter space Ω. To this end, we first select a (small) set of b points, $\boldsymbol{\phi}_1, \ldots, \boldsymbol{\phi}_b$, from Ω, and obtain the b corresponding locally optimal designs. This allows us to evaluate the RE-values of a given design $\mathbf{d}$ at the b selected $\boldsymbol{\phi}$-values. By using $W(\boldsymbol{\phi})$ to denote the RE-value at $\boldsymbol{\phi}$, we consider the following model:

$$W(\boldsymbol{\phi}) = \beta + Z(\boldsymbol{\phi}),$$

where β is an unknown parameter, and $Z(\boldsymbol{\phi})$ is a Gaussian process with mean 0 and $Cov(Z(\boldsymbol{\phi}), Z(\boldsymbol{\phi}')) = \sigma_z^2 R(\boldsymbol{\phi}, \boldsymbol{\phi}')$ for all $\boldsymbol{\phi}, \boldsymbol{\phi}' \in \Omega$, $\sigma_z^2 > 0$, and $R(\boldsymbol{\phi}, \boldsymbol{\phi}') = \exp\{-\theta\|\boldsymbol{\phi} - \boldsymbol{\phi}'\|_2^2\}$. Let $\mathbf{j}_b$ be the vector of b ones, $\mathbf{w} = (W(\boldsymbol{\phi}_1), \ldots, W(\boldsymbol{\phi}_b))^T$, $\mathbf{R} = ((R(\boldsymbol{\phi}_i, \boldsymbol{\phi}_j))_{i,j=1,\ldots,b}$ be the b-by-b correlation matrix of $\mathbf{w}$, and for given

$\boldsymbol{\phi}_0 \in \Omega$, $\mathbf{r}_0 = (R(\boldsymbol{\phi}_0, \boldsymbol{\phi}_1), R(\boldsymbol{\phi}_0, \boldsymbol{\phi}_2), \ldots, R(\boldsymbol{\phi}_0, \boldsymbol{\phi}_b))^T$. We then approximate the RE-value $W(\boldsymbol{\phi}_0)$ by the best linear unbiased predictor (BLUP):

$$\hat{W}(\boldsymbol{\phi}_0) = \hat{\beta} + \mathbf{r}_0^T \mathbf{R}^{-1}(\mathbf{w} - \mathbf{j}_b \hat{\beta}), \qquad (6.5)$$
$$\hat{\beta} = \mathbf{j}_b^T \mathbf{R}^{-1}\mathbf{w}/(\mathbf{j}_b^T \mathbf{R}^{-1}\mathbf{j}_b).$$

The θ parameter in $\mathbf{R}$ can be estimated by, e.g., the likelihood method; see Sect. 3.3.2 of Santner et al. (2003) for details. In our case study, we set $\theta = 10$, and find that other values of θ do not lead to a significant improvement in the final result.

We now are ready to present the steps of our proposed method for obtaining a maximin design. This method is referred to as **Method IA** in the next section.

Step 1 For a given b, select $\boldsymbol{\phi}_1, \ldots \boldsymbol{\phi}_b$ from Ω.

Step 2 For a specified optimality criterion $F(\mathbf{d}; \boldsymbol{\phi})$, obtain locally optimal designs $\mathbf{d}^*_{s,\phi_i} = argmax_d F_s(\mathbf{d}; \boldsymbol{\phi}_i)$ for $i = 1, \ldots, b$. Here, $F_s(\mathbf{d}; \boldsymbol{\phi})$ is calculated based on the $\mathbf{X}_s$ obtained from the previously described Steps X-1 to X-3.

Step 3 Obtain $\mathbf{d}^*_{Mm}$ that maximizes $\min_\phi RE_s(\mathbf{d}; \mathbf{d}^*_{s,\phi}) = \min_\phi F_s(\mathbf{d}; \boldsymbol{\phi}) / F_s(\mathbf{d}^*_{s,\phi}; \boldsymbol{\phi})$. Here, $\min_\phi RE_s(\mathbf{d}; \mathbf{d}^*_\phi)$ for a design $\mathbf{d}$ is approximated by using the following steps:

- Step 3-1 Set $W(\boldsymbol{\phi}) = RE_s(\mathbf{d}; \mathbf{d}^*_{s,\phi})$. Obtain $W(\boldsymbol{\phi}_1), \ldots, W(\boldsymbol{\phi}_b)$ by using the $\mathbf{d}^*_{s,\phi_i}$ obtained in Step 2.
- Step 3-2 Use the BLUP in (6.5) to predict $W(\boldsymbol{\phi})$ for each $\boldsymbol{\phi}$ in, e.g., a fine grid over Ω.
- Step 3-3 Approximate $\min_\phi W(\boldsymbol{\phi})$ with the minimum of the $\hat{W}(\boldsymbol{\phi})$'s obtained in Step 3-2.

For Step 1, it is often recommended to set $b = 10p$, where p is the dimension of Ω, and to consider a space-filling sampling plan for selecting $\boldsymbol{\phi}_1, \ldots \boldsymbol{\phi}_b$. The latter suggestion is partly because the surfaces of the RE-value are uncertain, and a space-filling sampling plan tends to give a representative sample from the different regions of Ω to facilitate the explorations of these uncertain surfaces. A sampling plan that possesses the space-filling property is the well-known Sobol sequence (Niederreiter 1988); we will consider this sampling plan in our case study (Sect. 6.4). When searching for the locally optimal designs in Step 2 and the maximin designs in Step 3, we choose to adopt the genetic algorithm of Kao et al. (2009) which is briefly described in the Appendix. Nevertheless, our proposed method is not restricted to this search algorithm and can be easily applied when another algorithm is considered. In the next section, we describe the implementation of our proposed method and demonstrate its performance with case studies.

6.4 Case Study

In this section, we obtain maximin designs by considering three different scenarios that, respectively, have $(Q, N) = (1, 127)$, $(2, 121)$, and $(3, 127)$. For all three cases, we set $(\tau_{ISI}, \tau_{TR}) = (4, 0.1)$; this means that a stimulus can possibly be presented to the experimental subject every $\tau_{ISI} = 4$ s, and the response sampling rate is 10 Hz. The lengths of the response vector $\mathbf{y}$ are then $T = 5080, 4840$, and 5080 for the three scenarios, respectively. Each stimulus is assumed to evoke an HRF that has a duration of $\tau_{dur} = 32$ s. With $\tau_{TR} = 0.1$ s, the HRF parameter vector $\mathbf{h}_q$ for the qth-type stimulus has a length of 321 $(= \lfloor \tau_{dur}/\tau_{TR} \rfloor + 1)$, and $\mathbf{h} = (\mathbf{h}_1, \ldots, \mathbf{h}_Q)$ of Model (6.1) has $321Q$ elements. The term $\mathbf{S}\boldsymbol{\gamma}$ of Model (6.1) is set to allow a second-order polynomial drift (Liu 2004), and the error term $\boldsymbol{\varepsilon} = ((\varepsilon_t))_{t=1,\ldots,T}$ is assumed to follow a stationary $AR(2)$ process. In particular, $\varepsilon_t = \phi_1\varepsilon_{t-1} + \phi_2\varepsilon_{t-2} + z_t$ where z_t is white noise, and ϕ_1, and ϕ_2 are unknown parameters satisfying

$$\phi_1 \in \begin{cases} [0, 0.5(1-\phi_2)], & \text{if } \phi_2 \in [0, 1/3]; \\ [0, \sqrt{0.5(1-2\phi_2)(1-\phi_2)}], & \text{if } \phi_2 \in (1/3, 0.5]. \end{cases} \tag{6.6}$$

The parameter space $\Omega \subset [0, 0.5]^2$ of $\boldsymbol{\phi} = (\phi_1, \phi_2)^T$ is presented as the unshaded area in Fig. 6.1. With this Ω, the first- and second-order autocorrelation coefficients of $\boldsymbol{\varepsilon}$ range between 0 and 0.5.

Four methods are considered for obtaining maximin designs. **Method IA** is our proposed method described in the previous section. In particular, we first obtain 27 Sobol points for $\boldsymbol{\phi} = (\phi_1, \phi_2)^T$ over $[0, 0.5]^2$ by using the statistics toolbox function `Sobolset` of MATLAB, and exclude those points that fall outside Ω. The remaining $b = 20$ Sobol points, $\boldsymbol{\phi}_1, \ldots, \boldsymbol{\phi}_{20}$, are presented in Fig. 6.1. We then adopt the genetic algorithm of Kao et al. (2009) to obtain a locally optimal design $\mathbf{d}^*_{s,\phi_i}$ that maximizes the A-optimality criterion $F_s(\mathbf{d}; \boldsymbol{\phi}_i) = m_s/trace\{\mathbf{M}_s^{-1}(\mathbf{d}; \boldsymbol{\phi}_i)\}$, where $\mathbf{M}_s(\mathbf{d}; \boldsymbol{\phi})$ is defined in (6.4), and m_s is the size of $\mathbf{M}_s(\mathbf{d}; \boldsymbol{\phi})$; $i = 1, \ldots, 20$. As described in Step 3 of our proposed approach, we then use the 20 $\mathbf{d}^*_{s,\phi_i}$s to approximate $\min_{\phi} RE_s(\mathbf{d}; \mathbf{d}^*_{s,\phi})$ for obtaining a maximin design $\mathbf{d}^*_{Mm}$. Specifically, this is done by considering a set $G_1 \subset \Omega$ with $G_1 = \tilde{G}_1 \cap \Omega$, and $\tilde{G}_1 = \{(\phi_1, \phi_2) \mid \phi_i = 0, 0.04, 0.08, \ldots, 0.48, i = 1, 2\}$ forms a grid on $[0, 0.5]^2$. We note that G_1 has a total of 124 (grid) points. For each $\mathbf{d}$, the BLUP of (6.5) is then used to approximate $W(\boldsymbol{\phi}) = RE_s(\mathbf{d}; \mathbf{d}^*_{s,\phi})$ for all $\boldsymbol{\phi} \in G_1$, and $\min_{\phi \in G_1} RE_s(\mathbf{d}; \mathbf{d}^*_{s,\phi})$ is obtained. We again use the genetic algorithm to obtain a $\mathbf{d}^*_{Mm}$ that maximizes $\min_{\phi \in G_1} RE_s(\mathbf{d}; \mathbf{d}^*_{s,\phi})$.

Method IB is similar to Method IA, but we directly use $\mathbf{X}$ instead of its subsampled version $\mathbf{X}_s$. Specifically, we first search for locally optimal designs, $\mathbf{d}^*_{\phi_1}$, $\ldots$, $\mathbf{d}^*_{\phi_{20}}$, for the 20 Sobol points that maximize $F(\mathbf{d}; \boldsymbol{\phi}) = m/trace\{\mathbf{M}^{-1}(\mathbf{d}; \boldsymbol{\phi})\}$, where $\mathbf{M}(\mathbf{d}; \boldsymbol{\phi})$ is the information matrix defined in (6.2), and m is its size. With these $\mathbf{d}^*_{\phi_i}$ s, we obtain the relative efficiencies $RE(\mathbf{d}; \mathbf{d}^*_{\phi_i}) = F(\mathbf{d}; \boldsymbol{\phi}_i)/F(\mathbf{d}^*_{\phi_i}; \boldsymbol{\phi}_i)$, $i = 1, \ldots, 20$, for given $\mathbf{d}$. By setting $W(\boldsymbol{\phi}) = RE(\mathbf{d}; \mathbf{d}^*_{\phi_i})$, we approximate $\min_{\phi \in G_1} RE(\mathbf{d}; \mathbf{d}^*_{\phi})$ by considering the same method as in Method IA. The genetic algorithm is used to search for a maximin design that maximizes the approximated $\min_{\phi \in G_1} RE(\mathbf{d}; \mathbf{d}^*_{\phi})$. We note that, with our current settings, the size of the information matrix $\mathbf{M}(\mathbf{d}; \boldsymbol{\phi})$ is $m = 321Q$ whereas that for the $\mathbf{M}_s(\mathbf{d}; \boldsymbol{\phi})$ used in Method IA is $m_s = 9Q$; $Q = 1, 2, 3$. The computing time for calculating $F(\mathbf{d}; \boldsymbol{\phi})$ in Method IB is expected to be much greater than that for calculating $F_s(\mathbf{d}; \boldsymbol{\phi})$ in Method IA.

For the remaining two methods that we consider, we obtain $\min_{\phi \in G_1} RE_s(\mathbf{d}; \mathbf{d}^*_{s,\phi})$ and $\min_{\phi \in G_1} RE(\mathbf{d}; \mathbf{d}^*_{\phi})$ without resorting to the BLUP of (6.5) used in Methods IA and IB. In particular, for **Method IIA**, we first obtain 124 locally optimal designs maximizing $F_s(\mathbf{d}; \boldsymbol{\phi})$ for the 124 grid points of $\boldsymbol{\phi}$ in G_1. This allows a direct calculation of $\min_{\phi \in G_1} RE_s(\mathbf{d}; \mathbf{d}^*_{s,\phi})$ for each $\mathbf{d}$. We then use the genetic algorithm to search for a maximin design maximizing $\min_{\phi \in G_1} RE_s(\mathbf{d}; \mathbf{d}^*_{s,\phi})$. For **Method IIB**, we consider the same grid-point method as Method IIA, but respectively replace $F_s(\mathbf{d}; \boldsymbol{\phi})$ and $RE_s(\mathbf{d}; \mathbf{d}^*_{s,\phi})$ with $F(\mathbf{d}; \boldsymbol{\phi})$ and $RE(\mathbf{d}; \mathbf{d}^*_{\phi})$; i.e., $\mathbf{X}_s$ is used in Method IIA, whereas the full design matrix $\mathbf{X}$ is considered in Method IIB. We note that a similar grid-point method is considered by Maus et al. (2010) and Kao and Mittelmann (2014) for obtaining fMRI designs although both previous works did not consider the use of the subsampled design matrix $\mathbf{X}_s$.

All the computations in this section are implemented on a desktop computer of a 3.7GHz Intel Core i7-8700k 6-core processor with 32GB RAM. We consider all the four methods for obtaining designs with $(Q, N) = (1, 127)$. Table 6.1 summarizes (1) the number of locally optimal designs needed for the four methods, (2) total computing time spent on obtaining all these locally optimal designs, and (3) the minimum of $RE(\mathbf{d}^*_{s,\phi}; \mathbf{d}^*_{\phi}) = F(\mathbf{d}^*_{s,\phi}, \boldsymbol{\phi})/F(\mathbf{d}^*_{\phi}, \boldsymbol{\phi})$ for the locally optimal designs $\mathbf{d}^*_{s,\phi}$ obtained by Methods IA and IIA. By using the subsampled design

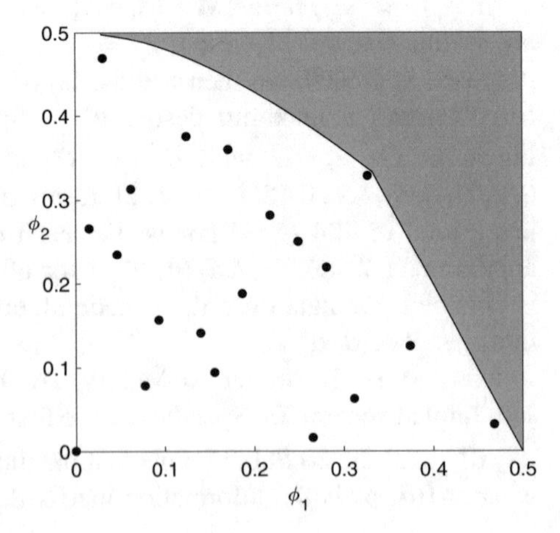

Fig. 6.1 The parameter space Ω and 20 Sobol points

matrix $\mathbf{X}_s$, the computing time needed for Method IA in obtaining locally optimal designs is about half of that of Method IB; the same is true when we compare the computing times of Methods IIA and IIB in Table 6.1. Nevertheless, all the $\mathbf{d}^*_{s,\phi}$ obtained by Methods IA and IIA still attain very high efficiencies under $F(\mathbf{d}; \boldsymbol{\phi})$. As presented in the last row of Table 6.1, the achieved $F(\mathbf{d}^*_{s,\phi}; \boldsymbol{\phi})$ is at least 98.7% of $F(\mathbf{d}^*_{\phi}; \boldsymbol{\phi})$. This result suggests that finding an optimal design by maximizing $F_s(\mathbf{d}; \boldsymbol{\phi})$ requires a much less computing time than optimizing $F(\mathbf{d}; \boldsymbol{\phi})$, but there is not much difference in the efficiencies of the obtained designs. $F_s(\mathbf{d}; \boldsymbol{\phi})$ thus serves as an efficient surrogate criterion for $F(\mathbf{d}; \boldsymbol{\phi})$. We consistently make the same observations in other case studies that will be reported elsewhere.

Table 6.1 Locally optimal designs (LODs) for the four methods with $(Q, N) = (1, 127)$

Method (LODs)	IA $(\mathbf{d}^*_{s,\phi})$	IB $(\mathbf{d}^*_{\phi})$	IIA $(\mathbf{d}^*_{s,\phi})$	IIB $(\mathbf{d}^*_{\phi})$
Number of LODs	20	20	124	124
Total time spent (min.)	30.55	59.13	192.65	392.33
$\min RE(\mathbf{d}^*_{s,\phi}; \mathbf{d}^*_{\phi})$	0.987	–	0.994	–

In Table 6.2, we compare the performance of the four methods in obtaining maximin designs, $\mathbf{d}^*_{Mm}$, for $(Q, N) = (1, 127)$. For comparison purposes, the reported computing time in the first row of that table does not include the time needed for obtaining the locally optimal designs, which can be found in Table 6.1. We note that the total times needed for obtaining $\mathbf{d}^*_{Mm}$ are 1.01, 1.73, 6.19 and 10.94 h for Methods IA, IB, IIA, and IIB, respectively. Our proposed approach (Method IA) is the most efficient while the times needed for Methods IIA and IIB are intimidating. For evaluating the performance of the obtained $\mathbf{d}^*_{Mm}$, we consider the set $G_2 = \tilde{G}_2 \cap \Omega$, where $\tilde{G}_2 = \{(\phi_1, \phi_2) \mid \phi_i = 0, 0.02, 0.04, \ldots, 0.5, i = 1, 2\}$ is a grid on $[0, 0.5]^2$ that is finer than $\tilde{G}_1$. A total of 464 grid points in $\tilde{G}_2$ fall inside Ω, and G_2 contains these 464 different values of $\boldsymbol{\phi}$. We obtain a locally optimal design $\mathbf{d}^*_{s,\phi}$ for each of these $\boldsymbol{\phi} \in G_2$, and use these $\mathbf{d}^*_{s,\phi}$s to calculate $\min_{\phi \in G_2} RE(\mathbf{d}^*_{Mm}; \mathbf{d}^*_{s,\phi}) = \min_{\phi \in G_2} F(\mathbf{d}^*_{Mm}; \boldsymbol{\phi})/F(\mathbf{d}^*_{s,\phi}; \boldsymbol{\phi})$ for the $\mathbf{d}^*_{Mm}$ obtained from each of the four methods. Note that $\min_{\phi \in G_2} RE(\mathbf{d}^*_{Mm}; \mathbf{d}^*_{s,\phi})$ is an approximation of $\min_{\phi \in G_2} RE(\mathbf{d}^*_{Mm}; \mathbf{d}^*_{\phi})$ since the former is calculated based on $\mathbf{d}^*_{s,\phi}$ instead of $\mathbf{d}^*_{\phi}$. Obtaining 464 $\mathbf{d}^*_{s,\phi}$s already requires much computing time, and the time needed for obtaining $\mathbf{d}^*_{\phi}$ is even more time consuming; see also the comparisons of Methods IA and IIA versus Methods IB and IIB in Table 6.1. For comparison purposes, we also use the genetic algorithm to search for these 464 $\mathbf{d}^*_{\phi}$ for $(Q, N) = (1, 127)$, and obtain $\min_{\phi \in G_2} RE(\mathbf{d}^*_{Mm}; \mathbf{d}^*_{\phi})$ as reported in the last row of Table 6.2. We see that $\min_{\phi \in G_2} RE(\mathbf{d}^*_{Mm}; \mathbf{d}^*_{s,\phi})$ provides a good approximation of $\min_{\phi \in G_2} RE(\mathbf{d}^*_{Mm}; \mathbf{d}^*_{\phi})$. As also shown in Table 6.2, the $\mathbf{d}^*_{Mm}$s obtained by the four methods have similar performances, and they all achieve at least

99% relative efficiency over G_2. These results indicate that our proposed Method IA can efficiently obtain a good maximin design.

Table 6.2 Maximin designs $\mathbf{d}^*_{Mm}$ from the four methods with $Q = 1$

Method	IA	IB	IIA	IIB
Time spent (min.)	30.26	44.53	178.52	263.80
$\min_{\phi \in G_2} RE(\mathbf{d}^*_{Mm}; \mathbf{d}^*_{s,\phi})$	0.997	0.995	0.994	0.997
$\min_{\phi \in G_2} RE(\mathbf{d}^*_{Mm}; \mathbf{d}^*_{\phi})$	0.994	0.992	0.991	0.994

For $Q = 2$ and $Q = 3$, the size of the full information matrix $\mathbf{M}(\mathbf{d}; \boldsymbol{\phi})$ becomes large; specifically, $m = 642$ for $Q = 2$, and $m = 963$ for $Q = 3$. By considering the full information matrix, Methods IB and IIB quickly become computationally very difficult. We thus consider only Methods IA and IIA and obtain locally optimal designs $\mathbf{d}^*_{s,\phi}$ and maximin designs $\mathbf{d}^*_{Mm}$ with $\mathbf{M}_s(\mathbf{d}; \boldsymbol{\phi})$ (and thus $\mathbf{X}_s$). In Table 6.3, we present the total computing time (i.e. the sum of the time spent on obtaining $\mathbf{d}^*_{s,\phi}$s and that on obtaining $\mathbf{d}^*_{Mm}$) for the two methods. The minimum relative efficiency $RE(\mathbf{d}^*_{Mm}; \mathbf{d}^*_{s,\phi})$ of the obtained $\mathbf{d}^*_{Mm}$ over G_2 is also reported there. Again, our proposed Method IA is more efficient than the grid method (Method IIA), and it achieves a maximin design that has a very high relative efficiency over G_2.

Table 6.3 Performance of methods IA and IIA for $Q = 2$ and $Q = 3$

	$Q = 2$		$Q = 3$	
Method	IA	IIA	IA	IIA
Total time (hr.)	1.44	8.37	2.46	14.87
$\min_{\phi \in G_2} RE(\mathbf{d}^*_{Mm}; \mathbf{d}^*_{s,\phi})$	0.998	0.996	0.992	0.994

Table 6.4 $\min_{\phi \in G_2} RE(\cdot; \mathbf{d}^*_{s,\phi})$ of $\mathbf{d}^*_{Mm}$ from Method IA versus some traditional designs

	$Q = 1$	$Q = 2$	$Q = 3$
$\mathbf{d}^*_{Mm}$	0.997	0.998	0.992
m-sequence-based design	0.916	0.886	0.938
Ten random designs	0.794–0.923	0.705–0.879	0.662–0.785

In Table 6.4, we compare the performance of the $\mathbf{d}^*_{Mm}$ obtained by Method IA with some traditional designs that are widely used in functional brain imaging studies. These traditional designs include an m-sequence-based design, and ten randomly generated designs. An m-sequence (or maximum length sift-register sequence) of length $N = (Q + 1)^r - 1$ exists for some integer r when $Q + 1$ is a prime power. It can be generated by, e.g., a primitive polynomial over the Galois field $GF(Q+1)$; see, Kao and Stufken (2015) for details. These sequences are advocated by Buračas and Boynton (2002) as efficient designs for estimating the HRF in fMRI studies (i.e. for estimating $\mathbf{h}$ of Model (6.1)). For $(Q, N) = (1, 127)$, an m-sequence exists with $r = 7$. As for $(Q, N) = (2, 121)$ and $(3, 127)$, we follow Liu and Frank (2004) to concatenate two identical m-sequences of a shorter length, and leave out the excessive elements in the tail of the concatenated sequence to generate m-sequence-based designs of length N. As shown in Table 6.4, these m-sequence-based designs yield relatively high $\min_{\phi \in G_2} RE(\cdot; \mathbf{d}^*_{s,\phi})$. However, they do not perform as well as the $\mathbf{d}^*_{Mm}$ obtained from Method IA. We also randomly generate ten design sequences for each of the three scenarios that we consider. The range of the $\min_{\phi \in G_2} RE(\cdot; \mathbf{d}^*_{s,\phi})$ of these random designs are also reported in Table 6.4. Again, they do not perform as well as the maximin design obtained from our proposed method. In Fig. 6.2, we provide these maximin designs by presenting the value of the nth element d_n of $\mathbf{d}^*_{Mm}$ versus n for $n = 1, \ldots, N$.

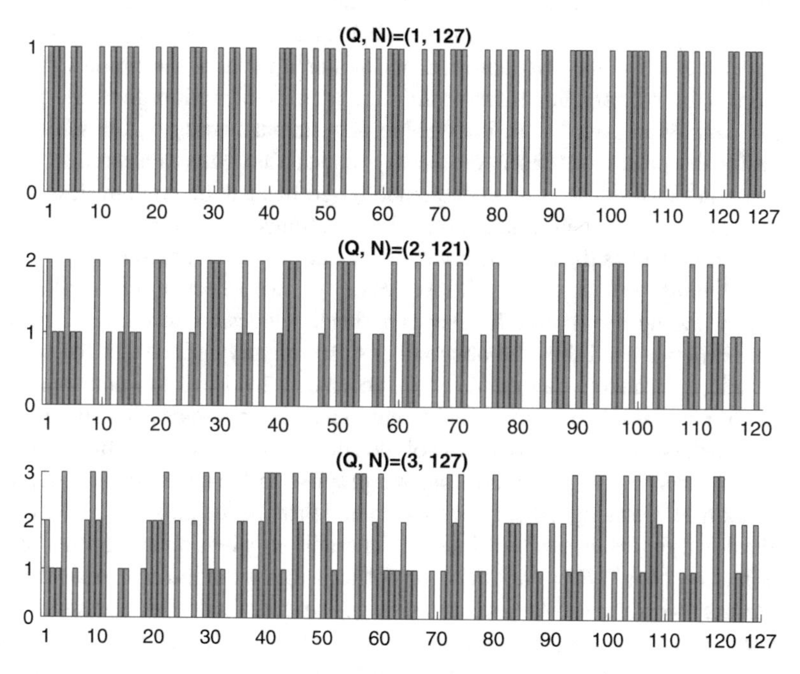

Fig. 6.2 The maximin designs $\mathbf{d}^*_{Mm} = (d_1, \ldots, d_N)$ obtained from Method IA; the y-axis is the value of d_n, and the x-axis is n

6.5 Conclusion and a Discussions

In this work, we propose an efficient approach for obtaining maximin designs for functional brain imaging studies where a pioneering brain mapping technology with high temporal resolution is utilized. Our proposed approach (Method IA) involves two components: (1) an easy-to-calculate surrogate criterion $F_s(\mathbf{d}; \boldsymbol{\phi})$ formed by the information matrix of a reduced size, and (2) a Kriging approximation of the minimum relative efficiency of each candidate design over the parameter space of the error correlation parameters.

In our case studies, we consider experiments of 8–9 min with $Q = 1, 2,$ and 3 stimulus types. These settings are not uncommon in practice, and our approach can easily be applied for a different setting. For comparison purposes, we consider three other methods as well as some existing designs. Specifically, Method IIB is essentially a direct application of the hitherto method used by Maus et al. (2010), and Kao and Mittelmann (2014). Although it tends to give good results in the traditional studies, this method quickly becomes infeasible in the present settings. The other two competing methods (IB and IIA) allow us to assess the improvement achieved by each of the two components of our approach. As presented in our case studies, both components are useful, and their combination gives the greatest improvement. Moreover, to our knowledge, a systematic study on the selection of designs for the present setting is previously unavailable. Many researchers tend to adopt existing designs such as m-sequence-based designs or random designs that are known to perform well in some traditional settings. However, these designs do not perform as well as the designs obtained by our proposed method.

For obtaining the surrogate criterion $F_s(\mathbf{d}; \boldsymbol{\phi})$, we select a subsample of the columns of the design matrix $\mathbf{X}$. Our experience suggests that the quality of the surrogate criterion, and thus the obtained designs, greatly depends on the columns that we select. A random subsample of the columns often leads to a poor result, and in the extreme case, the obtained design can have zero efficiency in estimating the HRF parameters. A judicious selection of the columns is thus crucial. In all the cases that we considered, the procedure that we propose in Sect. 6.3 consistently gives good outcomes. Deriving a theory to support and provide insights into our proposed procedure is a future research of interest. In addition, when obtaining the Kriging approximation of min RE, we adopt the settings that are popular in the computer experiments literature, and they tend to give good maximin designs for our study. A future research of interest includes the consideration of other approximations, and an extension of our approach by relaxing the assumption that τ_{ISI}/τ_{TR} is an integer.

Acknowledgements The research of Ming-Hung Kao was in part supported by the National Science Foundation grants DMS-13-52213, and CMMI-17-26445.

Appendix

Here, we present the steps of the genetic algorithm of Kao et al. (2009) for searching for an optimal design (i.e. optimal stimulus sequence of Q stimulus types).

Step GA-1 Generate $2G$ initial designs as the first generation, which include m-sequence-based designs, random designs, block designs of various block sizes, and mixed (block and random) designs. The fitness of these $2G$ designs is then evaluated by the objective function (e.g., the selected optimality criterion).

Step GA-2 With probability proportional to the fitness, G pairs of distinct designs are randomly selected with replacement from the current generation. Use the crossover operator to generate a pair of offspring designs from each selected paired design; i.e. to select a random cut-point and then generate offsprings by exchanging the corresponding subsequences before the cut-point of the selected paired designs.

Step GA-3 Randomly select $q\%$ of the elements of the $2G$ offspring designs, and replace each selected element by an integer randomly generated from the discrete uniform distribution over $0, 1, 2, \ldots, Q$. Obtain the fitness of the resulting $2G$ designs.

Step GA-4 Generate I immigrant designs from random designs, block designs, and their combinations. Obtain the fitness of these I immigrants.

Step GA-5 In the current pool of the parent, offspring and immigrant designs, select the best $2G$ designs to form the next generation.

Step GA-6 Repeat Steps GA-2 to GA-5 until a stopping rule is met.

In this work, we set $G = 10$, $q\% = 1\%$, and $I = 4$ as suggested by Kao et al. (2009). The second stopping rule presented in Kao (2009) is considered, and thus the search is terminated when no significant improvement can be expected in the next iterations. Specifically, the improvement in the objective function is evaluated every 200 generations. The algorithm stops when the current 200 generations result in an improvement that is no more than 10^{-7} of that of the first 200 generations.

References

Aguirre, G.K., Mattar, M.G., Magis-Weinberg, L.: De Bruijn cycles for neural decoding. Neuroimage **56**(3), 1293–1300 (2011)

Buračas, G.T., Boynton, G.M.: Efficient design of event-related fMRI experiments using m-sequences. NeuroImage **16**(3), 801–813 (2002)

Cheng, C.-S., Kao, M.-H.: Optimal experimental designs for fMRI via circulant biased weighing designs. Ann. Stat. **43**(6), 2565–2587 (2015)

Chernoff, H.: Locally optimal designs for estimating parameters. Ann. Math. Stat. **24**(4), 586–602 (1953)

Dale, A.M.: Optimal experimental design for event-related fMRI. Hum. Brain Mapp. **8**(2–3), 109–114 (1999)

Friston, K.J., Zarahn, E., Josephs, O., Henson, R.N.A., Dale, A.M.: Stochastic designs in event-related fMRI. NeuroImage **10**(5), 607–619 (1999)

Friston, K.J. Ashburner J.T., Kiebel, S.J., Nichols, T.E., Penny, W.D.: Statistical Parametric Mapping: The Analysis of Functional Brain Images. Amsterdam, Elsevier/Academic Press (2007)

Kao, M.H.: Multi-objective optimal experimental designs for ER-fMRI using MATLAB. J. Stat. Softw. **30**(11), 1–13 (2009)

Kao, M.-H.: A new type of experimental designs for event-related fMRI via Hadamard matrices. Stat. Prob. Lett. **84**(0), 108–112 (2014)

Kao, M.-H., Mittelmann, H.D.: A fast algorithm for constructing efficient event-related functional magnetic resonance imaging designs. J. Stat. Comput. Simul. **84**(11), 2391–2407 (2014)

Kao, M.-H., Stufken, J.: Optimal design for event-related fMRI studies. In: Dean, A., Morris M., Stufken, J., Bingham, D. (eds.) Handbook of Design and Analysis of Experiments, pp. 895–924 CRC Press, Boca Raton (2015)

Kao, M.-H., Mandal, A., Lazar, N., Stufken, J.: Multi-objective optimal experimental designs for event-related fMRI studies. NeuroImage **44**(3), 849–856 (2009)

Kao, M.-H., Mandal, A., Stufken, J.: Constrained multi-objective designs for functional MRI experiments via a modified nondominated sorting genetic algorithm. J. R. Stat. Soc. Ser. C (Appl. Stat.) **61**(4), 515–534 (2012)

Kao, M.-H., Majumdar, D., Mandal, A., Stufken, J.: Maximin and maximin-efficient event-related fMRI designs under a nonlinear model. Ann. Appl. Stat. **7**(4), 1940–1959 (2013)

Kiefer, J.: Optimum experimental designs. J. R. Stat. Soc. Ser. B (Methodol.) **21**(2), 272–319 (1959)

Liu, T.T.: Efficiency, power, and entropy in event-related fMRI with multiple trial types: part II: design of experiments. NeuroImage **21**(1), 401–413 (2004)

Liu, T.T., Frank, L.R.: Efficiency, power, and entropy in event-related fMRI with multiple trial types: part I: theory. NeuroImage **21**(1), 387–400 (2004)

Maus, B., van Breukelen, G.J.P., Goebel, R., Berger, M.P.F.: Robustness of optimal design of fMRI experiments with application of a genetic algorithm. Neuroimage **49**(3), 2433–2443 (2010)

Niederreiter, H.: Low-discrepancy and low-dispersion sequences. J. Number Theory **30**(1), 51–70 (1988)

Proulx, S., Safi-Harb, M., Levan, P., An, D., Watanabe, S., Gotman, J.: Increased sensitivity of fast bold fMRI with a subject-specific hemodynamic response function and application to epilepsy. NeuroImage **93**(1), 59–73 (2014)

Saleh, M., Kao, M.-H., Pan, R.: Design *D*-optimal event-related functional magnetic resonance imaging experiments. J. R. Stat. Soc. Ser. C (Appl. Stat.) **66**(1), 73–91 (2017)

Santner, T.J., Williams, B.J., Notz, W.: The Design and Analysis of Computer Experiments. Springer, New York (2003)

Scholkmann, F., Kleiser, S., Metz, A.J., Zimmermann, R., Mata Pavia, J., Wolf, U., Wolf, M.: A review on continuous wave functional near-infrared spectroscopy and imaging instrumentation and methodology. NeuroImage **85**, Part 1(0), 6–27 (2014)

Wager, T.D., Nichols, T.E.: Optimization of experimental design in fMRI: a general framework using a genetic algorithm. NeuroImage **18**(2), 293–309 (2003)

Worsley, K.J., Liao, C.H., Aston, J., Petre, V., Duncan, G.H., Morales, F., Evans, A.C.: A general statistical analysis for fMRI data. NeuroImage **15**(1), 1–15 (2002)

Chapter 7
A Global Optimization Algorithm for Sparse Mixed Membership Matrix Factorization

Fan Zhang, Chuangqi Wang, Andrew C. Trapp, and Patrick Flaherty

7.1 Introduction

Mixed membership matrix factorization (MMMF) has been used in document topic modeling (Blei et al. 2003), collaborative filtering (Mackey et al. 2010), population genetics (Pritchard et al. 2000), and social network analysis (Airoldi et al. 2008). The underlying assumption is that an observed feature for a given sample is a mixture of shared, underlying groups. These groups are called topics in document modeling, subpopulations in population genetics, and communities in social network analysis; in bioinformatics applications the groups are called subtypes and we adopt that terminology here. MMMF simultaneously identifies

F. Zhang
Center for Data Sciences at Brigham and Women's Hospital, Boston, MA, USA

Broad Institute of Massachusetts Institute of Technology and Harvard University, Boston, MA, USA

Department of Biomedical Informatics, Harvard Medical School, Boston, MA, USA
e-mail: fanzhang@broadinstitute.org

C. Wang
Department of Biomedical Engineering, Worcester Polytechnic Institute, Worcester, MA, USA
e-mail: cwang7@wpi.edu

A. C. Trapp
Robert A. Foisie Business School, Worcester Polytechnic Institute, Worcester, MA, USA
e-mail: atrapp@wpi.edu

P. Flaherty (✉)
Department of Mathematics & Statistics, University of Massachusetts Amherst, Amherst, MA, USA
e-mail: flaherty@math.umass.edu

L. Zhang et al. (eds.), *Contemporary Biostatistics with Biopharmaceutical Applications*, ICSA Book Series in Statistics,
https://doi.org/10.1007/978-3-030-15310-6_7

both the underlying subtypes and the distribution over those subtypes for each individual sample.

7.1.1 Mixed Membership Models

The MMMF problem can be viewed as inference in a particular statistical model (Singh and Gordon 2008). The model typically has a latent Dirichlet random variable that allows each sample to have its own distribution over subtypes and a latent variable for the feature weights that describe each subtype. The inferential goal is to estimate the joint posterior distribution over these latent variables and thus obtain the distribution over subtypes for each sample and the feature vector for each subtype. Non-negative matrix factorization techniques have been used in image analysis and collaborative filtering applications (Lee and Seung 1999; Mackey et al. 2010). Topic models for document clustering have also been cast as a matrix factorization problem (Xu et al. 2003).

The basic mixed membership model structure has been extended in various interesting ways. A hierarchical Dirichlet prior allows one to obtain a posterior distribution over the number of subtypes (Teh et al. 2005). A prior on the subtype variables allows one to impose specific sparsity constraints on the subtypes (Kabán 2007; MacKay 1992; Taddy 2013). Correlated information may be incorporated to improve the coherence of the subtypes (Blei and Lafferty 2006). Gaussian-Laplace-Dirichlet Model (GLAD) is hierarchical model that performs mixed membership matrix factorization with sparsity inducing Laplace prior on feature weights (Saddiki et al. 2015).

Sampling or variational inference methods are commonly used to estimate the posterior distribution of interest for mixed membership models, but these only provide local or approximate estimates. A mean-field variational algorithm (Blei et al. 2003) and a collapsed Gibbs sampling algorithm have been developed for Latent Dirichlet Allocation (Xiao and Stibor 2010). However, Gibbs sampling is approximate for finite chain lengths and variational inference is only guaranteed to converge to a local optimum (Blei et al. 2017).

7.1.2 Benders' Decomposition and Global OPtimization (GOP)

In many applications it is important to obtain a globally optimal solution rather than a local or approximate solution. Recently, there have been significant advances in deterministic optimization methods for general biconvex optimization problems (Floudas and Gounaris 2008; Horst and Tuy 2013). Here, we show that mixed membership matrix factorization can be cast as a biconvex optimization problem and the ϵ-global optimum can be obtained by these deterministic optimization methods.

Benders' decomposition exploits the idea that in a given optimization problem there are often *complicating variables*—variables that when held fixed yield a much simpler problem over the remaining variables (Benders 1962). Benders developed a cutting plane method for solving mixed integer optimization problems that can be so decomposed. Geoffrion later extended Benders' decomposition to situations where the primal problem (parametrized by fixed complicating variable values) no longer needs to be a linear program (Geoffrion 1972). The Global OPtimization (GOP) approach is an adaptation of the original Benders' decomposition that can handle a more general class of problems that includes mixed-integer biconvex optimization problems (Floudas 2013). Here, we exploit the GOP approach for solving a particular mixed membership matrix factorization problem.

7.1.3 Contributions

Our contribution is bringing the Global OPtimization (GOP) algorithm into contact with the mixed membership matrix factorization problem, computational improvements to the branch-and-bound GOP algorithm, and experimental results. Our discussion of the GOP algorithm here is necessarily brief. The details of problem conditions, convergence properties, and a full outline of the algorithm steps for the branch-and-bound version of the algorithm are found elsewhere (Floudas 2013).

We outline the general sparse mixed membership matrix factorization problem in Sect. 7.2. In Sect. 7.3, we use GOP to obtain an ϵ-global optimum solution for the mixed membership matrix factorization problem. In Sect. 7.4, we develop an A-star search algorithm that significantly improves the computational efficiency of our method. In Sect. 7.5, we show empirical accuracy and convergence time results on a synthetic data set. We also explore the performance of our algorithm on a small gene expression data set. Finally, we discuss further computational and statistical issues in Sect. 7.6.

7.2 Problem Formulation

The problem data is a matrix $y \in \mathbb{R}^{M \times N}$, where an element y_{ji} is an observation of feature j in sample i. We would like to represent each sample as a convex combination of K subtype vectors, $y_i = x\theta_i$, where $x \in \mathbb{R}^{M \times K}$ is a matrix of K subtype vectors and θ_i is the mixing proportion of each subtype. We would like x to be sparse because doing so makes interpreting the subtypes easier and often x is believed to be sparse *a priori* for many interesting problems. In the specific case of cancer subtyping, y_{ji} may be a normalized gene expression measurement for gene j in sample i. We write this matrix factorization problem as

$$\begin{aligned} &\underset{\theta,x}{\text{minimize}}\ \|y_i - x\theta_i\|_2^2 \\ &\text{subjectto}\ \|x\|_1 \leq P \\ &\qquad\qquad \theta_i \in \Delta^{K-1}\ \forall i, \end{aligned} \tag{7.1}$$

where Δ^{K-1} is a K-dimensional simplex.

Optimization problem (7.1) can be recast with a biconvex objective and a convex domain as

$$\begin{aligned} \underset{\theta,\,x,\,z}{\text{minimize}} \quad & \|y - x\theta\|_2^2 \\ \text{subject to} \quad & \sum_{j=1}^{M}\sum_{k=1}^{K} z_{jk} \leq P, \\ & -z_{jk} \leq x_{jk} \leq z_{jk} \qquad \forall (j,k), \\ & \theta_i \in \Delta^{K-1} \quad \forall i, \\ & z_{jk} \geq 0 \qquad \forall (j,k) \end{aligned} \tag{7.2}$$

If either x or θ is fixed then (7.2) reduces to a convex optimization problem. Indeed, if x is fixed, the optimization problem is a form of constrained linear regression. If θ is fixed, we have a form of LASSO regression. We prove that (7.1) is a biconvex problem in Appendix 2. Since both problems are computationally simple, we could take either x or θ to be the complicating variables in Benders' decomposition and we choose θ.

A common approach for solving an optimization problem with a nonconvex objective function is to alternate between fixing one variable and optimizing over the other. However, this approach only provides a local optimum (Gorski et al. 2007). A key to the GOP algorithm is the Benders'-based idea that feasibility and optimality information is shared between the primal problems in the form of constraints.

7.3 Algorithm

The Global OPtimization (GOP) algorithm, which we describe here, solves for ϵ-global optimum values of x and θ (Floudas and Visweswaran 1990; Floudas 2000, 2013). The algorithm proceeds by first partitioning the optimization problem decision variables into complicating and non-complicating variables. Then, the GOP algorithm alternates between solving a *primal problem* over θ for fixed x, and solving a *relaxed dual problem* over x for fixed θ. The primal problem provides an upper bound on the original optimization problem because it contains more constraints than the original problem (x is fixed). The relaxed dual problem contains fewer constraints and forms a valid global lower bound. The algorithm iteratively tightens the upper and lower bounds on the global optimum by alternating between the primal and relaxed dual problem.

7.3.1 Initialization

The algorithm starts by partitioning the problem into a relaxed dual problem and a primal problem. The solution of the relaxed dual problem is an optimal x for fixed values of the complicating variables θ and the solution of the primal problem is an optimal θ. An iteration counter $T = 1$ is initialized.

For each iteration, the relaxed dual problem is solved by forming a partition of the domain of x and solving a relaxed dual subproblem for each subset. A branch-and-bound tree data structure is used to store the solution of each of these relaxed dual subproblems and we initialize the root node $\mathtt{n}(0)$ where $T = 0$. The parents of $\mathtt{n}(T)$ is denoted $\mathtt{par}(\mathtt{n}(T))$, the set of ancestors of $\mathtt{n}(T)$ is denoted $\mathtt{anc}(\mathtt{n}(T))$, and the set of children of $\mathtt{n}(T)$ is denoted $\mathtt{ch}(\mathtt{n}(T))$. The root node is formed by initializing x at a random feasible point, $x^{\mathtt{n}(0)}$, and storing it in $\mathtt{n}(0)$.

7.3.2 Solve Primal Problem and Update Upper Bound

The primal problem (7.2) is constrained to a fixed value of x at $\mathtt{n}(T)$, $x^{(\mathtt{n}(T))}$,

Primal problem
(x fixed)

$$\begin{aligned} \underset{\theta}{\text{minimize}} \quad & \|y - x\theta\|_2^2 \\ \text{subject to} \quad & \theta_i^T 1_K = 1 \quad \text{for all } i, \\ & \theta_{ki} \geq 0 \quad \text{for all } k, i \end{aligned} \tag{7.3}$$

Since the primal problem is more constrained than (7.2), the solution, $S^{(\mathtt{n}(T))}$, is a global upper bound. The value of the upper bound is $\text{PUBD} \leftarrow \min\{\text{PUBD}, S^{(\mathtt{n}(T))}\}$, so PUBD holds the tightest upper bound across iterations.

7.3.3 Solve the Relaxed Dual Problem and Update Lower Bound

The relaxed dual problem is a relaxed version of (7.2) in that it contains fewer constraints than the original problem. Initially, at the root node, $\mathtt{n}(0)$, the domain of the relaxed dual problem is the entire domain of x, $\mathscr{X}$. Each node stores a set of linear constraints (cuts) such that when all of the constraints are satisfied, they define a region in $\mathscr{X}$. Sibling nodes form a partition of parent's region and a node deeper in the tree defines a smaller region than shallower nodes when incorporating the constraints of the node and all of its ancestors. These partitioning constraints are called *qualifying constraints*. Since the objective function is convex in θ for a fixed value of x, a Taylor series approximation of the Lagrangian with respect to θ provides a valid lower bound on the objective function. Since the objective function is convex in θ, the Taylor approximation is linear and the optimal objective is at a bound of θ. The GOP algorithm as outlined in (Floudas and Gounaris 2008) makes these ideas rigorous.

The relaxed dual problem for the mixed membership matrix factorization problem (7.2) for a node $\mathtt{n}(T)$ is below.

Relaxed Dual Problem

(θ fixed)

$$
\begin{aligned}
\underset{Q,x,z}{\text{minimize}} \quad & Q \\
\text{subject to} \quad & \sum_{j=1}^{M}\sum_{k=1}^{K} z_{jk} \le P, \\
& -z_{jk} \le x_{jk} \le z_{jk},\ z_{jk} \ge 0, \\
& L(x, \theta^B(t), y, \lambda^t, \mu^t)\big|_{x^t,\theta^t}^{\text{lin}} \le Q \quad \text{for } t \in \{\mathtt{anc}(\mathtt{n}(T)), \mathtt{n}(T)\}, \\
& g_{ki}^t\big|_{x^t}^{\text{lin}}(x) \le 0 \quad \text{if } \theta^B(t)_{ki} = 1, \\
& g_{ki}^t\big|_{x^t}^{\text{lin}}(x) \ge 0 \quad \text{if } \theta^B(t)_{ki} = 0
\end{aligned}
\tag{7.4}
$$

The function $L(x, \theta^B(t), y, \lambda^t, \mu^t)\big|_{x^t,\theta^t}^{\text{lin}}$ is the linearized Lagrangian of (7.2), $g_{ki}^t\big|_{x^t}^{\text{lin}}(x)$ is the ki-th qualifying constraint, and $\theta^B(t)$ is the value of θ at the bound such that the linearized Lagrangian is a valid lower bound in the region defined by the qualifying constraints at node t. We have taken a second Taylor approximation with respect to x to ensure the qualifying constraints are linear in x and thus valid cuts as recommended in (Floudas and Gounaris 2008).

The algorithm for solving the relaxed dual problem comprises five steps:

1. Construct a child node in the branch-and-bound tree
2. Populate the child node with the linearized Lagrange function and qualifying constraints

3. Solve the relaxed dual subproblem at the child nodes
4. Update the lower bound
5. Check convergence

7.3.3.1 Construct a Child Node in the Branch-and-Bound Tree

Recall, a unique region in $\mathscr{X}$ for the leaf node $\mathtt{ch}(\mathtt{n}(T))$ is defined by the t-th row of θ^B derived from the primal problem at node $\mathtt{n}(T)$. This region can be expressed as the qualifying constraint set,

$$g_{ki}^{\mathtt{ch}(\mathtt{n}(T))}\Big|_{x^{\mathtt{n}(T)}}^{\mathrm{lin}}(x) \leq 0 \text{ if } \theta_{ki}^B(t) = 1,$$

$$g_{ki}^{\mathtt{ch}(\mathtt{n}(T))}\Big|_{x^{\mathtt{n}(T)}}^{\mathrm{lin}}(x) \geq 0 \text{ if } \theta_{ki}^B(t) = 0.$$

To generate the tth child node of $\mathtt{n}(T)$ and populate it with this constraint set and $\theta^B(t)$ which will be used in the construction of the Lagrange function lower bound in the relaxed dual problem.

7.3.3.2 Populate the Child Node with the Linearized Lagrange Function and Qualifying Constraints

The qualifying constraint sets contained in each node along the path in the branch-and-bound tree from $\mathtt{ch}(\mathtt{n}(T))$ to the root, inclusively, are added to the relaxed dual subproblem at the newly constructed child node. For example, the qualifying constraint set for a node n' along the path is

$$g_{ki}^{\mathrm{n}'}\Big|_{x^{\mathrm{n}'}}^{\mathrm{lin}}(x) \leq 0 \text{ if } \theta^B(\mathrm{n}')_{ki} = 1$$

$$g_{ki}^{\mathrm{n}'}\Big|_{x^{\mathrm{n}'}}^{\mathrm{lin}}(x) \geq 0 \text{ if } \theta^B(\mathrm{n}')_{ki} = 0,$$

where $g_{ki}^{\mathrm{n}'}$ is the node's kith qualifying constraint, $x^{\mathrm{n}'}$ is the node's relaxed dual problem optimizer, and $\theta^B(\mathrm{n}')$ is a 0-1 vector defining the unique region for node n' since $\theta_{ki} \in [0, 1]$.

Then, the Lagrangian function lower bound constraints from each node along the path in the branch-and-bound tree from $\mathtt{ch}(\mathtt{n}(T))$ to the root, inclusively, are added to the relaxed dual subproblem. For example the linearized Lagrange function for node n',

$$L(x, \theta^B(\mathrm{n}'), y, \lambda^{(\mathrm{n}')}, \mu^{(\mathrm{n}')})\Big|_{x^{(\mathrm{n}')}, \theta^{(\mathrm{n}')}}^{\mathrm{lin}}.$$

The Lagrangian function for the primal problem is

$$
\begin{aligned}
L(x, \theta, \lambda, \mu) &= \sum_{i=1}^{N} L(x, \theta_i, \lambda_i, \mu_i) \\
&= \sum_{i=1}^{N} (y_i - x\theta_i)^\top (y_i - x\theta_i) \\
&\quad - \lambda_i (\theta_i^\top 1_K - 1) - \mu_i^\top \theta_i \\
&= \sum_{i=1}^{N} y_i^\top y_i - 2y_i^\top x\theta_i + \theta_i^\top x^\top x\theta_i \\
&\quad - \lambda_i (\theta_i^\top 1_K - 1) - \mu_i^\top \theta_i
\end{aligned}
\tag{7.5}
$$

with Lagrange multipliers $\mu \in \mathbb{R}_+^{K \times N}$ and $\lambda \in \mathbb{R}^N$.

The relaxed dual problem makes use of this Lagrangian function linearized about $\theta^{(t)}$ which we obtain through a Taylor series approximation,

$$
\begin{aligned}
L(x, \theta_i, \lambda_i, \mu_i)\big|_{\theta^{(t)}}^{\text{lin}} &\triangleq L\left(x, \theta_i^{(t)}, \lambda_i^{(t)}, \mu_i^{(t)}\right) \\
&\quad + \sum_{k=1}^{K} g_{ki}^{(t)}(x) \cdot \left(\theta_{ki} - \theta_{ki}^{(t)}\right),
\end{aligned}
\tag{7.6}
$$

where the qualifying constraint function is

$$
\begin{aligned}
g_i^{(t)}(x) &\triangleq \nabla_{\theta_i} L\left(\theta_i, x, \lambda_i^{(t)}, \mu_i^{(t)}\right)\big|_{\theta_i^{(t)}} \\
&= -2y_i^\top x + 2\theta_i^{(t)\top} x^\top x \\
&\quad - 1_K^\top \lambda_i^{(k)} - \mu_i^{(k)\top}.
\end{aligned}
\tag{7.7}
$$

The qualifying constraint $g_i^{(t)}(x)$ is quadratic in x. However, the qualifying constraints must be linear in x to yield a convex domain whether $g_i^{(t)}(x) \geq 0$ or $g_i^{(t)}(x) \leq 0$. So, the Lagrangian is linearized first with respect to x about $x^{(t)}$ then about θ_i at $\theta_i^{(t)}$. While the linearized Lagrangian is not a lower bound everywhere in x, it is a valid lower bound in the region bound by the qualifying constraints with θ_i set at the corresponding bounds in the Lagrangian function.

The Lagrangian function linearized about $x^{(t)}$ is

$$\begin{aligned} L(y_i, \theta_i, x, \lambda_i, \mu_i)\Big|_{x^{(t)}}^{\text{lin}} \triangleq & y_i^T y_i - \theta_i^\top x^{(t)\top} x^{(t)} \theta_i \\ & - 2y_i^\top x\theta_i + 2\theta_i^\top x^{(t)\top} x\theta_i \\ & - \lambda_i(\theta_i^\top 1_K - 1) - \mu_i^\top \theta_i. \end{aligned} \tag{7.8}$$

Subsequently, the Lagrangian function linearized about $(x^{(t)}, \theta_i^{(t)})$ is

$$\begin{aligned} L(y_i, \theta_i, x, \lambda_i, \mu_i)\Big|_{x^{(t)}, \theta_i^{(t)}}^{\text{lin}} \triangleq & \; y_i^\top y_i + \theta_i^{(t)\top} x^{(t)\top} x^{(t)} \theta_i^{(t)} \\ & - 2\theta_i^{(t)\top} x^{(t)\top} x^{(t)} \theta_i \\ & - \lambda_i \left(1_K^\top \theta_i - 1\right) - \mu_i^\top \theta_i \\ & - 2\theta_i^{(t)\top} x^\top x^{(t)} \theta_i^{(t)\top} - 2y_i^\top x\theta_i \\ & + 2\theta_i^{(t)\top} \left(x^{(t)\top} x + x^\top x^{(t)}\right) \theta_i \end{aligned} , \tag{7.9}$$

and the gradient used in the qualifying constraint is

$$\begin{aligned} g_i^{(t)}\Big|_{x^{(t)}}^{\text{lin}}(x) &\triangleq \nabla_{\theta_i} \left[L(y_i, \theta_i, x, \lambda_i, \mu_i)\Big|_{x_0}^{\text{lin}} \right]\Big|_{\theta_i^{(t)}} \\ &= -2x^{(t)\top} x^{(t)} \theta_i^{(t)} - 2x^\top y_i \\ &\quad + 2(x^{(t)\top} x + x^\top x^{(t)})\theta_i^{(t)} - \lambda_i 1_K - \mu_i. \end{aligned} \tag{7.10}$$

The qualifying constraints, Lagrange function constraints, and Lagrangian comprise the relaxed dual subproblem at child node $\mathtt{ch}(\mathtt{n}(T))$.

7.3.3.3 Solve the Relaxed Dual Subproblem at the Child Node

Once the valid constraints from the previous $t = 1, \ldots, T-1$ iterations have been identified and incorporated, the constraint for the current Tth iteration is

$$\begin{aligned} Q &\geq L(x, \theta^{B_T}, y, \lambda^{(t)}, \mu^{(t)})\big|_{x^{(t)}, \theta^{(t)}}^{\text{lin}} \\ g_{ki}^{(T)}\big|_{x^{(t)}}^{\text{lin}}(x) &\leq 0 \text{ if } \theta_{ki}^{B_T} = 1 \\ g_{ki}^{(T)}\big|_{x^{(t)}}^{\text{lin}}(x) &\geq 0 \text{ if } \theta_{ki}^{B_T} = 0. \end{aligned}$$

The resulting relaxed dual problem is a linear program and can be solved efficiently using the off-the-shelf LP solver Gurobi (Gurobi Optimization, Inc. 2018). We store the optimal objective function value and the optimizing decision variables in the node.

7.3.3.4 Update the Lower Bound

The global lower bound, RLBD, is provided by the lowest lower bound across all the leaf nodes in the branch-and-bound tree. Operationally, a hash table maintains a value that is a pointer to a branch-and-bound tree node whose key is the optimal value of the relaxed dual problem at that leaf node. Using this dictionary, branch-and-bound selects the smallest key and bounds to the node of the tree indicated by the value. This element is eliminated from the dictionary since at the end of the next iteration, it will be an interior node and not available for consideration. The iteration count is incremented, $T \leftarrow T + 1$, and the global lower bound is updated with the optimal value of the relaxed dual problem at the new node.

7.3.3.5 Check Convergence

Since RLBD maintains the lowest lower bound provided by the relaxed dual problem, the lower bound is non-decreasing. If the convergence criteria PUBD $-$ RLBD $\leq \epsilon$ has been met, then the algorithm is exited and the optimal θ from the node's primal problem and the optimal x from the node's relaxed dual problem is reported. Finite ϵ-convergence and ϵ-global optimality proofs can be found elsewhere (Floudas 2000).

7.4 Computational Improvements

In the relaxed dual problem branch-and-bound tree, a leaf node below the current node $\mathrm{n}(T)$ is constructed for each unique region defined by the hyperplane arrangement. In the GOP framework, there are KN hyperplanes, one for each connected variable and all of the KN elements of θ are connected variables. So, an upper bound on the number of regions defined by KN cuts is 2^{KN} because each region may be found by selecting a side of each cut. Thus we have the computationally complex situation of needing to solve a relaxed dual problem for each of the 2^{KN} possible regions.

Let an arrangement $\mathscr{A}$ denote a set of hyperplanes and $r(\mathscr{A})$ denote the set of unique regions defined by $\mathscr{A}$. In our particular situation, all of the hyperplanes pass through the unique point $x^{(\mathrm{n}(T))}$, so all of the regions are unbounded except by the constraints provided in $\mathscr{X}$. A recursive algorithm for counting the number of regions $|r(\mathscr{A})|$ known as Zaslavsky's Theorem, is outlined in (Zaslavsky 1975).

Indeed, $|r(\mathscr{A})|$ is often much less that $2^{|\mathscr{A}|}$. Due to its recursive nature, computing the number of hyperplanes using Zaslavsky's theorem can be computationally slow, though it can also be much better than the original 2^{KN} number of subproblems.

7.4.1 Cell Enumeration Algorithm

To address the computational complexity we have developed an A-star search algorithm for cell enumeration to simultaneously identify and count the set of unique regions defined by arrangement $\mathscr{A}$ with sign vectors. The algorithm proceeds as follows. First, preprocess the arrangement $\mathscr{A}$ to eliminate trivial and redundant hyperplanes. Next, eliminate a hyperplane from $\mathscr{A}$ if the coefficients are all zero and eliminate duplicate hyperplanes in $\mathscr{A}$ (see Appendix 3). What is left is a reduced arrangement, $\mathscr{A}'$.

Here, we define two concepts, *strict hyperplane* and *adjacent region*. A strict hyperplane is defined as non-redundant bounding hyperplane in a single region. If two regions exist that have sign vectors differing in only one hyperplane, then this hyperplane is a strict hyperplane. We define an adjacent region of region r as a neighbor region of r if they are separated by exactly one strict hyperplane. The general idea of the A-star algorithm uses ideas from partial order sets. We first initialize a root region using an interior point method and then determine all of its adjacent regions by identifying the set of strict hyperplanes. This process guarantees that we can enumerate all unique regions.

We define $\theta^B \in \{0, 1\}^{|r(\mathscr{A}')|\times KN}$. The rows are regions and there are KN columns. Each element of this matrix is either 0 or 1. The bth region in $r(\mathscr{A}')$ is uniquely identified by the zero-one vector in the bth row of θ^B. If the bth element of the kith row of θ^B is $+1$, then $g_{ki} \leq 0$. Similarly, if the bth element of the kith row of θ^B is 0, then $g_{ki} \geq 0$. The A-star search algorithm completes the θ^B matrix for the current node $\mathtt{n}(T)$ and a leaf node is generated for each row of θ^B. Thus each unique region defined by the qualifying constraint cuts provided by the Lagrange dual of the primal problem at the current node. The details of the A-star search algorithm are covered in Appendix 3.

7.4.2 Theoretical Time Complexity

The GOP algorithm has four main components: primal problem, preprocessing, unique region identification, and relaxed dual problems. We analyze the computational complexity of each in turn.

7.4.2.1 Primal Problem

The primal problem is a convex quadratic program with KN decision variables. The time complexity for the primal problem solving is then $O(K^3N^3)$ (Boyd and Vandenberghe 2004).

7.4.2.2 Preprocessing

We address the cases of overlapping qualifying constraint cuts by sorting the rows of the $KN \cdot M$ qualifying constraint coefficient matrix and comparing the coefficients of adjacent rows. We first sort the KN rows of the qualifying constraint coefficient matrix using heapsort which takes $O(KN \cdot \log(KN))$ time on average. The algorithm subsequently passes through the rows of the matrix to identify all-zero coefficients and duplicate cuts; each pass takes $O(KN)$ time. We define $|\mathscr{A}'|$ as the number of unique qualifying constraints.

7.4.2.3 Unique Region Identification

The interior point method that we used in the A-star search algorithm is a linear program of size $|\mathscr{A}'| \cdot MK$ with the time complexity of $O(|\mathscr{A}'| \cdot MK)$. The time complexity for enumerating the set of unique regions is $O(|\mathscr{A}'| \cdot (|\mathscr{A}'| \cdot MK))$, which exhibits polynomial behavior. The time complexity of the partial order A-star algorithm is polynomial in the best case and exponential in the worst case, depending on the heuristic. We define $|r(\mathscr{A}')|$ as the number of identified unique regions.

7.4.2.4 Relaxed Dual Problems

There are $2MK + 1$ decision variables for each relaxed dual problem, so the time complexity for each is $O(M^3K^3)$. The total time for solving the relaxed dual problems is $O(|r(\mathscr{A}')| \cdot M^3K^3)$, which depends on the number of relaxed dual problems.

7.5 Experiments

In this section, we present our experiments on synthetic data sets and show accuracy and convergence speed. Computational complexity is evaluated by both the theoretical and empirical time complexity.

7.5.1 Illustrative Example

We use a simple data set to show the operation of the algorithm in detail and facilitate visualization of the cut sets. The data set, y, and true decision variable values, (x^*, θ^*), are

$$x^* = \begin{bmatrix} 0, & -1 \end{bmatrix}, \theta^* = \begin{bmatrix} 1, & 0, & 0.5 \\ 0, & 1, & 0.5 \end{bmatrix},$$

$$y = \begin{bmatrix} 0, & -1, & -0.5 \end{bmatrix}.$$

We ran the GOP algorithm with sparsity constraint variable $P = 1$ and convergence tolerance $\epsilon = 0.01$. There are $KN = 6$ connected variables, so we solve at most $2^{KN} = 64$ relaxed dual problems at each iteration. These relaxed dual problems are independent and can be distributed to different computational threads or cores. The primal problem is a single optimization problem and will not be distributed. The optimal decision variables after 72 iterations are

$$\hat{x} = x^{(72)} = \begin{bmatrix} 0.080, & -0.920 \end{bmatrix}, \hat{\theta} = \theta^{(72)} = \begin{bmatrix} 1.00, & 0.080, & 0.580 \\ 0.00, & 0.920, & 0.420 \end{bmatrix}, \tag{7.11}$$

and the Lagrange multipliers are $\hat{\lambda} = [-0.147, 0, 0]$ and $\hat{\mu} = [0, 0, 0; 0.160, 0, 0]$.

Figure 7.1a shows the convergence of the upper and lower bounds by iteration. The upper bound converges quickly and the majority of the time in the algorithm is spent proving optimality. With each iteration regions of the solution space are tested until the lower bound is tightened sufficiently to meet the stopping criterion. Figure 7.1b shows the first ten x values considered by the algorithm with isoclines of the objective function with θ^* fixed. It is evident that the algorithm is not performing hill-climbing or any other gradient ascent algorithm during its search for the global optimum. Instead, the algorithm explores a region bound by the qualifying constraints to construct a lower bound on the objective function. We run it using 20 random initial values and the optimal objective functions for all random initializations are all 0, which shows that the GOP algorithm found the globally optimal solutions of this small instance. Furthermore, the algorithm does not search nested regions, but considers previously explored cut sets (Fig. 7.1b).

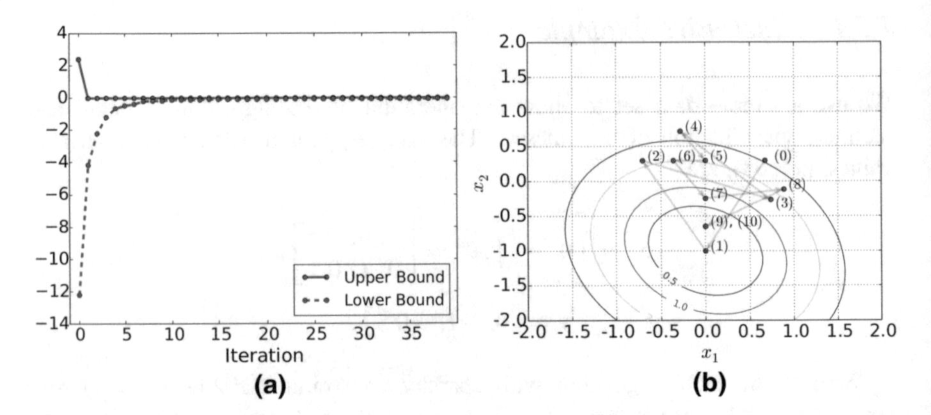

Fig. 7.1 (**a**) GOP optimal upper and lower bounds, (**b**) GOP optimal relaxed dual problem decision variables

Figure 7.2a and b shows the branch-and-bound tree and corresponding x-space region with the sequence of cut sets for the first three iterations of the algorithm. One cut in Fig. 7.2c–f is obtained for each of the KN qualifying constraints. We initialize the algorithm at $x^{(0)}$.

7.5.2 Accuracy and Convergence Speed

We ran our GOP algorithm using 64 processors on a synthetic data set which is randomly generated on the scale of one feature ($M = 1$), two subtyes ($K = 2$) and ten samples ($N = 10$). Figure 7.3a shows that our GOP algorithm converges very quickly to -0.17 duality gap (PUBD $-$ RLBD) in the first 89 iterations in 120 s. The optimal x (x_1, x_2) and θ (θ_1, θ_2) of each iteration are shown with a range of colors to represent corresponding RLBD in Fig. 7.3b,c. The dark blue represents low RLBD and the dark red represents high RLBD. The RLBD of the initial x, $x^{(0)}$, is -59.87; The RLBD of iteration 89, $x^{(89)}$, is -0.17. It demonstrates that the GOP algorithm can change modes very easily without getting stuck in local optima.

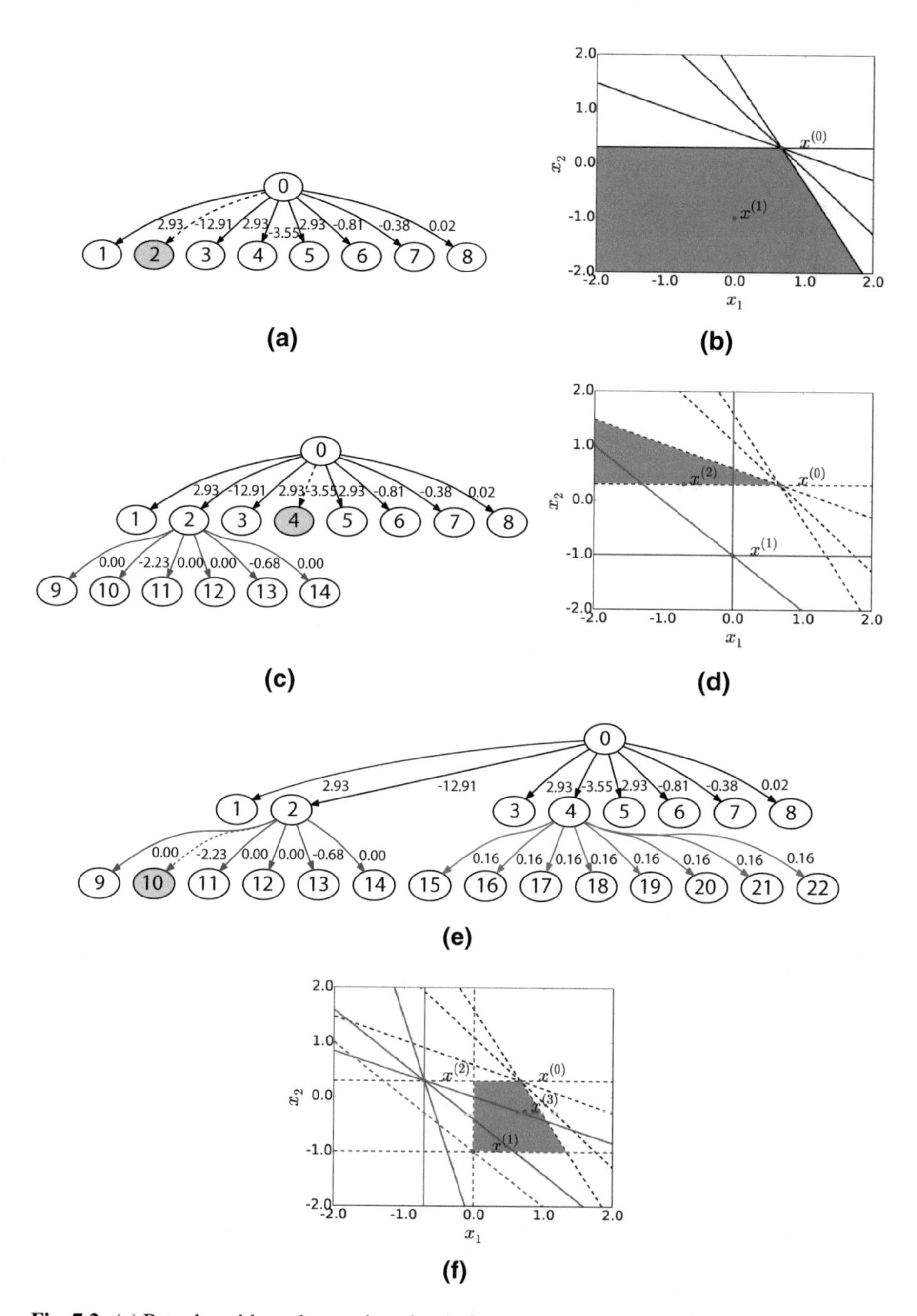

Fig. 7.2 (**a**) Branch-and-bound tree at iteration 1, (**b**) x-space region at iteration 1, (**c**) Branch-and-bound tree at iteration 2, (**d**) x-space region at iteration 2, (**e**) Branch-and-bound tree at iteration 3, (**f**) x-space region at iteration 3

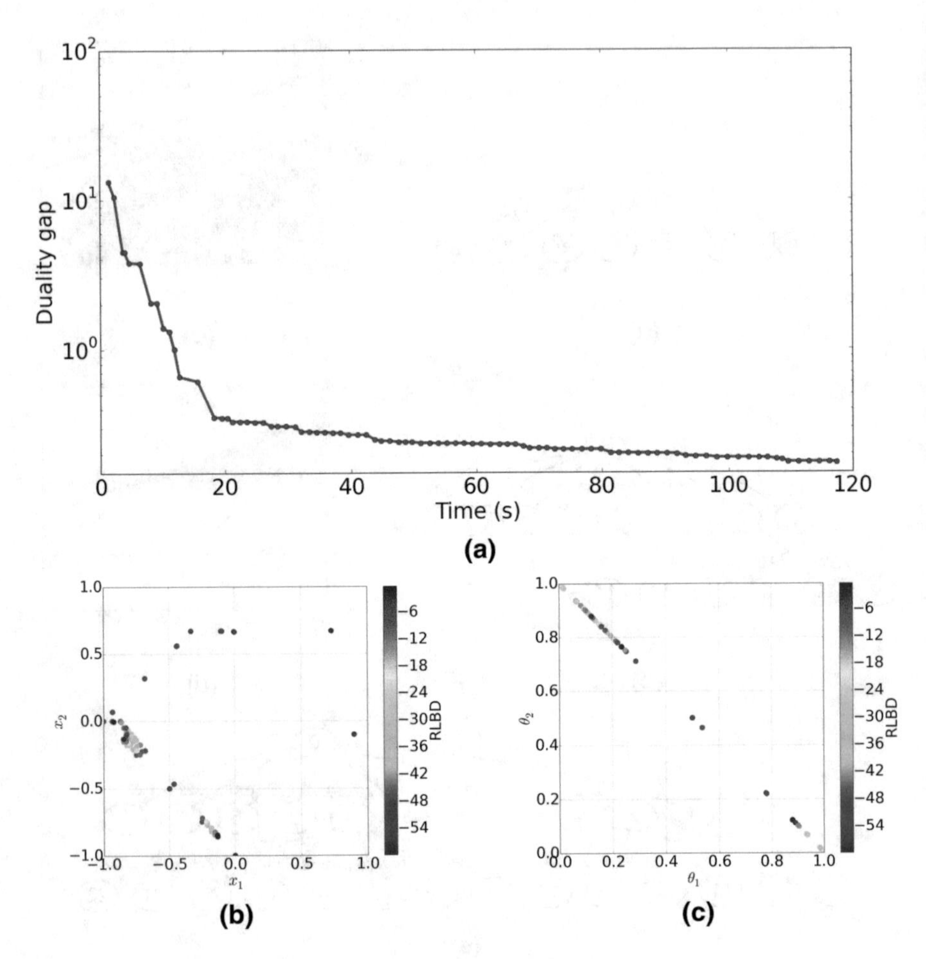

Fig. 7.3 (**a**) Duality gap through the first 120 s, (**b**) Optimal x of each iteration. The true x is $(0, -1)$, (**c**) Optimal θ of each iteration. The true θ is $(0.22, 0.78)$

7.5.3 Computational Complexity

We compare our theoretical complexity analysis with empirical measurements of the time complexity on simulated data sets.

We constructed 12 synthetic data sets in a full-factorial arrangement with $M \in \{20, 40, 60, 80\}$, $K \in \{2\}$, and $N \in \{4, 5, 6\}$ and measured CPU time for each component of one iteration. For each arrangement, each element of the true x^* is:

$$x^*_{mk} = \begin{cases} 1 & \text{if } 0 \leq m < M/4,\ k = 0 \\ -1 & \text{if } M/4 \leq m < M/2,\ k = 1 \\ \mathcal{N}(0, 0.5^2) & \text{if } M/2 \leq m < M, \forall k \\ 0 & \text{otherwise} \end{cases}$$

Here $\mathcal{N}(0, 0.5^2)$ is the sample from a Normal distribution by its mean 0 and standard deviation 0.5. For the true θ^*, θ^*_{kn} for $k = 0$ are n evenly spaced samples over the interval of [0, 1]; θ^*_{kn} for $k = 1$ are n evenly spaced samples over the interval of [1, 0].

Table 7.1 Timing profile (in seconds) of each component of the GOP algorithm for one iteration varying problem size

N	M	Primal problem	Preprocessing	Unique region ID	Relaxed dual problems	Total
4	20	0.10	1.69	1.29	1.54 (33%)	4.62
	40	0.12	1.91	1.72	1.69 (31%)	5.44
	60	0.12	2.03	1.11	1.77 (35%)	5.03
	80	0.13	2.39	2.05	3.70 (45%)	8.27
5	20	0.11	1.99	1.31	11.26 (77%)	14.67
	40	0.11	2.07	1.37	11.45 (76%)	15.00
	60	0.11	1.86	1.41	12.33 (78%)	15.71
	80	0.12	2.23	1.26	17.96 (83%)	21.57
6	20	0.14	2.21	2.50	65.71 (93%)	70.56
	40	0.13	2.83	2.49	67.08 (92%)	72.53
	60	0.12	3.45	2.80	69.00 (92%)	75.37
	80	0.12	3.15	2.80	77.62 (93%)	83.69

Table 7.1 shows that the time per iteration increases linearly with M when K and N are fixed. The time for solving all the relaxed dual problems increases as the number of samples increases. Even though the step of solving all the relaxed dual problems takes more than 90% of the total time per iteration when the number of samples is 6, our algorithm is easily parallelized to solve the relaxed dual problems, allowing the algorithm to scale nearly linearly with the size of the data set.

7.5.4 Real Data Analysis

To explore the performance of our algorithm on real data, we performed experiments on the TCGA pancancer high throughput DNA sequencing data set (Weinstein et al. 2013; Dheeru and Karra Taniskidou 2017). The original data was subsetted to the top two most variable genes and the top ten most variable samples by standard deviation. Then it was log transformed and centered across genes. The number of clusters was set to $K = 2$, the sparsity constraint was set to $P = 1$.

At early iterations, the optimal θ is a nearly 0–1 matrix, so we report the samples associated with each of the $K = 2$ subtypes. Samples 1, 4, 5, and 7 were assigned to subtype A and samples 2, 3, 6, 8, 9 and 10 were assigned to subtype B; subtypes are labeled arbitrarily with letters. The optimal x values were $x_A = [-0.204, 0]$ and $x_B = [0.561, 0.234]$. The inference algorithm enforced the L_1 penalty—the sum of the absolute values of x are at $P = 1$. And, the L_1 penalty clearly enforced sparsity in that one of the elements is exactly equal to zero.

The data set provides the anatomical regions associated with each of the cancer samples, and we explored those assignments to see if there is an association between the subtypes and the anatomical site of the cancer. Subtype A contains three colon adenocarcinomas and one prostate adenocarcinomas; subtype B contains four breast invasive carcinomas, one lung adenocarcinoma, and one kidney adenocarcinoma. Clearly, the algorithm is effectively clustering colon adenocarcinomas and cancers that are genomically more like that type from breast adenocarcinomas and cancers that are genomically more like that type.

At later iterations, when the duality gap had narrowed to 3.65, the optimal θ is more mixed. Still, the majority of the colorectal adenocarcinomas had subtype A as their largest component, and the majority of breast invasive carcinomas had subtype B as their largest component. These results indicate that this globally optimal inference algorithm performs well on a real data set. Since the algorithm provides both upper and lower bounds, a proof of ϵ-optimality is provided. Within this tolerance, the algorithm provides confidence that the provided estimates are globally optimal and not merely an artifact of local convergence.

7.6 Discussion

We have presented a global optimization algorithm for a mixed membership matrix factorization problem. Our algorithm brings ideas from the global optimization community (Benders' decomposition and the GOP method) into contact with statistical inference problems for the first time. The naïve computational cost of the global optimal solution is the need to solve a number of linear programs that grows exponentially in the number of connected variables in the worst case—in this case the KN elements of θ. Many of these linear programs are redundant or yield optimal solutions that are greater than the current upper bound and thus not useful. A branch-and-bound framework (Floudas 2000) reduces the need to solve all possible relaxed dual problems by fathoming parts of the solution space We further mitigate this cost by developing an search algorithm for identifying and enumerating the true number of unique linear programs.

Finally, we have derived an algorithm for particular loss functions for the sparsity constraint and objective function. The GOP framework can handle integer variables and thus may be used with an ℓ_0 counting "norm" rather than the ℓ_1 norm to induce sparsity. This would give us a mixed-integer biconvex program, but the conditions for the framework. Structured sparsity constraints can also be defined as is done for elastic-net extensions of LASSO regression. It may be useful to consider other loss functions for the objective function depending on the application.

We are exploring the connections between GOP and the other alternating optimization algorithms such as the expectation maximization (EM) and variational EM algorithm. Since the complexity of GOP only depends on the connected variables, the graphical model structure connecting the complicating and non-complicating variables may be used to identify the worst-case complexity of the algorithm prior to running the algorithm. A factorized graph structure may provide an approximate, but computationally efficient algorithm based on GOP. Additionally, because the Lagrangian function factorizes into the sum of Lagrangian functions for each sample in the data set, we may be able to update the parameters based on GOP for a selected subset of the data in an iterative or sequential algorithm. We are exploring the statistical consistency properties of such an update procedure.

Acknowledgements We acknowledge Hachem Saddiki for valuable discussions and comments on the manuscript.

Appendix 1: Derivation of Relaxed Dual Problem Constraints

The Lagrange function is the sum of the Lagrange functions for each sample,

$$L(y, \theta, x, \lambda) = \sum_{i=1}^{n} L(y_i, \theta_i, x, \lambda_i, \mu_i), \tag{7.12}$$

and the Lagrange function for a single sample is

$$L(y_i, \theta_i, x, \lambda_i, \mu_i) = y_i^T y_i - 2y_i^T x\theta_i + \theta_i^T x^T x\theta_i - \lambda_i(\theta_i^T 1_K - 1) - \mu_i^T \theta_i. \tag{7.13}$$

We see that the Lagrange function is biconvex in x and θ_i. We develop the constraints for a single sample for the remainder.

Linearized Lagrange Function with Respect to x

Casting x as a vector and rewriting the Lagrange function gives

$$L(y_i, \theta_i, \bar{x}, \lambda_i, \mu_i) = a_i - 2b_i^T \bar{x} + \bar{x}^T C_i \bar{x} - \lambda_i(\theta_i^T 1_K - 1) - \mu_i^T \theta_i, \tag{7.14}$$

where $\bar{x}$ is formed by stacking the columns of x in order. The coefficients are formed such that

$$a = y_i^T y_i,$$
$$b_i^T \bar{x} = y_i^T x \theta_i,$$
$$\bar{x}^T C_i \bar{x} = \theta_i^T x^T x \theta_i.$$

The linear coefficient matrix is the $KM \times 1$ vector,

$$b_i = [y_i \theta_{1i}, \cdots, y_i \theta_{Ki}]$$

The quadratic coefficient is the $KM \times KM$ and block matrix

$$C_i = \begin{bmatrix} \theta_{1i}^2 I_M & \cdots & \theta_{1i}\theta_{Ki} I_M \\ \vdots & \ddots & \vdots \\ \theta_{Ki}\theta_{1i} I_M & \cdots & \theta_{Ki}^2 I_M \end{bmatrix}$$

The Taylor series approximation about x_0 is

$$L(y_i, \theta_i, \bar{x}, \lambda_i, \mu_i)\Big|_{\bar{x}_0}^{\text{lin}} = L(y_i, x_0, \theta_i, \lambda_i, \mu_i) + (\nabla_x L|_{x_0})^T (x - x_0). \tag{7.15}$$

The gradient with respect to x is

$$\nabla_x L(y_i, \theta_i, \bar{x}, \lambda_i, \mu_i) = -2b_i + 2C_i \bar{x}. \tag{7.16}$$

Plugging the gradient into the Taylor series approximation gives

$$L(y_i, \theta_i, \bar{x}, \lambda_i)\Big|_{\bar{x}_0}^{\text{lin}} = a_i - 2b_i^T \bar{x}_0 + \bar{x}_0^T C_i \bar{x}_0 - \lambda_i \left(\theta_i^T 1_K - 1\right) - \mu_i^T \theta_i + (-2b_i + 2C_i \bar{x}_0)^T (\bar{x} - \bar{x}_0). \tag{7.17}$$

Simplifying the linearized Lagrange function gives

$$L(y_i, \theta_i, \bar{x}, \lambda_i, \mu_i)\Big|_{\bar{x}_0}^{\text{lin}} = \left(y_i^T y_i - \bar{x}_0^T C_i \bar{x}_0 - \lambda_i \left(\theta_i^T 1_K - 1\right) - \mu_i^T \theta_i\right) - 2b_i^T \bar{x} + 2\bar{x}_0^T C_i \bar{x} \tag{7.18}$$

Finally, we write the linearized Lagrangian using the matrix form of x_0,

$$L(y_i, \theta_i, x, \lambda_i, \mu_i)\Big|_{x_0}^{\text{lin}} = y_i^T y_i^T - \theta_i^T x_0^T x_0 \theta_i - 2y_i^T x \theta_i + 2\theta_i^T x_0^T x \theta_i - \lambda_i \left(\theta_i^T 1_K - 1\right) - \mu_i^T \theta_i \tag{7.19}$$

While the original Lagrange function is convex in θ_i for a fixed x, the linearized Lagrange function is not necessarily convex in θ_i. This can be seen by collecting the quadratic, linear and constant terms with respect to θ_i,

$$L(y_i, \theta_i, x, \lambda_i, \mu_i)\Big|_{x_0}^{\text{lin}} = \left(y_i^T y_i^T + \lambda_i\right) + \left(-2y_i^T x - \lambda_i 1_K^T - \mu_i^T\right)\theta_i + \theta_i^T \left(2x_0^T x - x_0^T x_0\right)\theta_i. \tag{7.20}$$

Now, if and only if $2x_0^T x - x_0^T x_0 \succeq 0$ is positive semidefinite, then $L(y_i, \theta_i, x, \lambda_i, \mu_i)\Big|_{x_0}^{\text{lin}}$ is convex. The condition is satisfied at $x = x_0$ but may be violated at some other value of x.

Linearized Lagrange Function with Respect to θ_i

Now, we linearize (7.18) with respect to θ_i. Using the Taylor series approximation with respect to θ_{0i} gives

$$L(y_i, \theta_i, x, \lambda_i, \mu_i)\Big|_{x_0,\theta_{0i}}^{\text{lin}} = L(y_i, \theta_{0i}, x, \lambda_i, \mu_i)\Big|_{x_0}^{\text{lin}} + \left(\nabla_{\theta_i} L(y_i, \theta_i, x, \lambda_i, \mu_i)\Big|_{x_0}^{\text{lin}}\Big|_{\theta_{0i}}\right)^T (\theta_i - \theta_{0i}) \tag{7.21}$$

The gradient for this Taylor series approximation is

$$g_i(x) \triangleq \nabla_{\theta_i} L(y_i, \theta_i, x, \lambda_i, \mu_i)\Big|_{x_0}^{\text{lin}}\Big|_{\theta_{0i}} = -2x_0^T x_0\theta_{0i} - 2x^T y_i + 2\left(x_0^T x + x^T x_0\right)\theta_{0i} - \lambda_i 1_K - \mu_i, \tag{7.22}$$

where $g_i(x)$ is the vector of K qualifying constraints associated with the Lagrange function. The qualifying constraint is linear in x. Plugging the gradient into the approximation gives

$$L(y_i, \theta_i, x, \lambda_i, \mu_i)\Big|_{x_0,\theta_{0i}}^{\text{lin}} = y_i^T y_i^T - \theta_{0i}^T x_0^T x_0\theta_{0i} - 2y_i^T x\theta_{0i} + 2\theta_{0i}^T x_0^T x\theta_{0i} - \lambda_i\left(\theta_{0i}^T 1_K - 1\right)$$
$$-\mu_i^T\theta_{0i} + \left(-2x_0^T x_0\theta_{0i} - 2x^T y_i + 2(x_0^T x + x^T x_0)\theta_{0i} - \lambda_i 1_K - \mu_i\right)^T (\theta_i - \theta_{0i}) \tag{7.23}$$

The linearized Lagrange function is bi-linear in x and θ_i. Finally, simplifying the linearized Lagrange function gives

$$L(y_i, \theta_i, x, \lambda_i, \mu_i)\Big|_{x_0,\theta_{0i}}^{\text{lin}} = y_i^T y_i^T + \theta_{0i}^T x_0^T x_0 \theta_{0i} - 2\theta_{0i}^T x_0^T x_0 \theta_i - \lambda_i (1_K^T \theta_i - 1) - \mu_i^T \theta_i$$
$$- 2\theta_{0i}^T x^T x_0 \theta_{0i} - 2y_i^T x \theta_i + 2\theta_{0i}^T (x_0^T x + x^T x_0)\theta_i \tag{7.24}$$

Appendix 2: Proof of Biconvexity

To prove the optimization problem is biconvex, first we show the feasible region over which we are optimizing is biconvex. Then, we show the objective function is biconvex by fixing θ and showing convexity with respect to x, and then vice versa.

The Constraints Form a Biconvex Feasible Region

Our constraints can be written as

$$||x||_1 \leqslant P \tag{7.25}$$

$$\sum_{k=1}^{K} \theta_{ki} = 1 \ \forall i \tag{7.26}$$

$$0 \leqslant \theta_{ki} \leqslant 1 \ \forall (k, i). \tag{7.27}$$

The inequality constraint (7.25) is convex if either x or θ is fixed, because any norm is convex. The equality constraints (7.26) is an affine combination that is still affine if either x or θ is fixed. Every affine set is convex. The inequality constraint (7.27) is convex if either x or θ is fixed, because θ is a linear function.

The Objective Is Convex with Respect to θ

We prove the objective is a biconvex function using the following two theorems.

Theorem 1 *Let $A \subseteq \mathbb{R}^n$ be a convex open set and let $f : A \to \mathbb{R}$ be twice differentiable. Write $H(x)$ for the Hessian matrix of f at $x \in A$. If $H(x)$ is positive semidefinite for all $x \in A$, then f is convex (Boyd and Vandenberghe 2004).*

Theorem 2 *A symmetric matrix A is positive semidefinite (PSD) if and only if there exists B such that $A = B^T B$ (Lancaster et al. 1985).*

The objective of our problem is,

$$f(y, x, \theta) = ||y - x\theta||_2^2 = (y - x\theta)^T (y - x\theta) \tag{7.28}$$

$$= (y^T - \theta^T x^T)(y - x\theta) \tag{7.29}$$

$$= y^T y - y^T x\theta - \theta^T x^T y + \theta^T x^T x\theta. \tag{7.30}$$

The objective function is the sum of the objective functions for each sample.

$$f(y, x, \theta) = \sum_{i=1}^{N} f(y_i, x, \theta_i) \tag{7.31}$$

$$= \sum_{i=1}^{N} y_i^T y_i - 2y_i^T x\theta_i + \theta_i^T x^T x\theta_i. \tag{7.32}$$

The gradient with respect to θ_i,

$$\nabla_{\theta_i} f(y_i, x, \theta_i) = -2y_i^T x + \left(x^T x + \left(x^T x\right)^T\right)\theta_i \tag{7.33}$$

$$= -2x^T y_i + 2x^T x\theta_i. \tag{7.34}$$

Take second derivative with respect to θ_i to get Hessian matrix,

$$\nabla_{\theta_i}^2 f(y_i, x, \theta_i) = \nabla_{\theta_i} \left(-2x^T y_i + 2x^T x\theta_i\right) \tag{7.35}$$

$$= 2\nabla_{\theta_i} \left(x^T x\theta_i\right) \tag{7.36}$$

$$= 2\left(x^T x\right)^T \tag{7.37}$$

$$= 2x^T x. \tag{7.38}$$

The Hessian matrix $\nabla_{\theta_i}^2 f(y_i, x, \theta_i)$ is positive semidefinite based on Theorem 2. Then, we have $f(y_i, x, \theta_i)$ is convex in θ_i based on Theorem 1. The objective $f(y, x, \theta)$ is convex with respect to θ, because the sum of convex functions, $\sum_{i=1}^{N} f(y_i, x, \theta_i)$, is still a convex function.

The Objective Is Convex with Respect to x

The objective function for sample i is

$$f(y_i, x, \theta_i) = y_i^T y_i - 2y_i^T x\theta_i + \theta_i^T x^T x\theta_i. \tag{7.39}$$

We cast x as a vector $\bar{x}$, which is formed by stacking the columns of x in order. We rewrite the objective function as

$$f(y_i, \bar{x}, \theta_i) = a_i - 2b_i^T \bar{x} + \bar{x}^T C_i \bar{x}. \tag{7.40}$$

The coefficients are formed such that

$$a = y_i^T y_i, \tag{7.41}$$

$$b_i^T \bar{x} = y_i^T x\theta_i, \tag{7.42}$$

$$\bar{x}^T C_i \bar{x} = \theta_i^T x^T x\theta_i. \tag{7.43}$$

The linear coefficient matrix is the $KM \times 1$ vector

$$b_i = [y_i\theta_{1i}, \ldots, y_i\theta_{Ki}] \tag{7.44}$$

The quadratic coefficient is the $KM \times KM$ and block matrix

$$C_i = \begin{bmatrix} \theta_{1i}^2 I_M & \cdots & \theta_{1i}\theta_{Ki} I_M \\ \vdots & \ddots & \vdots \\ \theta_{Ki}\theta_{1i} I_M & \cdots & \theta_{Ki}^2 I_M \end{bmatrix} \tag{7.45}$$

The gradient with respect to $\bar{x}$

$$\nabla_{\bar{x}} f(y_i, \bar{x}, \theta_i) = -2b_i + 2C_i\bar{x}. \tag{7.46}$$

Take second derivative to get Hessian matrix,

$$\nabla_{\bar{x}^2} f(y_i, \bar{x}, \theta_i) = 2C_i^T \tag{7.47}$$

$$= 2\left(\theta_i\theta_i^T\right)^T \tag{7.48}$$

$$= 2\left(\theta_i^T\right)^T\left(\theta_i^T\right). \tag{7.49}$$

The Hessian matrix $\nabla^2_{\bar{x}} f(y_i, \bar{x}, \theta_i)$ is positive semidefinite based on Theorem 2. Then, we have $f(y_i, \bar{x}, \theta_i)$ is convex in $\bar{x}$ based on Theorem 1. The objective $f(y, x, \theta)$ is convex with respect to x, because the sum of convex functions, $\sum_{i=1}^{N} f(y_i, x, \theta_i)$, is still a convex function.

The objective is biconvex with respect to both x and θ. Thus, we have a biconvex optimization problem based on the proof of biconvexity of the constraints and the objective.

Appendix 3: A-Star Search Algorithm

In this procedure, first we remove all the duplicate and all-zero coefficients hyperplanes to get unique hyperplanes. Then we start from a specific region r and put it into a open set. Open set is used to maintain a region list which need to be explored. Each time we pick one region from the open set to find adjacent regions. Once finishing the step of finding adjacent regions, region r will be moved into a closed set. Closed set is used to maintain a region list which already be explored. Also, if the adjacent region is a newly found one, it also need to be put into the open set for exploring. Finally, once the open set is empty, regions in the closed set are all the unique regions, and the number of the unique regions is the length of the closed set. This procedure begins from one region and expands to all the neighbors until no new neighbor is existed.

The overview of the A-star search algorithm to identify unique regions is shown in Algorithm 1.

Algorithm 1 A-star Search Algorithm

1: Sort the rows of the KN x M qualifying constraint coefficient matrix.
2: Compare adjacent rows of the qualifying constraint coefficient matrix and eliminate duplicate rows.
3: Eliminate rows of the qualifying constraint coefficient matrix with all-zero coefficients.
4: Determine the list of unique qualifying constraints by pairwise test.
5: Set S and $|\mathscr{A}'|$ to the set of unique, non-trivial qualifying constraints and the number of them.
6: Initialize a region $root$ using an interior point method (Component 1).
7: Put region $root$ into the open set.
8: **if** open set is not empty **then**
9: Get a region R from the open set.
10: Calculate the adjacent regions set $Radj$ (Component 2).
11: Put region R into the closed set.
12: **for** each region r in $Radj$ **do**
13: **if** r is not in the open set *and* not in the closed set **then**
14: Put region r into the open set.
15: Reflect the sign of the regions in the close set.
16: Get all the regions represented by string of 0 and 1.

Hyperplane Filtering

Assuming there are two different hyperplanes H_i and H_j represented by $A_i = \{a_{i,0}, \ldots, a_{i,MK}\}$ and $A_j = \{a_{j,0}, \ldots, a_{j,MK}\}$. We take these two hyperplanes duplicated when

$$\frac{a_{i,0}}{a_{j,0}} = \frac{a_{i,1}}{a_{j,1}} = \ldots = \frac{a_{i,MK}}{a_{j,MK}} = \frac{\sum_{l=0}^{MK} a_{i,l}}{\sum_{l=0}^{MK} a_{j,l}}, a_{j,l}! = 0 \tag{7.50}$$

This can be converted to

$$\left| \sum_{l=0}^{MK} a_{i,l} \cdot a_{j,n} - \sum_{l=0}^{MK} a_{j,l} \cdot a_{i,n} \right| \leq \tau, \forall\, n \epsilon [0, MK] \tag{7.51}$$

where threshold τ is a very small positive value.

We eliminate a hyperplane H_i represented by $A_i = \{a_{i,0}, \ldots, a_{i,MK}\}$ from hyperplane arrangement $\mathscr{A}$ if the coefficients of A_i are all zero,

$$|a_{i,j}| \leqslant \tau \text{ forall } a_{i,j} \in A_i \text{ and } j \in [0, MK]$$

The arrangement $\mathscr{A}'$ is the reduced arrangement and $A'x = b$ are the equations of unique hyperplanes.

Interior Point Method

An interior point is found by solving the following optimization problem:

$$\begin{aligned} &\text{maximize} && z \\ &\text{subject to} && -A'_i x + z \leq b_i \quad \text{if } \theta_i^B = 0, \\ & && A'_i x + z \leq -b_i \quad \text{if } \theta_i^B = 1, \\ & && z > 0 \end{aligned} \tag{7.52}$$

Algorithm 2 Interior Point Method (Component 1)

1: Generate $2^{|\mathscr{A}'|}$ different strings using 0 and 1.
2: **for** each s in the strings **do**
3: Solve an optimization problem to get an interior point.
4: **if** Get a interior point **then**
5: Get the $root$ region represented by 0 and 1.

Algorithm 3 Get Adjacent Regions (Component 2)

1: Initialize an empty set SH for strict hyperplanes.
2: Initialize an adjacent region set ADJ.
3: # Find out all the strict hyperplanes for region R.
4: **for** each hyperplane H of $|\mathscr{A}'|$ hyperplanes **do**
5: Pick one hyperplane H from all the hyperplanes defining region R.
6: Flip the sign of H to get $\neg H$.
7: Form a new hyperplane arrangement $\neg\mathscr{A}'$ with $\neg H$.
8: Solve the problem to get an interior point constrained by $\neg\mathscr{A}'$.
9: **if** the interior point is not Non **then**
10: H is a strict hyperplane and put into set SH.
11: **else**
12: H is a redundant hyperplane.
13: # Find out all the adjacent regions for region R.
14: **for** each strict hyperplane sh in set SH **do**
15: Take the opposite sign $\neg sh$ of sh.
16: Form a adjacent region adj based on $\neg sh$ and all the else hyperplanes.
17: Put adj into set ADJ.

References

Airoldi, E.M., Blei, D.M., Fienberg, S.E., Xing, E.P.: Mixed membership stochastic blockmodels. J. Mach. Learn. Res. **9**, 1981–2014 (2008)

Benders, J.F.: Partitioning procedures for solving mixed-variables programming problems. Numer. Math. **4**(1), 238–252 (1962)

Blei, D.M., Lafferty, J.D.: Correlated topic models. In: Proceedings of the International Conference on Machine Learning, pp 113–120 (2006)

Blei, D.M., Ng, A.Y., Jordan, M.I.: Latent dirichlet allocation. J. Mach. Learn. Res. **3**, 993–1022 (2003)

Blei, D.M., Kucukelbir, A., McAuliffe, J.D.: Variational inference: a review for statisticians. J. Am. Stat. Assoc. **112**(518), 859–877 (2017)

Boyd, S., Vandenberghe, L.: Convex Optimization. Cambridge University Press, Cambridge (2004)

Dheeru, D., Karra T.E.: UCI machine learning repository. URL UCI machine learning repository (2017). http://archive.ics.uci.edu/ml

Floudas, C.A.: Deterministic Global Optimization, Nonconvex Optimization and Its Applications, vol 37. Springer, Boston (2000)

Floudas, C.A.: Deterministic Global Optimization: Theory, Methods and Applications, vol. 37. Springer, Berlin (2013)

Floudas, C.A., Gounaris, C.E.: A review of recent advances in global optimization. J. Glob. Optim. **45**, 3–38 (2008)

Floudas, C.A., Visweswaran, V.: A global optimization algorithm (GOP) for certain classes of nonconvex NLPS. Comput. Chem. Eng. **14**(12), 1–34 (1990)

Geoffrion, A.M.: Generalized benders decomposition. J. Optim. Theory Appl. **10**, 237–260 (1972)

Gorski, J., Pfeuffer, F., Klamroth, K.: Biconvex sets and optimization with biconvex functions: a survey and extensions. Math. Methods Oper. Res. **66**, 373–407 (2007)

Gurobi Optimization, Inc (2018) Gurobi optimizer version 8.0

Horst, R., Tuy, H.: Global Optimization: Deterministic Approaches. Springer, Berlin (2013)

Kabán, A.: On Bayesian classification with laplace priors. Pattern Recognit. Lett. **28**(10), 1271–1282 (2007)

Lancaster, P., Tismenetsky, M., et al.: The theory of matrices: with applications. Elsevier, San Diego (1985)

Lee, D.D., Seung, H.S.: Learning the parts of objects by non-negative matrix factorization. Nature **401**, 788–791 (1999)

MacKay, D.J.C.: Bayesian interpolation. Neural Comput. **4**(3), 415–447 (1992)

Mackey, L., Weiss, D., Jordan, M.I.: Mixed membership matrix factorization. In: International Conference on Machine Learning, pp 1–8 (2010)

Pritchard, J.K., Stephens, M., Donnelly, P.: Inference of population structure using multilocus genotype data. Genetics **155**, 945–959 (2000)

Saddiki, H., McAuliffe, J., Flaherty, P.: GLAD: a mixed-membership model for heterogeneous tumor subtype classification. Bioinformatics **31**(2), 225–232 (2015)

Singh, A.P., Gordon, G.J.: A unified view of matrix factorization models. In: Lecture Notes in Computer Science, vol. 5212, pp. 358–373, Springer, Berlin (2008)

Taddy, M.: Multinomial inverse regression for text analysis. J. Am. Stat. Assoc. **108**(503), 755–770, (2013). https://doi.org/10.1080/01621459.2012.734168

Teh, Y.W., Jordan, M.I., Beal, M.J., Blei, D.M.: Sharing clusters among related groups: hierarchical Dirichlet processes. In: Advances in Neural Information Processing Systems, vol. 1, MIT Press, Cambridge (2005)

Weinstein, J.N., Collisson, E.A., Mills, G.B., Shaw, K.R.M., Ozenberger, B.A., Ellrott, K., Shmulevich, I., Sander, C., Stuart, J.M., Network CGAR, et al.: The cancer genome atlas pan-cancer analysis project. Nat. Genet. **45**(10), 1113 (2013)

Xiao, H., Stibor, T.: Efficient collapsed Gibbs sampling for latent Dirichlet allocation. In: Sugiyama, M., Yang, Q. (eds.) Proceedings of 2nd Asian Conference on Machine Learning, vol. 13, pp. 63–78 (2010)

Xu, W., Liu, X., Gong, Y.: Document clustering based on non-negative matrix factorization. In: Proceedings of the 26th Annual International ACM SIGIR Conference on Research and Development in Information Retrieval–SIGIR '03, p. 267 (2003)

Zaslavsky, T.: Facing Up to Arrangements: Face-Count Formulas for Partitions of Space by Hyperplanes: Face-Count Formulas for Partitions of Space by Hyperplanes, vol. 154. American Mathematical Society (1975)

Chapter 8
A Nonnegative Robust Linear Model for Deconvolution of Proportions

Hyonho Chun and Hyuna Yang

8.1 Introduction

There have been many linear deconvolution methods that estimate mixing rates of diverse cell types from gene expression measurements. With given set of gene expression measurements from pure cells, Abbas et al. (2009) formulated this deconvolution problem as a least squares problem. The approach was then extended to a quadratic programming (QP) method by imposing non-negativity constraints (Gong et al. 2011). Next, Qiao et al. (2012) proposed the PERT method that extends latent Dirichlet allocation (LDA). They took into account the non-Gaussian nature of expression data while performing linear deconvolution by using a Multinomial mixture model to gene expression data. However, the model is over-parametrized, and hence needs a strong prior specification to be an identifiable model.

Recently, CIBERSORT (Newman et al. 2015) was proposed and applied to estimate immune cell infiltration in tumor samples. The role of immune cells in tumor samples is not yet clearly understood. Yet, accurately finding the presence of immune cells is an important scientific question. CIBERSORT estimated the mixing ratios robustly by using support vector regression (SVR). The SVR was advocated due to the non-Gaussian nature of gene expression data that is manifested by the too

H. Chun (✉)
Department of Mathematics and Statistics, Boston University, Boston, MA, USA
e-mail: chunh@bu.edu

H. Yang
IBM Watson Health, Cambridge, MA, USA

L. Zhang et al. (eds.), *Contemporary Biostatistics with Biopharmaceutical Applications*, ICSA Book Series in Statistics,
https://doi.org/10.1007/978-3-030-15310-6_8

high or too small expression values. Despite the excellent performance, it produces only relative abundance estimates conditional on the given cell types. Hence the absolute level of abundances in the presence of other unknown cell types or noises is hard to be gauged. It was also reported that spurious negative correlations are observed among estimated mixing ratios (Li et al. 2017). Also, CIBERSORT does not impose non-negativity constraints and thus requires post-processing of making negatives to zeros. Due to these problems, Li et al. (2017) used QP with non-negativity constraints in estimating immune infiltration from tumor samples. Li et al. (2017) needed careful selection of genes because their QP estimates were heavily influenced by high values of gene expression.

For this reason, we propose a non-negative robust linear regression (NRLM) approach for a linear deconvolution problem. Our approach adopts ϵ-insensitive loss function with non-negativity constraints as well as L_1 norm constraint to handle non-Gaussian nature of expression while producing interpretable mixing ratio estimates. Details of our approach are in the next section.

8.2 Method

Unlike other regression-based approaches, we model the proportions, rather than gene expression values. We assume that the observed proportion is in part the mixture of specified cell profiles (e.g., immune cell profiles). The other part of proportion is related to noise or other non-specified cell contents (e.g., tumors). Given a signature proportion matrix of N genes and K specified cell types ($\mathbf{A}_{N\times K}$), we assume that the observed proportion $\mathbf{p}$ is decomposed into $\mathbf{p} = \mathbf{A}\mathbf{w} + \mathbf{v}$ where $\mathbf{v}$ captures noise or non-specified cells proportions, and $\mathbf{w}$ is K dimensional vector describing the mixing ratios of the specified cell types.

Under this model, our parameter of interest is $\mathbf{w}$. We propose to estimate $\mathbf{w}$ by minimizing the following ϵ-insensitive loss function.

$$\min_{\mathbf{w}} \sum_{i=1}^{N} \max(|p_i - \mathbf{A}_{i\cdot}\mathbf{w}| - \epsilon, 0)$$

such that $0 \leq w_i \leq 1$ and $\sum w_i \leq 1$. Here $\epsilon > 0$ is a fixed constant and $\mathbf{A}_{i\cdot}$ is the ith row of the matrix. The ϵ-insensitive loss function is used in SVR to promote the robustness of estimates. The difference between SVR and our NRLM is in the use of different regularization on $\mathbf{w}$. SVR uses L_2 regularization to resolve an ill-posedness problem due to correlated predictors in a linear model or over-parametrization in a highly non-linear model. However, the given deconvolution problem is inherently linear and it becomes less crucial to use L_2 regularization as long as the features are not highly correlated. Instead, NRLM uses L_1 regularization with non-negativity constraints to make the estimate interpretable. In addition, we do not use the intercept term to keep the probabilistic interpretation of cell mixtures.

By introducing non-negative slack variables ξ_i^+ and ξ_i^-, the objective function is written as $\min \sum(\xi_i^+ + \xi_i^-)$ s.t. $-\xi_i^- - \epsilon \le p_i - \sum_j A_{i,j} w_j \le \xi_i^+ + \epsilon$; $\sum w_i \le 1$; $\xi_i^+ > 0$; $\xi_i^- > 0$; and $w_i > 0$. This linear programming (LP) problem is solved by using the LPSolve package in **R**. The only tuning parameter is ϵ and is selected from (10, 25, 50, 75, 90)% percentiles of the entire observed proportions that minimizes L_1 norm of $|\mathbf{p} - \mathbf{A}\mathbf{w}|$.

8.3 Results

8.3.1 Simulation Studies

8.3.1.1 No Unwanted Negative Correlations

We perform a simulation study that is similar to as in Li et al. (2017). Using the example, they illustrated the unwanted negative correlations among mixing rate estimates from CIBERSORT. We show that our NLRM gives accurate estimates without any unwanted correlations using two simulated datasets. The first data is generated from two unrelated cell types CD8 T cells (X_1) and neutrophils (X_2) of the LM22 matrix (Newman et al. 2015). The second data is using two highly related cell types, naive (X_1) and memory B-cells (X_2).

The mixing proportions f_1^i and f_2^i are simulated from Uniform(0, 0.5). $\tilde{\mathbf{v}}^i$ are random proportion vectors from Dirichlet(1, (0.1, …, 0.1)) after multiplication of the total expression of CD8 T cells. The model is written as $\mathbf{Y}^i = f_1^i \mathbf{X}_1 + f_2^i \mathbf{X}_2 + \tilde{\mathbf{v}}^i$, where $i = 1, \ldots, 500$.

The model is re-written with proportions as follows:

$$
\begin{aligned}
\mathbf{p}^i &= f_1^i \frac{|\mathbf{X}_1|}{|\mathbf{Y}^i|}\mathbf{A}_1 + f_2^i \frac{|\mathbf{X}_2|}{|\mathbf{Y}^i|}\mathbf{A}_2 + \frac{1}{|\mathbf{Y}^i|}\tilde{\mathbf{v}}^i \\
&= w_1^i \mathbf{A}_1 + w_2^i \mathbf{A}_2 + \mathbf{v}^i,
\end{aligned}
$$

where $|\mathbf{X}|$ represents L_1 the norm of the vector $\mathbf{X}$. Our approach yields the estimates of w_j^i. We convert w_j^i to f_j^i by adjusting with totals as follows: $f_j^i = w_j^i \frac{|\mathbf{Y}^i|}{|\mathbf{X}_j|}$. As seen in Fig. 8.1, our proposed approach estimates the true proportions accurately and

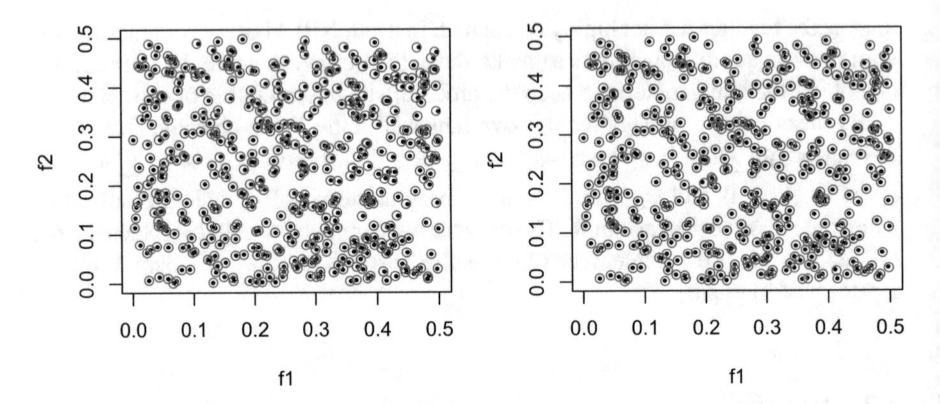

Fig. 8.1 Proportion estimations from cells mixture. Left: Simulation Scenario 1 (independent cell mixture), Right: Simulation Scenario 2 (dependent cell mixture). Solid black dots and circles represent the true and estimates, respectively

the estimates from our method do not expose any unwanted negative correlations in both related (the first simulation) and unrelated (the second simulation) signatures.

8.3.1.2 Deconvolution of *In-Vitro* Cell Mixtures

In this subsection, we present the performance of our deconvolution methods from *in-vitro* mixtures. Four pure cell types (Jurkat, IM9, Raji, and THP1) were mixed *in-vitro* and true mixing ratios were computed by using flow cytometry by Abbas et al. (2009). We download datasets from CIBERSORT's website.

Since our model uses proportions, the gene selection affects the estimates. For a selected gene set S, let $\mathbf{p}_S$ be the sub-vector of $\mathbf{p}$ and $\mathbf{A}_S$ bs the sub-row matrix of $\mathbf{A}$. The model can be written as

$$\begin{aligned}\frac{\mathbf{p}_S^i}{|\mathbf{p}_S^i|} &= \alpha_1^i \frac{\mathbf{A}_{S,1}}{|\mathbf{A}_{S,1}|} + \ldots + \alpha_K^i \frac{\mathbf{A}_{S,1}}{|\mathbf{A}_{S,1}|} + \frac{\mathbf{v}_S^i}{|\mathbf{p}_S^i|} \\ &= f_1^i \frac{|\mathbf{X}_j|}{|\mathbf{p}_S^i||\mathbf{Y}^i|}\mathbf{A}_{S,1} + \ldots + f_1^i \frac{|\mathbf{X}_j|}{|\mathbf{p}_S^i||\mathbf{Y}^i|}\mathbf{A}_{S,1} + + \frac{\mathbf{v}_S^i}{|\mathbf{p}_S^i|}.\end{aligned}$$

Hence, we recover $f_j^i = \frac{|\mathbf{Y}^i|}{|\mathbf{X}_j|}\frac{|\mathbf{p}_S^i|}{|\mathbf{A}_{S,j}|}\alpha_j^i = \frac{|\mathbf{Y}_S^i|}{|\mathbf{X}_{S,j}|}\alpha_j^i$.

The estimated and true proportions are presented in Fig. 8.2. The correlation between true and estimates of our NRLM is 0.91 which is close to 0.92 from CIBERSORT. We notice that the bias of each cell type estimate has the same direction and suggests that the discrepancy may come from the inaccuracy of total estimates due to microarray normalization.

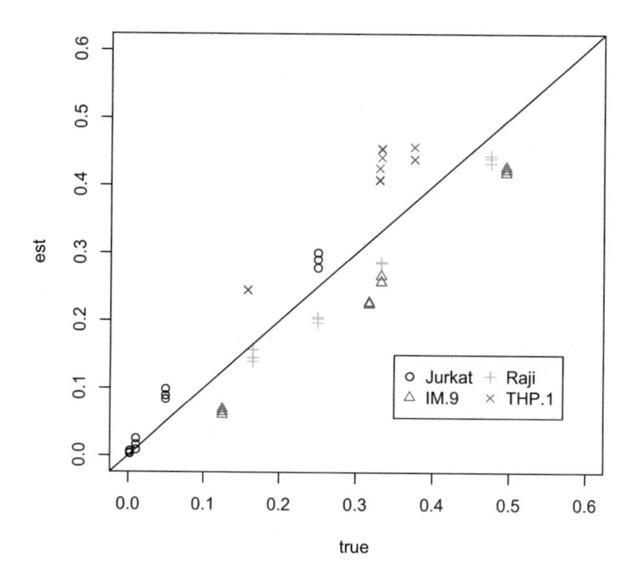

Fig. 8.2 The estimated and true proportions. The correlation between true and estimate is 0.91

8.3.1.3 Performance Comparison Cell with Tumor Contents or Noises

In this subsection, we compare the performance of our NRLM to CIBERSORT and robust linear model (RLM) by adding tumor contents and noises to the 12 *in-vitro* mixture data of Abbas et al. (2009). We remark that Newman et al. (2015) reported the superior performance of CIBERSORT to other methods such as QP and PERT. Hence, we do not include QP or PERT in the simulation. We add RLM because the proportion based RLM has not been compared in the previous study. We also remark that RLM uses the Huber loss function and hence it produces robust estimates. However, it does not impose non-negative constraints.

We download the tumor contents from the colon cancer cell line (HCT116) and use the mean value reported in Newman et al. (2015) as the mean of two datasets (GSM269529 and GSM269530). The MAS5 and quantile normalization are used. The noise is simulated by using log normal model with mean 0 and standard deviation $f \cdot \sigma$ where $\sigma = 11.4$ and f varies as 0, 0.3, 0.6, and 0.9. We compare the correlation between estimates and the truth. As seen in Fig. 8.3, our NRLM yields the highest correlation and performs much better when a large amount of noise is added.

8.3.1.4 Estimating Actual Mixing Rates

As an added advantage, our NLRM estimates the absolute amount of a specific cell content. We follow the spike-in *in-silico* simulation of Newman et al. (2015).

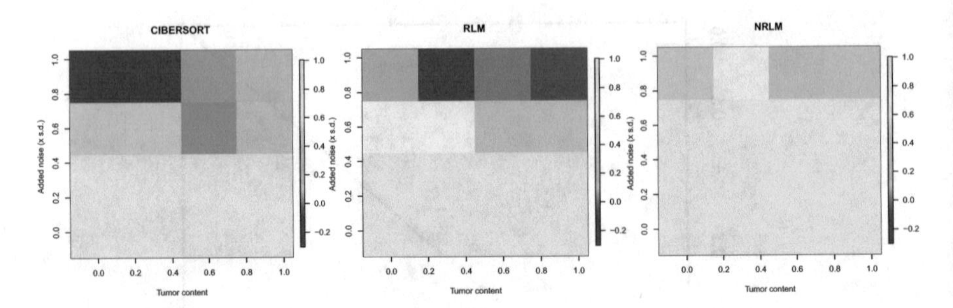

Fig. 8.3 For each combination of tumor content and noise, correlation between true and estimated proportions are presented with colors

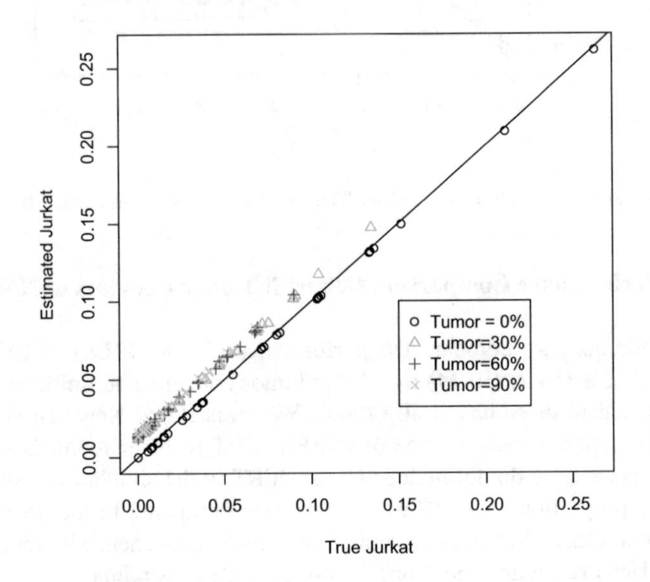

Fig. 8.4 Accuracy of absolute content estimates in *in-silico* Jurkat spike-in experiment

We mix a varying proportion of Jurkat cells from Abbas et al. (2009) to other five different type of mixture cells consisting of the other three cells. We then add tumor cells that are described in the previous subsection with varying total amount. When compared to the truth, we find that the estimate is almost identical to the truth (Fig. 8.4) when there is no tumor content. But, with tumor contents, the Jurkat content is slightly over-estimated although the difference is less than 0.016.

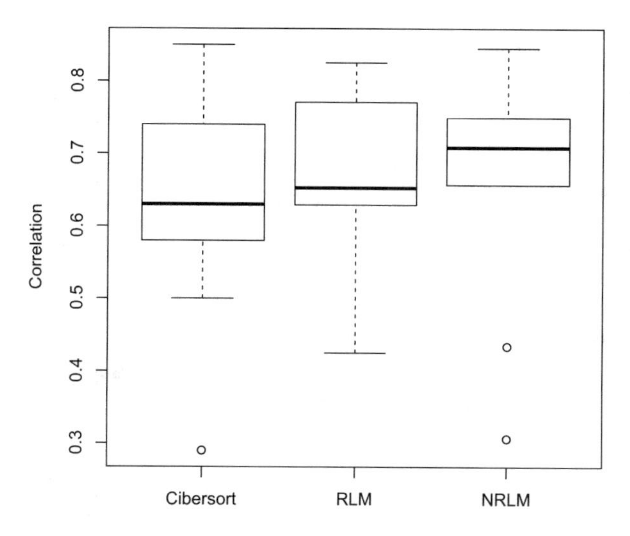

Fig. 8.5 Performance comparison in deep deconvolution of PBMC samples

8.3.2 *Real Data Analysis*

We compare the performance of CIBERSORT, RLM, and NRLM with a deep convolution problem discussed in Newman et al. (2015). In their study, peripheral blood mononuclear cells (PBMCs) were collected from 20 adults of varying ages. Gene expression levels were measured by using Illumina BeadChip arrays. To find ground truths, they used flow cytometry and computed the relative abundance of nine immune cell subsets.

The data is obtained from CIBERSORT's website. Here we do not apply any total/marker correction to the RLM and NRLM estimates because only marker gene expression data is downloaded. As seen in Fig. 8.5, our NRLM performance is comparable to the other methods. We remark that, in this example, some of the estimates from CIBERSORT or RLM are negative and these negative coefficients are modified to zeros. Since these estimates are conditional on the negative estimates, care must be taken when the coefficients are interpreted. We also present the estimated relative abundance compared with the truth at Fig. 8.6. Although NRLM performs better than other methods and shows clear positive correlations, it appears that the estimates from NRLM are not so accurate. In fact, the absolute level of proportions explained by this LM22 matrix is 0.48 from our method. Hence, these blood samples still have a large component that cannot be explained by the signature matrix. When we check the histogram of proportions from randomly selected subjects, their frequency distributions are very different from the frequency distributions of the LM22 matrix (Fig. 8.7). This suggests that we may still need a better signature matrix for deep deconvolution.

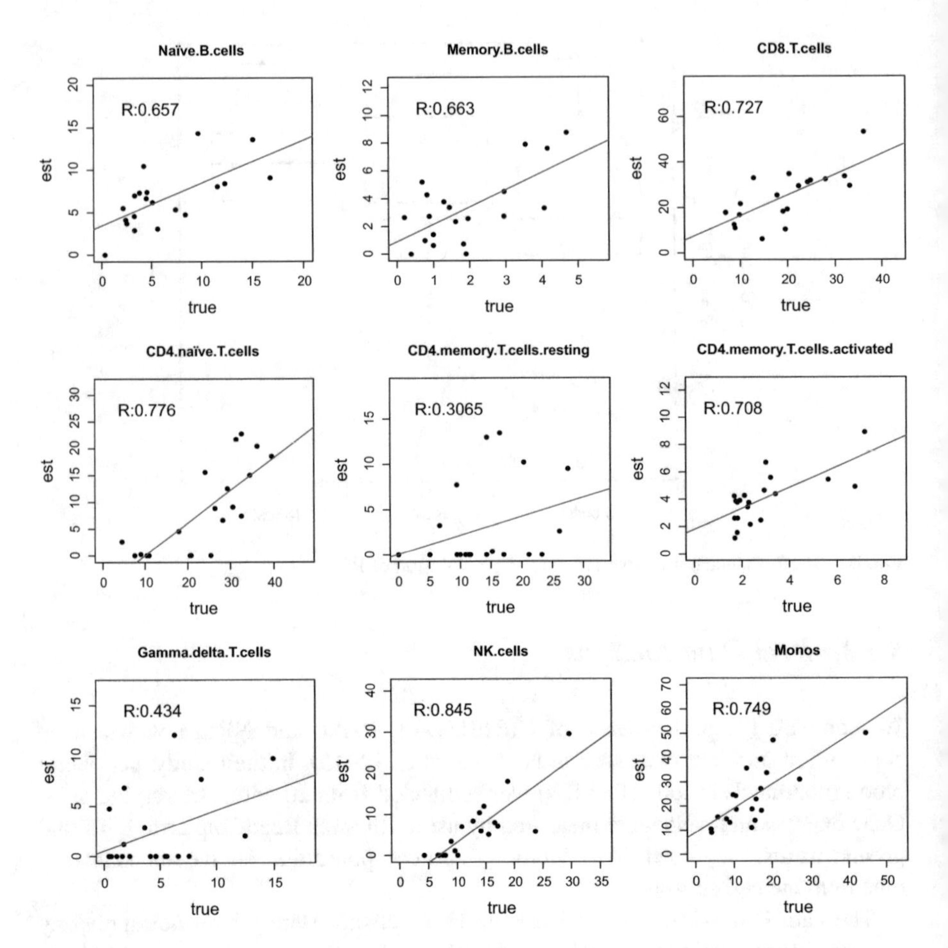

Fig. 8.6 NRLM estimates vs true proportion from flow cytometry by using PBMC samples

8.4 Discussion

The most important factor in estimating mixing rates of given cell types is the choice of the signature matrix. If the signature matrix **A** is not accurate, no method would perform well. In this regard, we highly appreciate the importance of the LM22 matrix (Newman et al. 2015) in estimating immune content. Our contribution is to provide interpretable estimates of mixing ratios robustly. To promote the robustness, we adopted the ϵ- insensitive loss function just like SVR. However, we imposed non-negativity as well as the L_1 constraint for interpretability. Also, we normalized each column of data to have the sum of one and did not use the intercept term to keep the probabilistic interpretation. From these changes, our method showed a great performance in estimating mixing rates. However, the current method only produced point estimates without uncertainty quantification. Our future work includes deriving a large sample property and producing interval estimates.

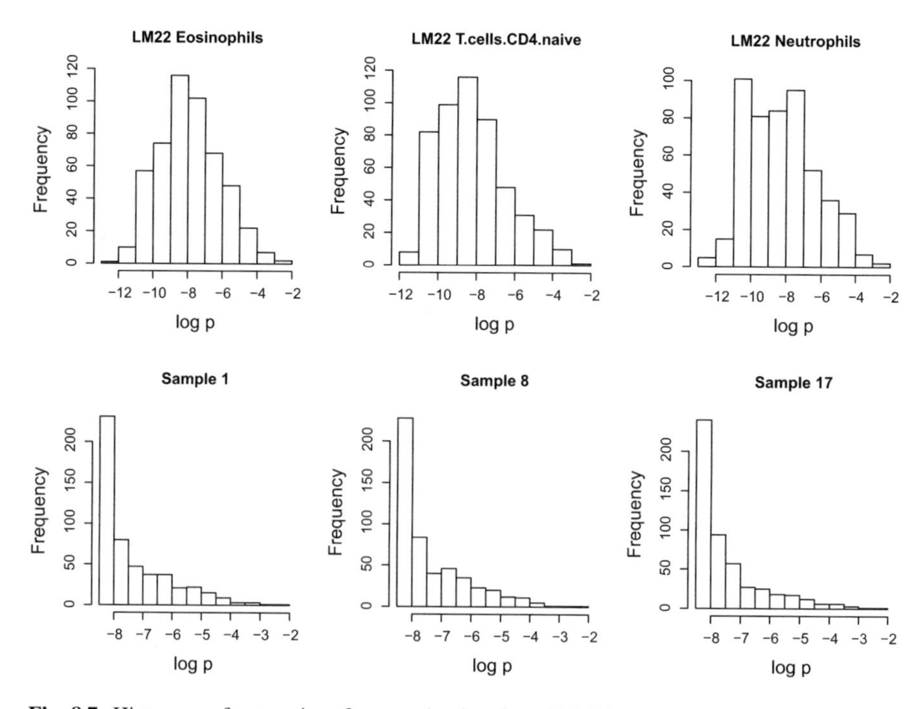

Fig. 8.7 Histogram of proportions from randomly selected LM22 matrix and PBMC samples

References

Abbas, A.R., et al.: Deconvolution of blood microarray data identifies cellular activation patterns in systemic lupus erythematosus. PloS One **4**(7), e6098 (2009). https://doi.org/10.1371/journal.pone.0006098

Gong, T., et al.: Optimal deconvolution of transcriptional profiling data using quadratic programming with application to complex clinical blood samples. PloS One **6**(11), e27156 (2011). https://doi.org/10.1371/journal.pone.0027156

Li, B., Liu, J.S., Liu, S.L.: Revisit linear regression-based deconvolution methods for tumor gene expression data. Genome Biol. **18**, 127 (2017). https://doi.org/10.1186/s13059-017-1258-3

Neweman, A.M., et al.: Robust enumeration of cell subsets from tissue expression profiles. Nat. Methods **12**, 453–457 (2015)

Qiao, W., et al.: PERT: a method for expression deconvolution of human blood samples from varied microenvironmental and developmental conditions. PloS Comput. Biol. **8**(12), e1002838 (2012). https://doi.org/10.1371/journal.pcbi.1002838

References

Chapter 9
Practical Design Approaches for Assessing Parallelism in Dose Response Modelling

Timothy E. O'Brien and Jack Silcox

9.1 Introduction

Scientific researchers in biomedicine, pharmaceutical science and toxicology often face situations in which binary logistic regression model fits are used to compare two drugs or substances, often by means of comparisons of median doses or concentrations (EC50 or LD50). Applications are given in works spanning early bioassay findings of Finney (1971, 1978) to more recent results in Rich (2013) and Gupta and Vale (2017). Furthermore, Wheeler et al. (2006) underscores the caution that instead of examining for overlap in separate EC50 confidence intervals, testing is best based on estimation and confidence intervals associated with the relative potency parameter. Notably, before fitting such curves and testing for differing potencies, an important requirement is that these dose response curves be parallel. As such, various works have introduced meaningful means to assess parallelism in logistic regression settings, including Gottschalk and Dunn (2005), Jonkman and Sidak (2009), Novick et al. (2012), Yang and Zhang (2012), Yang et al. (2012), Fleetwood et al. (2015) and Sidak and Jonkman (2016).

So as to efficiently test for common slopes of drug curves, our focus here is on developing robust, efficient and practical design strategies in the assessment of dose response curve parallelism.

T. E. O'Brien (✉)
Department of Mathematics and Statistics, Loyola University Chicago, Chicago, IL, USA

Institute of Environmental Sustainability, Loyola University Chicago, Chicago, IL, USA

J. Silcox
Department of Mathematics and Statistics, Loyola University Chicago, Chicago, IL, USA

University of Utah, Cognitive and Neural Science PhD Program, Salt Lake City, UT, USA
e-mail: tobrie1@luc.edu

L. Zhang et al. (eds.), *Contemporary Biostatistics with Biopharmaceutical Applications*, ICSA Book Series in Statistics,
https://doi.org/10.1007/978-3-030-15310-6_9

Overviews of optimal design theory and applications are given in O'Brien and Funk (2003) and Atkinson et al. (2007). Additionally, classical experimental design strategies for binary logistic regression models are given in Abdelbasit and Plackett (1983), Minkin (1987) and Kalish (1990), and model-robust design approaches are given and explored in Atkinson (1972), O'Brien (2005), O'Brien (2016), O'Brien (2018), O'Brien and Rawlings (1996), O'Brien and Lim (2018), O'Brien et al. (2009) and O'Brien et al. (2010).

9.2 Assessing Parallelism in Dose Response

In situations where the outcome variable is a percentage derived from binary outcomes—such as percentage mortality in a microbiology or toxicology experiment—the two-parameter (binomial, logit-link) logistic model is often used to model the dose-response data. This generalized nonlinear model is written

$$\log\left(\frac{\pi}{1-\pi}\right) = \eta = \beta\,(x-\gamma) \tag{9.1}$$

Here, π is the probability of outcome (e.g., mortality), β is the slope, x is the concentration or dose of the drug or compound, and γ is the EC50/LD50 parameter so that $x = \gamma$ coincides with $\pi = \frac{1}{2}$ (or 50% chance of death). Equivalent to (9.1) is the expression $\pi = \frac{e^{\beta(x-\gamma)}}{1+e^{\beta(x-\gamma)}}$. This model can be extended to simultaneously model two curves (such as corresponding to two viruses or drugs, labelled "A" and "B"), as graphed in Fig. 9.1, by modifying the right-hand side in (9.1) to be

$$\eta = \begin{cases} \beta\,(x-\gamma_A)\,, & drug\ A \\ (\beta+\delta)\,(x-\gamma_B)\,, & drug\ B \end{cases} \tag{9.2}$$

In (9.2), β is the slope of the drug A curve, $(\beta + \delta)$ is the slope of the drug B curve, and the respective EC50's are γ_A and γ_B for drugs A and B. Our goal in fitting this model is to fit the respective curves with particular focus on the difference-of-slopes parameter δ. Both curves can be written in a single model as $\pi = \frac{e^{\eta}}{1+e^{\eta}}$ with η here given in (9.2). It is important to point out that both curves are fit—and indeed designed—simultaneously since they share the joint parameter β. These curves are plotted in Fig. 9.1—where π is the percent response—using parameter values $\beta = 0.30$, $\delta = 0$, $\gamma_A = 11.0$, $\gamma_B = 14.5$.

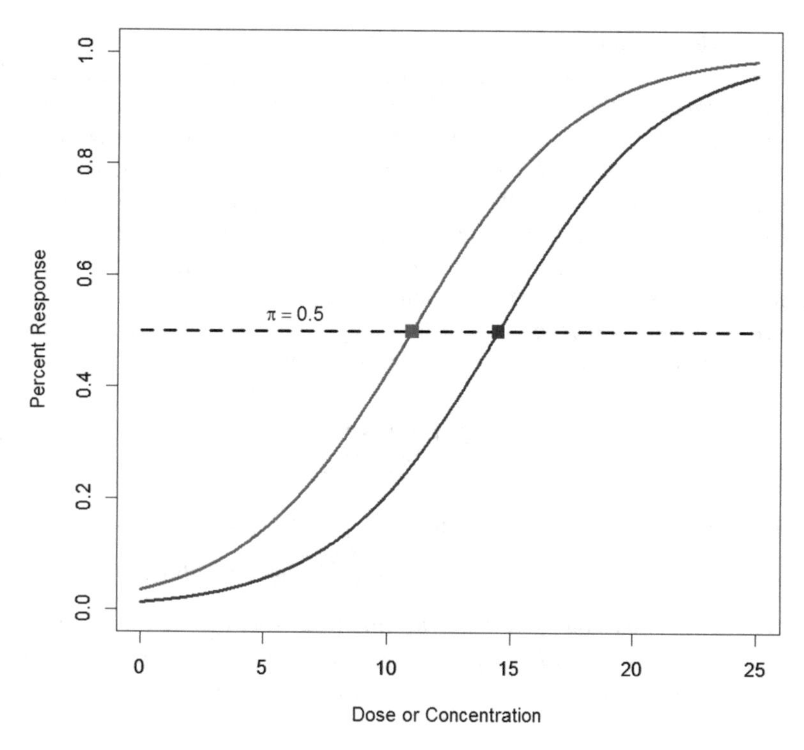

Fig. 9.1 Plot of parallel ($\delta = 0$) binary logistic curves with vertical axis corresponding to probability of outcome or mortality, slope $\beta = 0.30$ and EC50's $\gamma_A = 11.0$ (left curve) and $\gamma_B = 14.5$ (right curve); here, relative potency is then estimated to be 14.5/11.0 = 1.32. Solid squares indicate EC50's as points where the respective curves cross cut-line, $\pi = 1/2$

9.3 Optimal Design Background

Approximate designs, denoted ξ, are written

$$\xi = \begin{Bmatrix} x_1 & x_2 & \dots & x_n \\ \omega_1 & \omega_2 & \dots & \omega_n \end{Bmatrix} \tag{9.3}$$

The ω_i are non-negative design weights which sum to one, and the x_i are design points (i.e., concentrations) that belong to the design space and are not necessarily distinct. Further, the p model parameters are stacked into the p-vector $\boldsymbol{\theta}^T = (\beta, \gamma_A, \gamma_B, \delta)$. In the constant-variance Normal/Gaussian setting with linear or nonlinear normal model function $\eta(x, \boldsymbol{\theta})$, the $n \times p$ Jacobian matrix is $\boldsymbol{V} = \frac{\partial \eta}{\partial \theta}$ and with $\boldsymbol{\Omega} = diag\ \{\omega_1, \omega_2, \ \dots\ , \omega_n\}$, the $p \times p$ (Fisher) information matrix is written

$$\boldsymbol{M}(\xi, \boldsymbol{\theta}) = \boldsymbol{V}^T \boldsymbol{\Omega} \boldsymbol{V} \tag{9.4}$$

Atkinson et al. (2007) demonstrates that for binomial logistic models in general, the information matrix for the relative-potency logistic model considered here has the same form as in (9.4) with a modification of the weight matrix $\boldsymbol{\Omega}$. Specifically, $\boldsymbol{\Omega}$ in this case is a diagonal matrix with i^{th} diagonal element $\omega_i\pi_i(1 - \pi_i)$, where π_i is the success probability. In regression settings, since the (asymptotic) variance of the maximum-likelihood estimator $\hat{\boldsymbol{\theta}}_{MLE}$ is proportional to $\boldsymbol{M}^{-1}(\xi, \boldsymbol{\theta})$, designs are often chosen to minimize some (convex) function of $\boldsymbol{M}^{-1}(\xi, \boldsymbol{\theta})$. For example, designs which minimize the determinant of $\boldsymbol{M}^{-1}$—and equivalently which maximize the determinant of $\boldsymbol{M}$—are called D-optimal.

Since our focus is on the difference-of-slopes parameter (δ) more so than the other parameters, we partition the Fisher information matrix as

$$\boldsymbol{M} = \begin{bmatrix} \boldsymbol{M}_{11} & \boldsymbol{M}_{12} \\ \boldsymbol{M}_{21} & \boldsymbol{M}_{22} \end{bmatrix} \tag{9.5}$$

Each sub-matrix $\boldsymbol{M}_{ij}$ is of dimension $p_i \times p_j$ for $i, j = 1, 2$, and $p_1 + p_2 = p$. The parameter vector is also partitioned, $\boldsymbol{\theta} = \begin{pmatrix} \boldsymbol{\theta}_1 \\ \boldsymbol{\theta}_2 \end{pmatrix}$ with $\boldsymbol{\theta}_1$ (the so-called nuisance parameters) of dimension $p_1 \times 1$ and $\boldsymbol{\theta}_2$ (the parameter of interest) of dimension $p_2 \times 1$. In the current situation, $\boldsymbol{\theta}_1^T = (\beta, \gamma_A, \gamma_B)$ so $p_1 = 3$ and $\theta_2 = \delta$ so $p_2 = 1$. As outlined in Atkinson et al. (2007), $D_s(\boldsymbol{\theta}_2)$ subset designs maximize

$$\left| \boldsymbol{M}_{22} - \boldsymbol{M}_{21}\boldsymbol{M}_{11}^{-1}\boldsymbol{M}_{12} \right| = \frac{|\boldsymbol{M}|}{|\boldsymbol{M}_{11}|} \tag{9.6}$$

Because of problems associated with subset designs, some authors suggest combining the subset and full parameter criteria so that for a given $\alpha \in \left[\frac{p_2}{p}, 1\right]$, designs be chosen to maximize the compound objective function (see O'Brien (2005) and Atkinson et al. (2007)),

$$\Phi_\alpha(\xi, \boldsymbol{\theta}) = \frac{1-\alpha}{p_1}\log|\boldsymbol{M}_{11}| + \frac{\alpha}{p_2}\log\left| \boldsymbol{M}_{22} - \boldsymbol{M}_{21}\boldsymbol{M}_{11}^{-1}\boldsymbol{M}_{12} \right| \tag{9.7}$$

A generalized inverse is used in (9.7) when $\boldsymbol{M}_{11}$ is not invertible. This objective function ranges from the D-optimal criterion for $\alpha = \frac{p_2}{p}$ to the subset design criterion (for δ) in (9.6) for $\alpha = 1$. For a given choice of $\alpha \in \left[\frac{p_2}{p}, 1\right]$, we call designs that maximize (9.7) D_α-optimal designs.

Extending the results given in O'Brien (2016), our results here for the two-logistic situation in (9.2) and Fig. 9.1 validate that the optimal values of $s = e^\eta$ for D_α-optimal designs satisfy the expression,

$$(1 + s) + A\,(1 - s)\,\log(s) = 0 \tag{9.8}$$

In (9.8), $A = \frac{\alpha p_1 + 3(1-\alpha)p_2}{2(\alpha p_1 + (1-\alpha)p_2)}$. The s-values which solve this expression are observed to be reciprocals, and this has been demonstrated to be the case in other situations when working with logistic regression (see O'Brien (2016), O'Brien and Lim (2018), and O'Brien et al. (2009)). As in these other situations, Eq. (9.8) has exactly two roots for $\alpha \in \left[\frac{p_2}{p}, 1\right]$. For example, for $\alpha = \frac{p_2}{p} = 0.25$ (i.e., the D-optimality criterion), $A = 1$ and solution of (9.8) gives $s_1 = 0.2137$ (and $\pi_1 = \frac{s_1}{1+s_1} = 0.1760$) and $s_2 = 4.6805$ (and $\pi_2 = 0.8240$). As α approaches unity (i.e., the $D_s(\delta)$ subset design), we obtain $A = \frac{1}{2}$, $s_1 = 0.0908$ (and $\pi_1 = 0.0832$) and $s_2 = 11.0161$ (and $\pi_2 = 0.9168$). Since the s values here are reciprocals, the corresponding values of π necessarily sum to one.

Figure 9.2 shows the D-optimal design points along with the corresponding cut-lines at $\pi_1 = 0.1760$ and $\pi_2 = 0.8240$. The values of the design support points are obtained using the relations $x = \gamma_A \pm \frac{1}{\beta} \log(s_1)$ for drug A (left curve in Fig. 9.2) and $x = \gamma_B \pm \frac{1}{(\beta+\delta)} \log(s_1)$ for drug B (right curve in Fig. 9.2). Thus, for the specific parameter values used here, the support points are $x = 5.8554, 16.1447$ for drug A (left curve) and $x = 9.3554, 19.6447$ for drug B (right curve).

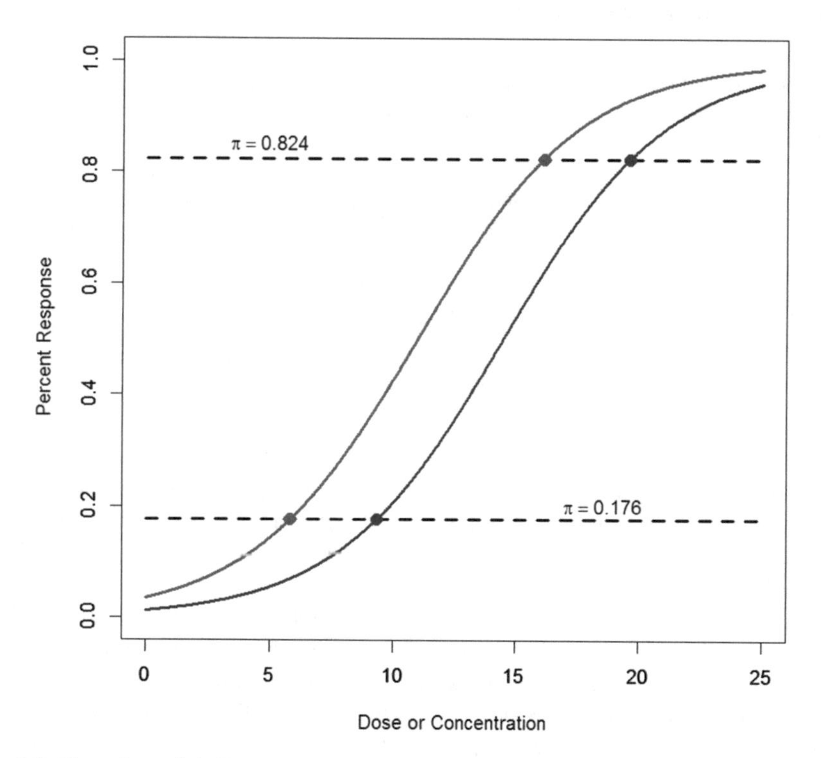

Fig. 9.2 Plot of parallel ($\delta = 0$) binary logistic curves with vertical axis corresponding to probability of outcome/mortality, slope $\beta = 0.30$ and EC50's $\gamma_A = 11.0$ (left curve) and $\gamma_B = 14.5$ (right curve), along with respective D-optimal design points (circles on respective curves). Cut-lines at $\pi_1 = 0.1760$ and $\pi_2 = 0.8240$ demonstrate D-optimality of these design support points

A measure of the distance between an arbitrary design ξ and a D-optimal design ξ_D^* is the D-efficiency (*DEFF*) given by the expression $\left(\frac{|M(\xi)|}{|M(\xi_D^*)|}\right)^{1/p}$ (see O'Brien and Funk (2003) and Atkinson et al. (2007)). A similar D-efficiency expression ($DEFF_s$) can be given for subset efficiency in (9.6) but using $\left|M_{22} - M_{21}M_{11}^{-1}M_{12}\right|$ in place of $|M|$ in both numerator and denominator and raised to the power $(1/p_2)$ instead of $(1/p)$. For the current situation, full and subset efficiencies of the D-optimal and the $D_s(\delta)$ subset design are given in Table 9.1. Note that as one shifts from the D-optimal to the $D_s(\delta)$ subset design (i.e., as α increases from 0.25 to 1.0), the design support points spread out away from the EC50s since the proportion cut lines (see Fig. 9.2) drop from $\pi = 0.1760$ to $\pi = 0.0832$ and increase from $\pi = 0.8240$ to $\pi = 0.9168$. Also, as we shift from the D-optimal to the subset design, as expected we note the decrease in the variance term (diagonal term in M^{-1}) associated with δ of over 20% (from 1.0419 to 0.8196) but also the approximately 90% increase in the variance terms associated with the γ (EC50) terms (from 153.2 to 291.3). Finally, in noting the efficiency values of these two designs, one readily sees the trade-off nature in that as one efficiency increases, the other decreases, and vice versa.

Table 9.1 *D*- and $D_s(\delta)$-optimal designs in the parallel logistic setting: design support points and corresponding proportions, associated variance estimates for parameter values, and *D*- and subset-efficiencies

α	Design	Drug A support points	Drug B support points	Proportion values	Diagonal elements of M^{-1} corresponding to $\theta^T = (\beta, \gamma_A, \gamma_B, \delta)$	*DEFF*	$DEFF_s$
0.25	*D*-optimal	$x_1 = 5.8554$ $x_2 = 16.1447$	$x_1 = 9.3554$ $x_2 = 19.6447$	$\pi_1 = 0.1760$ $\pi_2 = 0.8240$	0.5210, 153.2, 153.2, 1.0419	1.00	0.787
1.0	$D_s(\delta)$ subset-optimal	$x_1 = 3.0024$ $x_2 = 18.9977$	$x_1 = 6.5019$ $x_2 = 22.4977$	$\pi_1 = 0.0832$ $\pi_2 = 0.9168$	0.4098, 291.3, 291.3, 0.8196	0.818	1.00

Translating efficiencies into sample size requirements, note that an efficiency of 0.80 of a given design relative to an optimal design translates to $1/0.80 = 1.25$. So, a sample size 25% higher for the less-efficient design is needed (compared with the optimal design) to yield equivalent information.

The above advantages in terms of efficiency notwithstanding, optimal designs are often only used as a starting point since they often have shortcomings. In most practical situations, optimal designs for p-parameter model functions comprise only p support points; this is observed here since the D-optimal designs for the two-parameter logistic curves graphed in Fig. 9.2 have only two support points for each curve/drug. As such, these designs provide little or no ability to test for lack of fit of the assumed model. As a result, researchers often desire near-optimal, so-called "robust", designs which have extra support points that can then be used to test for model adequacy. Another important disadvantage of the optimal designs plotted in Fig. 9.2 is that practitioners typically wish to use the same concentration values for both drugs. In the next section, we introduce and explore very useful strategies to obtain near-optimal, robust designs which address these shortcomings.

9.4 Robust Design Approaches

In this section, we propose practical experimental design strategies addressing the lack-of-fit and same-concentration considerations raised in the previous section. The requirement that the same concentrations be used for both dose-response curves leads us to consider finding the corresponding same-concentration-restricted D- and $D_s(\delta)$-optimal designs, and this is done in the following section. In the subsequent sections, we introduce and examine so-called reflection designs based upon the original optimal designs, as well as geometric and uniform designs.

9.4.1 Same-Concentration Designs

In the current illustration—as well as in other situations with the two-drug logistic model in (9.2)—the same-concentration-restricted D-optimal design comprises two support points (for both drugs), and this is demonstrated in Fig. 9.3 and Table 9.2. As demonstrated in Fig. 9.3, the optimal proportions of $\pi_{1A} = 0.2583$ and $\pi_{2A} = 0.8914$ for drug A and $\pi_{1B} = 0.1086$ and $\pi_{2B} = 0.7417$ for drug B are such that (1) the lower proportions (π_{1A} and π_{1B}) straddle the cut-line of $\pi = 0.176$ from Fig. 9.2 and the upper proportions (π_{2A} and π_{2B}) straddle the cut-line of $\pi = 0.824$ from Fig. 9.2, and (2) are reciprocally related via $\pi_{2A} = 1 - \pi_{1B}$ and $\pi_{2B} = 1 - \pi_{1A}$.

In examining the first row of Table 9.2 (i.e. for $\alpha = 0.25$, i.e., the D-optimal design), note that the same-concentration D-optimal design yields a D-efficiency (relative to the best design plotted in Fig. 9.2) of 96.1%; with a modest 3.9% information loss, this design is thus deemed to be highly efficient. Table 9.2 also gives analogous results for the $D_s(\delta)$-optimal ($\alpha = 1$) situation, where it is noted that in shifting to subset optimality, the design support points and optimal proportions again shift outward away from the EC50s. The subset efficiency of this subset design is 89.3%. Our findings bear out that the patterns and observations noted here generalize to other parameter choices.

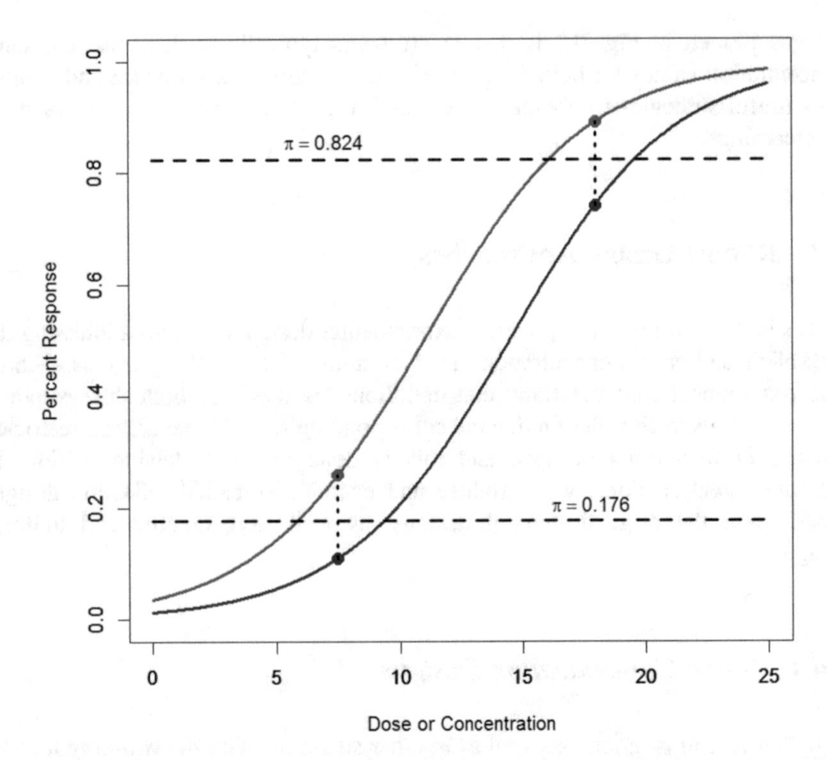

Fig. 9.3 Plot of parallel ($\delta = 0$) binary logistic curves with vertical axis corresponding to probability of outcome/mortality, slope $\beta = 0.30$ and EC50's $\gamma_A = 11.0$ (left curve) and $\gamma_B = 14.5$ (right curve), along with same-concentration D-optimal design points (circles on the respective curves). Cut-lines at $\pi_1 = 0.1760$ and $\pi_2 = 0.8240$ correspond to unrestricted D-optimal design points represented in Fig. 9.2

Table 9.2 Same-concentration-restricted *D*- and $D_s(\delta)$ designs in the parallel logistic setting: design support points and corresponding proportions, associated variance estimates for parameter values, and *D*- and subset-efficiencies

α	Design	Drug A and B support points	Drug A proportion values	Drug B proportion values	Diagonal elements of $\boldsymbol{M}^{-1}$ corresponding to $\boldsymbol{\theta}^T = (\beta, \gamma_A, \gamma_B, \delta)$	$DEFF$	$DEFF_s$
0.25	*D*-optimal	$x_1 = 7.4844$ $x_2 = 18.0156$	$\pi_1 = 0.2583$ $\pi_2 = 0.8914$	$\pi_1 = 0.1086$ $\pi_2 = 0.7417$	0.5607, 154.1, 154.1, 1.1214	0.961	0.731
1.0	$D_s(\delta)$ subset-optimal	$x_1 = 4.8600$ $x_2 = 20.6404$	$\pi_1 = 0.1368$ $\pi_2 = 0.9475$	$\pi_1 = 0.0526$ $\pi_2 = 0.8632$	0.4587, 275.6, 275.6, 0.9173	0.811	0.893

A major disadvantage of the same-concentration restriction is that the resulting near-optimal designs are observed to comprise only two support points, and thus provide no ability to check for model misspecification. As such, we now consider the following robust modification of the original unrestricted D-optimal design approach illustrated in Fig. 9.2 and Table 9.1.

9.4.2 *Reflection Designs*

One means of obtaining robust near-optimal designs (i.e., designs with reasonably high efficiency and additional support points) in the two-logistic situation of (9.2) is to simply use for both drugs each of the four distinct concentrations in Fig. 9.2—that is, the four concentrations where—for at least one of the curves—$\pi = 0.1760$ and $\pi = 0.8240$. We call such designs reflection designs. For the parameter values used here, this situation is illustrated in Fig. 9.4. The additional reflection π values use $ex = e^{\beta(\gamma_B - \gamma_A)}$, $\pi_1 = \frac{s_1}{1+s_1} = \pi_7, \pi_2 = \frac{s_2}{1+s_2} = 1 - \pi_1 = \pi_8$ and the relations for all such parallel-curve (i.e., $\delta = 0$) reflection design situations are as follows:

$$\pi_3 = \frac{ex\ s_1}{1+ex\ s_1}; \quad \pi_4 = \frac{ex\ s_2}{1+ex\ s_2}; \quad \pi_5 = \frac{s_1}{ex+s_1} = 1 - \pi_4; \quad \pi_6 = \frac{s_2}{ex+s_2} = 1 - \pi_3 \tag{9.9}$$

For the parameter values used here, we obtain $s_1 = 0.2137$, $s_2 = 4.6805$, $ex = 2.858$, $\pi_1 = \pi_7 = 0.1760$, $\pi_2 = \pi_8 = 0.8240$, $\pi_3 = 0.3791$, $\pi_4 = 0.9304$, $\pi_5 = 0.0696$, $\pi_6 = 0.6209$.

For the chosen parameter values, designs, support points, proportion values and summary statistics are given in Table 9.3 for reflection designs for α values in Eq. (9.7) of $\alpha = 0.25$ (i.e. *D*-optimality), $\alpha = 0.87$ (equal-efficiency), and $\alpha = 1.0$ (i.e., $D_s(\delta)$-optimality); note that the equal-efficiency design has been chosen so the two efficiency values, *DEFF* and $DEFF_s$, are approximately equal. It is important to note that, with a *D*-efficiency of 93.2%, the *D*-optimal reflection design is observed to be highly efficient—i.e., with an efficiency loss of less than 7%. This design is also robust in that it provides additional support points to test for model lack-of-fit, and it is very practical to use in that scientific researchers merely need to sketch the anticipated dose-response curves for the two drugs/substances, and obtain design support points resulting from the cut lines at $\pi_1 = 0.1760$ and $\pi_2 = 0.8240$ (or thereabouts). The subset efficiency of this design is 67.1%; should a researcher desire higher subset efficiency, the value of α could be increased to meet the researcher's objectives. For example, the equal-efficiency design ($\alpha = 0.87$) results in efficiencies for the full parameter vector and for the δ (difference of slopes) parameter of about 84%; in this case, the proportion cut values, at $\pi_1 = 0.0951$ and

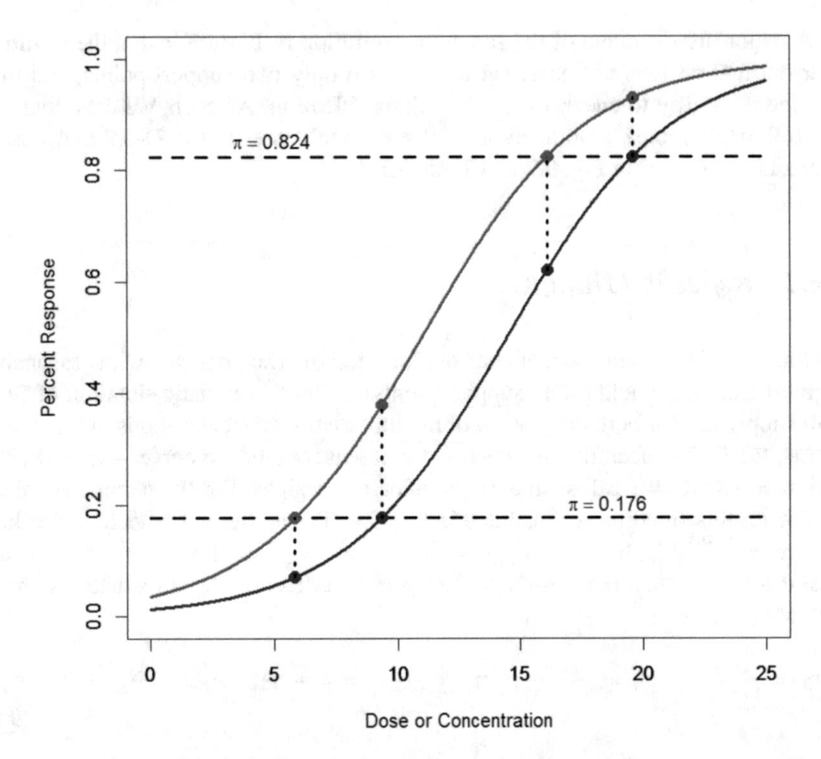

Fig. 9.4 Plot of parallel ($\delta = 0$) binary logistic curves with vertical axis corresponding to probability of outcome/mortality, slope $\beta = 0.30$ and EC50's $\gamma_A = 11.0$ (left curve) and $\gamma_B = 14.5$ (right curve), cut-lines at $\pi_1 = 0.1760$ and $\pi_2 = 0.8240$ correspond to unrestricted D-optimal design points, and reflection design points (four circles for each of the drugs)

$\pi_2 = 0.9049$—or very nearly $\pi_1 = 0.10$ and $\pi_2 = 0.90$—are also very practical to implement by researchers. As has been observed above for unconstrained and same-concentration designs, as α values are increased in (9.7) to emphasize efficient estimation of the difference-in-slopes parameter (δ), the design support points shift out (i.e., away from the EC50's). As noted in Table 9.3, one down-side of this shift is that the variability associated with these EC50 values increase here by 56% (from 151.2 to 235.7) for $\alpha = 0.87$ and by 74% (from 151.2 to 263.8) for $\alpha = 1.0$. We discuss implications of these results in terms of design performance, as well as overall recommendations, in Sects. 9.5 and 9.6.

Table 9.3 Reflection D-optimal, equal-efficiencies, and $D_s(\delta)$ designs in the parallel logistic setting with parameter values given above: design support points and corresponding proportions, associated variance estimates for parameter values, and D- and subset-efficiencies

α	Design	Drug A and B support points	Drug A proportion values	Drug B proportion values	Diagonal elements of $\boldsymbol{M}^{-1}$ corresponding to $\boldsymbol{\theta}^T = (\beta, \gamma_A, \gamma_B, \delta)$	$DEFF$	$DEFF_s$
0.25	D-optimal	$x_1 = 5.8554$ $x_2 = 9.3554$ $x_3 = 16.1447$ $x_4 = 19.6447$	$\pi_1 = 0.1760$ $\pi_2 = 0.3791$ $\pi_3 = 0.8240$ $\pi_4 = 0.9304$	$\pi_1 = 0.0696$ $\pi_2 = 0.1760$ $\pi_3 = 0.6209$ $\pi_4 = 0.8240$	0.6106, 151.2, 151.2, 1.2212	0.932	0.671
0.87	Equal efficiency	$x_1 = 3.4889$ $x_2 = 6.9888$ $x_3 = 18.5111$ $x_4 = 22.0111$	$\pi_1 = 0.0951$ $\pi_2 = 0.2309$ $\pi_3 = 0.9049$ $\pi_4 = 0.9645$	$\pi_1 = 0.0355$ $\pi_2 = 0.0951$ $\pi_3 = 0.7691$ $\pi_4 = 0.9049$	0.4873, 235.7, 235.7, 0.9746	0.841	0.841
1.0	$D_s(\delta)$ subset-optimal	$x_1 = 3.0021$ $x_2 = 6.5023$ $x_3 = 18.9977$ $x_4 = 22.4980$	$\pi_1 = 0.0832$ $\pi_2 = 0.2060$ $\pi_3 = 0.9168$ $\pi_4 = 0.9692$	$\pi_1 = 0.0308$ $\pi_2 = 0.0832$ $\pi_3 = 0.7940$ $\pi_4 = 0.9168$	0.4828, 263.8, 263.8, 0.9657	0.802	0.849

9.4.3 Geometric and Uniform Designs

Noting the common-usage of geometric and uniform designs in practical settings, O'Brien (2016) and O'Brien et al. (2009) combined the D-optimality criterion with the geometric design structure of the form $x = a, ab, ab^2 \ldots ab^K$ and the uniform design structure of the form $x = A, A + B, A + 2B \ldots A + KB$, where K is an adjustment value chosen by the researcher to provide a sufficient number of design support points. We adopt a similar approach here for the two-logistic model and, using the chosen parameter values, focus here only on the D-optimality criterion ($\alpha = 0.25$). (Nonetheless, these results generalize to other choices of α as well.) Results are given in Table 9.4 for the optimal geometric and uniform designs for $K = 3$; this choice of K makes these designs comparable with the 4-support-point designs given above.

As noted in Table 9.4, the geometric design yields slightly lower efficiencies compared with the uniform design, as well as diverse proportions (i.e., values of π) and unequal variance terms for the two EC50 values. The uniform design, on the other hand, results in a recognizable pattern of proportions (viz, $\pi_{1A} + \pi_{4B} = \pi_{2A} + \pi_{3B} = \pi_{3A} + \pi_{2B} = \pi_{4A} + \pi_{1B} = 1$), and somewhat lower variance values. Both of these designs, however, perform slightly worse than the D-optimal ($\alpha = 0.25$) reflection design in the previous section in terms of the efficiency measures. These designs are further examined using simulations in the following section.

Table 9.4 Optimal geometric and uniform D-optimal designs in the parallel logistic setting with parameter values given above: design support points and corresponding proportions, associated variance estimates for parameter values, and D- and subset-efficiencies

α	Design	Drug A and B support points	Drug A proportion values	Drug B proportion values	Diagonal elements of $\boldsymbol{M}^{-1}$ corresponding to $\boldsymbol{\theta}^T = (\beta, \gamma_A, \gamma_B, \delta)$	*DEFF*	*DEFF$_s$*
0.25	D-optimal geometric $a = 6.216$ $b = 1.471$	$x_1 = 6.2162$ $x_2 = 9.1428$ $x_3 = 13.4473$ $x_4 = 19.7783$	$\pi_1 = 0.1923$ $\pi_2 = 0.3642$ $\pi_3 = 0.6757$ $\pi_4 = 0.9330$	$\pi_1 = 0.0769$ $\pi_2 = 0.1670$ $\pi_3 = 0.4217$ $\pi_4 = 0.8297$	0.7640, 133.1, 164.4, 1.4377	0.889	0.570
0.25	D-optimal uniform $A = 5.575$ $B = 4.783$	$x_1 = 5.5755$ $x_2 = 10.3585$ $x_3 = 15.1415$ $x_4 = 19.9245$	$\pi_1 = 0.1642$ $\pi_2 = 0.4520$ $\pi_3 = 0.7760$ $\pi_4 = 0.9357$	$\pi_1 = 0.0643$ $\pi_2 = 0.2240$ $\pi_3 = 0.5480$ $\pi_4 = 0.8358$	0.6830, 146.1, 146.1, 1.3660	0.902	0.600

9.5 Simulation Results

The highlighted designs from the previous sections—and listed below for ease of comparison—were each evaluated using simulation methods with S=5000 simulations to assess their performance in practice and using key measures discussed below.

Table 9.5 Designs examined in simulations: each design comprised $4n$ observations for both drugs

Design	Description	Text table	Replicates	Support points
1	Reflection design: $\alpha = 0.87$ (equal efficiency)	3	n observations for each drug and each support point	$x = 3.4889, 6.9888$ 18.5111, 22.0111
2	Reflection design: $\alpha = 0.25$ (D-optimal)	3	n observations for each drug and each support point	$x = 5.8554, 9.3554$ 16.1447, 19.6447
3	Geometric design: $\alpha = 0.25$ (D-optimal)	4	n observations for each drug and each support point	$x = 6.2162, 9.1428$ 13.4473, 19.7783
4	Uniform design: $\alpha = 0.25$ (D-optimal)	4	n observations for each drug and each support point	$x = 5.5755, 10.3585$ 15.1415, 19.9245
5	Same-concentration design: $\alpha = 0.25$ (D-optimal)	2	$2n$ observations for each drug and each support point	$x = 7.4844, 18.0156$
6	Same-concentration design: $\alpha = 1.0$ (D-subset)	2	$2n$ observations for each drug and each support point	$x = 4.8600, 20.6404$

Case I (small-to-moderate size) with $n = 15$ (i.e., 60 observations for both drugs) and Case II (moderate-to-large size) with $n = 30$ (i.e., 120 observations for both drugs)

As noted in Table 9.5, our first simulation, called “Case I: 15/30”, involved the small-to-moderate-sample situation of 15 replicates of each of the 4-point designs (designs 1–4) and 30 replicates of each of the 2-point designs (designs 5 and 6) for each drug—so total sample size of 60. This process was subsequently repeated for the moderate-to-large sample case, called “Case II: 30/60”, which entailed 30 replicates of each of the 4-point designs and 60 replicates of each of the 2-point designs for each drug—i.e., total sample size of 120. In each case, independent binary data were generated from the two-logistic situation in (9.2) using parameter values used above, viz, $\beta = 0.30$, $\delta = 0$, $\gamma_A = 11.0$, $\gamma_B = 14.5$. Simulation results for Cases I and II are given in Tables 9.6 and 9.7 respectively.

Table 9.6 Simulations results for Case I (15/30 study), i.e., with $n = 15$ in Table 9.5

Design	Average of $\hat{\delta}$ estimates	Variance of $\hat{\delta}$ estimates	Proportion of simulations with $\left\|\hat{\delta}\right\| > 1$	Proportion of simulations with abs-value of $\|M^{-1}\| > 1$
1	0.0317	2.3625	0.0548(≈274/5000)	0.0557(≈278/5000)
2	0.0058	0.0940	0.0014(≈7/5000)	0.0014(≈7/5000)
3	0.0014	0.0251	0.0002(≈1/5000)	0.0002(≈1/5000)
4	−0.0013	0.0435	0.0012(≈6/5000)	0.0012(≈6/5000)
5	−0.0040	0.2951	0.0510(≈255/5000)	0.0500(≈250/5000)
6	0.0435	0.8514	0.3325(≈1662/5000)	0.3564(≈1782/5000)

The results in Table 9.6 confirm—as do additional unreported findings—that as α is chosen other than the D-optimal case ($\alpha = 0.25$), estimation of the difference-of-slopes parameter δ becomes unstable; this is clearly demonstrated by Designs 1 and 6 in Table 9.6. As noted above, in these cases, the design support points move away from the EC50 values, and thus estimation of the full parameter vector is also unstable; this is demonstrated above in large values of the generalized variance, $|M^{-1}|$. An important empirical result observed here is that although our ultimate focus is on assessing the difference-of-slopes parameter, designs must be chosen to estimate all model parameters in order to be viable. Also, given the weak performance of the same-concentration D-optimal (two-support point) design in Design 5, this design is also dismissed. As such, reasonable contenders include only the reflection, geometric and uniform D-optimal designs: i.e., Designs 2, 3 and 4. As samples sizes are doubled for the Case II “30/60” (i.e., Table 9.7), the situation improves as expected for all designs, but clearly the most viable robust design approaches are seen to be the reflection, geometric and uniform D-optimal designs.

Table 9.7 Simulations results for Case II (30/60 study), i.e., with $n = 30$ in Table 9.5

Design	Average of $\hat{\delta}$ estimates	Variance of $\hat{\delta}$ estimates	Proportion of simulations with $\left|\hat{\delta}\right| > 1$	Proportion of simulations with abs-value of $|M^{-1}| > 1$
1	−0.0033	0.0733	0.0016(≈8/5000)	0.0016(≈8/5000)
2	0.0011	0.0059	0.0000(≈0/5000)	0.0000(≈0/5000)
3	3.6*e*−06	0.0069	0.0000(≈0/5000)	0.0000(≈0/5000)
4	−0.0014	0.0069	0.0000(≈0/5000)	0.0000(≈0/5000)
5	0.0020	0.0186	0.0022(≈11/5000)	0.0020(≈10/5000)
6	0.0033	0.0186	0.0762(≈381/5000)	0.0770(≈385/5000)

9.6 Discussion

The theoretical results in Sect. 9.4 as well as the empirical results in Sect. 9.5 lead us to advocate for the reflection, geometric and uniform designs introduced and illustrated in Sect. 9.4 in designing for the assessment of parallelism for the two-logistic model in (9.2). In terms of efficiencies, reflection designs are preferred. Furthermore, in terms of straightforward ease-of-implementation, these same reflection designs are highly favorable since as noted above the researcher merely sketches the drug/compound logistic curves and reads off design support points at the intersections with cut lines at $\pi_1 = 0.1760$ and $\pi_2 = 0.8240$. (Geometric and uniform designs generally require optimal design software.) The above simulation results confirm the favorable performance of these reflection designs even in small-to-moderate sample size situations as described above in Table 9.6. The addition of design support points when using the reflection design over the theoretical two-point D-optimal designs in Fig. 9.2 cannot be overstressed since researchers typically wish to both efficiently estimate model parameters and check for model adequacy.

References

Abdelbasit, K.M., Plackett, R.L.: Experimental design for binary data. J. Am. Stat. Assoc. **78**, 90–98 (1983)

Atkinson, A.C.: Planning experiments to detect inadequate regression models. Biometrika. **59**, 275–293 (1972)

Atkinson, A.C., Donev, A.N., Tobias, R.D.: Optimum Experimental Designs, with SAS. Oxford University Press, New York (2007)

Finney, D.J.: Probit Analysis, 3rd edn. Cambridge University Press, London (1971)

Finney, D.J.: Statistical Method in Biological Assay, 3rd edn. Charles Griffin, London (1978)

Fleetwood, K., Bursa, F., Yellowlees, A.: Parallelism in practice: approaches to parallelism in bioassays. PDA J. Pharm. Sci. Technol. **69**, 248–263 (2015)

Gottschalk, P.G., Dunn, J.R.: Measuring parallelism, linearity, and relative potency in bioassay and immunoassay data. J. Biopharm. Stat. **15**, 437–463 (2005)

Gupta, V., Vale, P.F.: Nonlinear disease tolerance curves reveal distinct components of host responses to viral infection. R. Soc. Open Sci. **4**, 170342 (2017)
Jonkman, J.N., Sidak, K.: Equivalence testing for parallelism in the four-parameter logistic model. J. Biopharm. Stat. **19**, 818–837 (2009)
Kalish, L.A.: Efficient design for estimation of median lethal dose and quantal dose-response curves. Biometrics. **46**, 737–748 (1990)
Minkin, S.: Optimal designs for binary data. J. Am. Stat. Assoc. **82**, 1098–1103 (1987)
Novick, S.J., Yang, H., Peterson, J.J.: A Bayesian approach to parallelism testing in bioassay. Stat. Biopharm. Res. **4**, 357–374 (2012)
O'Brien, T.E.: Designing for parameter subsets in Gaussian nonlinear regression models. J. Data Sci. **3**, 179–197 (2005)
O'Brien, T.E.: Efficient experimental design strategies in toxicology and bioassay. Stat. Optim. Inf. Comput. **4**, 99–106 (2016)
O'Brien, T.E.: Contemporary robust optimal design strategies. In: Tez, M., von Rosen, D. (eds.) Trends and Perspectives in Linear Statistical Inference, pp. 165–180. Springer, Cham (2018)
O'Brien, T.E., Funk, G.M.: A gentle introduction to optimal design for regression models. Am. Stat. **57**, 265–267 (2003)
O'Brien, T.E., Lim, C.: New challenges and strategies in robust optimal design for multicategory logit modelling. In: Chen, D., Jin, Z., Li, G., Li, Y., Liu, A., Zhao, Y. (eds.) New Advances in Statistics and Data Science, pp. 61–74. Springer, Cham (2018)
O'Brien, T.E., Chooprateep, S., Homkham, N.: Efficient geometric and uniform design strategies for sigmoidal regression models. S. Afr. Stat. J. **43**, 49–83 (2009)
O'Brien, T.E., Jamroenpinyo, S., Bumrungsup, C.: Curvature measures for nonlinear regression models using continuous designs with applications to optimal design. Involve J. Math. **3**, 317–332 (2010)
O'Brien, T.E., Rawlings, J.O.: A non-sequential design procedure for parameter estimation and model discrimination in nonlinear regression models. J. Stat. Plann. Inference. **55**, 77–93 (1996)
Rich, I.N.: Potency, proliferation and engraftment potential of stem cell therapeutics: the relationship between potency and clinical outcome for hematopoietic stem cell products. J. Cell Sci. Ther. **S13**, 001 (2013). https://doi.org/10.4172/2157-7013.S13-001
Sidak, K., Jonkman, J.N.: Testing for parallelism in the heteroskedastic four-parameter logistic model. J. Biopharm. Stat. **26**, 250–268 (2016)
Wheeler, M.W., Park, R.M., Bailer, A.J.: Comparing median lethal concentration values using confidence interval overlap or ratio tests. Environ. Toxicol. Chem. **25**, 1441–1444 (2006)
Yang, H., Zhang, L.: Evaluations of parallelism testing methods using ROC analysis. Stat. Biopharm. Res. **4**, 162–173 (2012)
Yang, H., Kim, H.J., Zhang, L., Strouse, R.J., Schenerman, M., Jiang, X.-R.: Implementation of parallelism testing for four-parameter logistic model in bioassays. PDA J. Pharm. Sci. Technol. **66**, 262–269 (2012)

Part II
Biopharmaceutical Applications

Chapter 10
Optimal Adaptive Phase III Design with Interim Sample Size and Dose Determination

Lanju Zhang, Lu Cui, and Yaoyao Xu

10.1 Introduction

Randomized and controlled clinical trials (RCT) are the gold standard method for confirmatory studies (phase III trials) in drug developments. Traditionally, fixed sample size designs are employed with the sample size, doses, and other parameters determined at the planning stage. The trial is then executed without major modifications. Therefore the probability of success of the trial heavily depends on the accuracy of the projection of the assumed relevant parameters, such as the effect size of treatment. Often, these assumptions are based on limited data from phase II studies or literature, and can be of significant variability (Liu et al. 2008). For example, Gan et al. (2012) reviewed 253 phase III oncology trials and found that "Investigators consistently make overly optimistic assumptions regarding treatment benefits when designing RCTs." Inaccurate specification of these parameters often leads to an under-powered or over-powered study, thus failure of a potentially successful program or a waste of time and resources.

To remedy, adaptive designs have been proposed to modify the study design and improve the probability of the success based on available interim data of the same trial. A vast literature exists for sample size adapted trial designs allowing the

L. Zhang (✉) · L. Cui
Data and Statistical Sciences, AbbVie Inc., North Chicago, IL, USA
e-mail: Lanju.zhang@abbvie.com

Y. Xu
PAREXEL International, Durham, NC, USA

L. Zhang et al. (eds.), *Contemporary Biostatistics with Biopharmaceutical Applications*, ICSA Book Series in Statistics,
https://doi.org/10.1007/978-3-030-15310-6_10

sample size determination using interim data with strict type I error rate control (e.g., Cui et al. 1997, 1999; Lehmacher and Wassmer 1999; Muller and Schafer 2001; Bauer and Köhne 1994; Proschan and Hunsberger 1995; Denne 2001; Posch et al. 2011). A recent comprehensive review is conducted by Bauer et al. (2016). These designs were discussed in the FDA guidance (FDA 2010) and increasingly used and adopted by the pharmaceutical industry and regulatory agencies. Another way to improve the trial efficiency is to allow dropping ineffective doses or treatments using interim data (Thall et al. 1988; Stallard and Todd 2003, 2011). This approach, sometimes in the context of a seamless phases II/III trial design, is a variation of interim analysis for futility. The latter results in the termination of entire study while the former terminates a few arms. All data collected before and after the interim treatment selection in the remaining treatment arms will be used in the final analysis. Examples with dose selections include INHANCE trial (Lawrence et al. 2014; Lawrence and Bretz 2014) and ADVENT trial (Chaturvedi et al. 2014).

In this paper, we consider a two-stage adaptive phase III trial design with both dose selection AND sample size determination using the first stage data, which has not been addressed in the literature. By doing this we can not only avoid unnecessary patient enrollment to a worse performing arm, but also determine an accurate final sample size based on in-trial data. We further optimize the trial design with respect to the timing of the interim analysis and the weights in combining two stages data. Optimization with respect to these parameters has been demonstrated to be able to increase robustness of the study power against treatment effect size misspecification at the design stage (Liu et al. 2008; Bretz et al. 2009; Zhang et al. 2016).

In summary, we present an optimized two-stage adaptive phase III clinical trial design with final sample size determination and dose selection based on the first stage data. Strict type I error rate control is achieved using a weighted test statistic (Cui et al. 1997, 1999), adjusting for potential inflation from multiple treatment arms and sample size determination based on interim data. We provide explicit formulae so that optimization can be conducted in an economical manner without time-consuming simulation. The rest of the paper is organized as follows. Section 10.2 describes statistical methodology for the trial design and formulae for design performance measurements such as power, type I error rate, and average sample size. Section 10.3 discusses how to optimize the design using method in Sect. 10.2. Section 10.4 presents a real example. We conclude with a discussion in Sect. 10.5.

10.2 Methodology

We consider a two-stage phase III trial including two treatment arms and a control arm. These two treatments can be two doses of the same compound, or two compounds for the same disease. The trial will include an interim analysis with the purpose to determine a better performing treatment arm that will advance to the second stage and determine the total sample size for the selected treatment arm and control arm based on the interim data. No testing is conducted at the interim. The final analysis uses stage 1 and stage 2 data of the selected treatment arm and control arm.

Suppose responses of two treatment arms are X_1, X_2 and that of control is X_0 and $X_i \sim N(\mu_i, \sigma^2)$. We assume σ^2 is known. Otherwise, it can be estimated from data and all our following discussions should hold asymptotically. Let $\Delta_1 = \mu_1 - \mu_0$, $\Delta_2 = \mu_2 - \mu_0$ be the treatment effect of treatments 1 and 2 compared to control group 0. The hypotheses of interest are,

$$H_0 : \Delta_1 = \Delta_2 = 0 \quad vs \qquad H_1 : any\ \Delta_i > 0. \tag{10.1}$$

More discussions on null hypothesis configurations are in Sect. 10.2.4 and Appendix 4. Suppose at the interim with n subjects per treatment group, we have data of a random sample $\{X_{ij}, i = 0, 1, 2, \ j = 1, \ \ldots, n\}$ from each group and define $Y_i^{(n)} = \sum_{j=1}^{n} \left(X_{ij} - X_{0j}\right)/n, i = 1, \ 2$. Then $Y_i^{(n)}$ is an estimate of Δ_i. Let A be an event generated by $Y_1^{(n)}$ and $Y_2^{(n)}$. Event A occurs if treatment 1 is selected and otherwise if treatment 2 is selected. Let $I_1 = I_A$ be the indicator of selecting treatment 1 and $I_2 = 1 - I_A$. At the interim, if treatment i is selected, the total sample size per arm is determined to be M_i based on its observed treatment effect, thus M_i is a random variable depending on $Y_i^{(n)}$. Let w_1, $w_2 > 0$ such that $w_1^2 + w_2^2 = 1$. At the final analysis, the test statistic (Cui et al. 1997, 1999) for treatment effect of treatment i ($i = 1, 2$) is defined as,

$$Z_i = w_1 \frac{Y_i^{(n)}}{\sigma\sqrt{2/n}} + w_2 \frac{Y_i^{(M_i - n)}}{\sigma\sqrt{2/\left(M_i - n\right)}}, \tag{10.2}$$

where $Y_i^{(M_i-n)}$ is defined similarly as $Y_i^{(n)}$ using the second stage data only. Since only one treatment is selected to move to the second stage, the final test statistic should be,

$$T_S = Z_1 I_1 + Z_2 I_2, \tag{10.3}$$

based on the selected treatment. In the following, we considered the property of this test statistic.

10.2.1 Conditional Power

At the interim, the observed treatment effect is $Y_i^{(n)}$ for treatment i. Conditional on $Y_i^{(n)}, i = 1, \ 2, I_i$ and M_i are fixed, and the random quantity in the definition of Z_i is $Y_i^{(M_i-n)}$. So the conditional power is determined as (Appendix 1),

$$P\left(Z_i > C|Y_i^{(n)}\right) = 1 - \Phi\left(\frac{1}{w_2}\left(C - \frac{w_1 Y_i^{(n)}}{\sigma\sqrt{2/n}}\right) - \frac{\Delta_i\sqrt{M_i - n}}{\sqrt{2}\sigma}\right),$$

where C is the critical value for final test statistic to control the type I error rate, to be determined in a later section. In practice, Δ_i, $i = 1, 2$, can be replaced with their estimates based on interim data or other values deemed appropriate. In this paper, we use conditional power to determine the total sample size based on the first stage data. Other methods for total sample size determination can also be used.

10.2.2 Total Sample Size M_i

Now we consider the total sample size M_i which can be determined such that the conditional power based on the observed effect size of the selected treatment is at least $(1 - \beta)$. Using the formula in the last section, and replacing Δ_i's with their estimates based on interim data, we can find M_i to be (Appendix 2),

$$M_i = n + \frac{2\sigma^2}{w_2^2}\left(\frac{C + w_2 z_\beta}{Y_i^{(n)}} - \frac{n w_1}{\sqrt{2n}\sigma}\right)^2. \tag{10.4}$$

M_i is not a continuous function of $Y_i^{(n)}$ and can approach infinity if $Y_i^{(n)}$ is close to zero. In practice, an upper limit is used to cap the sample size increase (Mehta and Pocock 2011). Another method to avoid this is to use a futility criterion, eg, stopping the trial if both $Y_1^{(n)}$ and $Y_2^{(n)}$ are less than the minimal clinically meaningful effect.

10.2.3 Absolute Power

Now we consider the absolute power, taking into account the variability of $Y_1^{(n)}$, $Y_2^{(n)}$ and the fact that M_i depends on $Y_i^{(n)}$. We assume the treatment with the larger estimated effect difference will be selected at the interim. A closed formula involving integration can be derived (Appendix 3),

$$
\begin{aligned}
P\left(T_S > C\right) = 1 &- \int_{-\infty}^{\infty} \Phi\left(\frac{1}{w_2}\left(C - w_1\left(v_1 + \frac{n\Delta_1}{\sqrt{2n}\sigma}\right)\right)\right. \\
&\left. - \frac{\Delta_1}{w_2}\min\left(\left|\frac{n\left(C + w_2 z_\beta\right)}{n\Delta_1 + \sqrt{2n}\sigma v_1} - \frac{n w_1}{\sqrt{2n}\sigma}\right|, \frac{w_2\sqrt{L-n}}{\sigma\sqrt{2}}\right)\right) \\
&\times \Phi\left(\frac{(1-\rho)\, v_1 - \frac{n(\Delta_2 - \Delta_1)}{\sqrt{2n}\sigma}}{\sqrt{1-\rho^2}}\right) \phi\left(v_1\right) dv_1 \\
&- \int_{-\infty}^{\infty} \Phi\left(\frac{1}{w_2}\left(C - w_1\left(v_2 + \frac{n\Delta_2}{\sqrt{2n}\sigma}\right)\right)\right. \\
&\left. - \frac{\Delta_2}{w_2}\min\left(\left|\frac{n\left(C + w_2 z_\beta\right)}{n\Delta_2 + \sqrt{2n}\sigma v_2} - \frac{n w_1}{\sqrt{2n}\sigma}\right|, \frac{w_2\sqrt{L-n}}{\sigma\sqrt{2}}\right)\right) \\
&\times \Phi\left(\frac{(1-\rho)\, v_2 + \frac{n(\Delta_2 - \Delta_1)}{\sqrt{2n}\sigma}}{\sqrt{1-\rho^2}}\right) \phi\left(v_2\right) dv_2,
\end{aligned}
\tag{10.5}
$$

where ρ is the correlation coefficient between $Y_1^{(n)}$ and $Y_2^{(n)}$ because they share the same control group data. In our setting, ρ=1/2, but we will use ρ to make our formulae general. When the equation is evaluated under the alternative hypothesis in (10.1), it gives the power of the final test statistic, which depends on both Δ_1 and Δ_2. When it is evaluated under null hypothesis, it gives the type I error rate, to be considered in the next section. This formula is appropriate when M_i is determined based on condition power according to Eq. (10.4). However, similar formula can be obtained if M_i is determined based on other methods.

10.2.4 Type I Error Rate

When $\Delta_1 = \Delta_2 = 0$, the formula in Sect. 10.2.3 for the type I error rate reduces to,

$$
\begin{aligned}
P\left(T_S > C\right) &= 1 - \int_{-\infty}^{\infty} \Phi\left(\frac{C - w_1 v_1}{w_2}\right) \Phi\left(\frac{(1-\rho)\, v_1}{\sqrt{1-\rho^2}}\right) \phi\left(v_1\right) dv_1 \\
&\quad - \int_{-\infty}^{\infty} \Phi\left(\frac{C - w_1 v_2}{w_2}\right) \Phi\left(\frac{(1-\rho)\, v_2}{\sqrt{1-\rho^2}}\right) \phi\left(v_2\right) dv_2 \\
&= 1 - 2\int_{-\infty}^{\infty} \Phi\left(\frac{C - w_1 v}{w_2}\right) \Phi\left(\frac{(1-\rho)\, v}{\sqrt{1-\rho^2}}\right) \phi(v) dv,
\end{aligned}
\tag{10.6}
$$

which doesn't depend on n; however, it depends on w_1, just as in group sequential designs in which the type I error rate depends on the information fraction corresponding to the interim analyses (Mehta and Pocock 2011). Since ρ=1/2 in our setting, letting (10.6) equal α, we can determine the corresponding critical value C, for any given w_1. Table 10.1 gives critical values for one-sided type I error rate $\alpha = 0.025$ with different weights w_1. All critical values are larger than 1.96 and increases as w_1 increases.

A few observations are in order. First, there are other null hypothesis configurations. For example, $\Delta_1 = 0, \quad \Delta_2 > 0$. A type I error would occur if treatment 1 is selected and rejected. But the type I error rate in this case would be much smaller than that when $\Delta_1 = \Delta_2 = 0$. For details, refer to Appendix 4. Secondly, when $w_1 = 0$ (so $w_2 = 1$), which means we choose one treatment arm at the beginning of the trial, Eq. (10.6) reduces to (Appendix 4),

$$1 - 2\int_{-\infty}^{\infty} \Phi(C)\Phi\left(\frac{(1-\rho)\,v}{\sqrt{1-\rho^2}}\right)\phi(v)dv = 1 - \Phi(C).$$

If $C = 1.96$, the type I error rate is exactly 0.025. In this case, we only use the second stage data with one comparison, so there is no need of adjustment of the type I error rate.

Thirdly, when $w_1 = 1$ (so $w_2 = 0$), which means two treatment arms are carried forward to the end of the trial, and Eq. (10.6) reduces to

$$1 - 2\int_{-\infty}^{\infty} \Phi\left(\frac{1}{w_2}(C - w_1 v)\right)\Phi\left(\frac{(1-\rho)\,v}{\sqrt{1-\rho^2}}\right)\phi(v)dv$$
$$= 1 - 2\int_{-\infty}^{C} \Phi\left(\frac{(1-\rho)\,v}{\sqrt{1-\rho^2}}\right)\phi(v)dv.$$

The type I error rate is controlled with a more conservative critical value C = 2.211.

Finally, in Appendix 4, we point out that there are more than one configuration of null hypotheses (e.g., $\Delta_1 = \Delta_2 = 0$; $\Delta_1 = 0, \quad \Delta_2 > 0$; $\Delta_1 > 0, \Delta_2 = 0$), leading to more than one type I error, but the type I error rate is dominated by the hypothesis $\Delta_1 = \Delta_2 = 0$. Figure 10.1 depicts type I error rates under different hypothesis configurations and confirms this dominance. For the purpose of validating the formula, we run a simulation under the null hypothesis $\Delta_1 = \Delta_2 = 0$ with critical values C determined as in Table 10.1 and plot the results in Fig. 10.2, which demonstrates that type I error rate is well controlled and agrees with our formula.

10.2.5 Average Sample Size

Since the final sample size is a random quantity, it is interesting to know the average sample size, which is determined by the following formula,

$$
\begin{aligned}
\mu_M &= \iint_A M_1 dF_n(y_1, y_2) + \iint_{A'} M_2 dF_n(y_1, y_2) \\
&= \int_{-\infty}^{\infty} \min\left(n + \frac{2\sigma^2}{w_2^2}\left(\frac{C + w_2 z_\beta}{\Delta_1 + \sqrt{2/n}\sigma v_1} - \frac{n w_1}{\sqrt{2n}\sigma}\right)^2, \ L\right) \\
&\quad \times \Phi\left(\frac{(1-\rho)\, v_1 - \frac{n(\Delta_2 - \Delta_1)}{\sqrt{2n}\sigma}}{\sqrt{1-\rho^2}}\right) \o(v_1)\, dv_1 \\
&\quad + \int_{-\infty}^{\infty} \min\left[n + \frac{2\sigma^2}{w_2^2}\left(\frac{C + w_2 z_\beta}{\Delta_2 + \sqrt{2/n}\sigma v_2} - \frac{n w_1}{\sqrt{2n}\sigma}\right)^2, \ L\right] \\
&\quad \times \Phi\left(\frac{(1-\rho)\, v_2 + \frac{n(\Delta_2 - \Delta_1)}{\sqrt{2n}\sigma}}{\sqrt{1-\rho^2}}\right) \o(v_2)\, dv_2
\end{aligned}
\tag{10.7}
$$

where L is the upper limit for sample size increase.

10.2.6 Probability to Select a Treatment Arm

The probability to select treatment arm 1 is,

$$
\iint_A 1 dF_n(y_1, y_2) = \int_{-\infty}^{\infty} \Phi\left(\frac{(1-\rho)\, v_1 - \frac{n(\Delta_2 - \Delta_1)}{\sqrt{2n}\sigma}}{\sqrt{1-\rho^2}}\right) \o(v_1)\, dv_1 \tag{10.8}
$$

10.3 Design Evaluation

With the proposed Dose and Sample size Adaptive (DSA) design and the formulae obtained, the design evaluation and optimization become possible. The goal here is to find a better DSA design which can select the right treatment with a high probability, achieve a stable and satisfactory statistical power, and attain a small average final sample size. This goal can be achieved by varying n, the sample size of the interim analysis, and w_1 , the weight for the first stage data, to generate a set of candidate designs. Each of the candidate design corresponding to a pair of values of n and w_1 is evaluated based on the power and average total sample size against the configuration of the underlying true treatment differences Δ_1 and Δ_2. The overall winner from the comparisons is the one achieving good balance between stable high statistical power and low average total sample size. This optimally chosen design is to be implemented in the trial.

The following computational examples are used to illustrate this process.

In the examples, we target a statistical power of $1 - \beta = 0.9$ at one-sided significance level $\alpha = 0.025$. The total sample size of the trial is capped by $L = 2000$. We further assume $\sigma = 1$ and $\rho = 1/2$ by the design. To evaluate individual designs, the average total sample size can be calculated as $2\mu_M$+n using (10.7) and the design power can be obtained using (10.5).

10.3.1 Simple Configuration of Δ_1 and Δ_2

In this example, we consider simple pointwise configuration of Δ_1 and Δ_2 of $\Delta_1 = 0.2$ and $\Delta_2 = 0.4$. A set of 792 candidate designs are generated by varying n from 40 to 320 by 40 and w_1 from 0.01 to 0.99 by 0.01. For each design the average total sample size and design power are calculated. The results are plotted (Fig. 10.3) for each given n and varying w_1 from 0.01 to 0.99. As shown in Fig. 10.3, the design with n = 120 achieves the smallest average total sample size. This optimal design is corresponding to $w_1 = 0.8$. It has about 93.9% of chance to correctly select treatment arm 2 to move forward. The average total sample size of this design is 544 and its statistical power is about 0.94.

Consider a traditional program with a three-arm phase 2 trial to choose a better treatment to move forward in a subsequent two-arm phase 3 trial. In the phase 2 part, with a maximum effect size 0.4, two-sided $\alpha = 0.025$ after adjustment of multiplicity, 80% power, the required sample size is 121 per arm, totaling 363 for phase 2. After phase 2, there is 94% chance to select the arm of better effect size 0.4 and 6% to select the arm of effect size 0.2. The former requires 133 per arm and latter 527 per arm for two-sided $\alpha = 0.05$ and 90% power in phase 3. So on average, phase 3 requires 313 patients. Combining phase 2 and phase 3, 676 patients are required. The advantage of the proposed adaptive design with 544 subjects and 94% of statistical power is obvious.

10.3.2 Complex Configuration of Δ_1 and Δ_2

This example considers a complex but more realistic configuration of Δ_1 and Δ_2 with Δ_1 unknown but believed to be within the range, say, from 0.1 to 0.3, and $\Delta_2 = \Delta_1 + 0.2$. Practically, the range of Δ_1 can be obtained as a confidence interval of the parameter and the treatment difference of the two active treatment arms can be projected as its point estimate from historical data or based on dose response relationship.

In this case, a set of 1980 candidate designs is generated by varying n from 40 to 800 by 40 and w_1 from 0.01 to 0.99 by 0.01. For each design with a fixed n and w_1 the total average sample size and statistical power of the design are calculated against the aforementioned two-dimensional configuration of the underlying parameters Δ_1 and Δ_2 with Δ_1 moving from 0.1 to 0.3 by 0.01. Denote this power as $Power(n, w_1, \Delta_1)$. The total *power deviation* from the targeted 0.9 is defined as $d(n, w_1) = \int_{0.1}^{0.3} |\, Power(n, w_1, \Delta_1) - 0.9| d\Delta_1$ and calculated numerically. The values of $d(n, w_1)$ are grouped by n with w_1 ranging from 0.01 to 0.09 and displayed in Fig. 10.4. The designs with $(n, w_1) = (80,0.7), (120,0.56), (160, 0.47), (200,0.39), (240, 0.34), (280, 0.3)$ achieve small power deviation and average total sample size. The power and average total sample size of these designs are further plotted against Δ_1 in Fig. 10.5. It can be seen that the design with n = 120 and $w_1 = 0.56$ achieves a stable power around 90%.

Again compare to a traditional program with a three-arm phase 2 trial to choose a better treatment to move forward in a subsequent two-arm phase 3 trial. In the ideal case, if we know the effect size and do similar calculation as in Sect. 10.3.1, a total of 1184 patients are needed for two phases using effect size 0.1 and 0.3 for two treatment arms, and a total of 439 patients using effect size 0.3 and 0.5 for two treatment arms, both resulting in larger sample sizes than the design we chose with n = 120 and $w_1 = 0.56$. So the optimized design takes into account the uncertainty of effect size and provides robust power with average sample size even smaller than those achieved if effect sizes are assumed to be known.

More general configuration of Δ_1 and Δ_2 is when we know a range for Δ_1 and another range forΔ_2. The design optimization in this case can be done in a similar way via global search the best design from all candidate designs.

10.4 An Example

In this section, we consider a real application of the proposed method to designing a clinical study of a new drug treatment of rheumatoid arthritis (RA). Assume that two doses of the new drug are of interest. The primary efficacy measurement is change from baseline in DAS28 at Week 26, which is commonly used in RA studies. Assume that the primary outcome in each treatment arm follow a normal distribution: N(-1.9, 1.36^2) for placebo (PBO), N(-2.5, 1.36^2) for low dose, and N(-2.9, 1.36^2) for high dose. The assumptions are based on Lee et al. (2014). We apply our two-stage design with dose selection and sample size determination using the first stage data. Taking variability into account, with our notation, assume $\Delta_1 \in [-0.8, -0.4]$, $\Delta_2 = \Delta_1 - 0.4$,and $\sigma = 1.36$. A fixed sample size trial comparing two groups with treatment effect -0.4 and -0.8 requires 61 and 243 per arm, respectively. Following the optimization process in Sect. 10.3.2, we search candidate designs with n from 20 to 320 by 20 and w_1 from 0.01 to 0.99 by 0.01. A design with n=100, $w_1 = 0.68$ results in a trial with almost 90% power over the whole interesting region for $\Delta_1$with a reasonable mean final sample size.

10.5 Conclusion and Discussions

Adaptive designs have provided abundant opportunities to improve the chance of success and likely the efficiency of conducting clinical trials. Their value has been increasingly recognized by regulatory agencies and industry. Such designs can be used to select a treatment arm using interim data followed with a confirmatory stage in seamless phase II/III designs or phase III designs (Stallard and Todd 2003; Lawrence and Bretz 2014) or to determine a final sample size based on interim data (Cui et al. 1997, 1999). It is recognized that the timing of interim analysis is important (Chaturvedi et al. 2014) and optimization can further improve the robustness of such designs when there is uncertainty in treatment effect size assumption at trial planning stage (Zhang et al. 2016). In this paper, we combine these three design techniques to provide an efficient and robust trial design for phase III clinical trials, which offers a strict type I error rate control, an opportunity to select a better performing treatment arm, and robust power over an effect size window through optimization. All relevant formulae and their derivation have been provided in the appendices, which help to facilitate readers to design and optimize their trials without time-consuming simulation.

During the interim, one can also apply some futility rule to stop the whole trial if neither treatment arm is promising. This should not change our design and optimization process. This same design can be also used in seamless phase II/phase III trials. In that situation, there might be more than two treatment (dose) groups. Our design can be readily generalized to more than two treatment groups, e.g., selecting the treatment arm with the largest observed effect size, as in Stallard and Todd (2003, 2011), but computation may rely on statistical simulations.

We optimize the trials with respect to interim sample size n and w_1, which take values independently. We discard the concept of "planned total sample size" and therefore there is no "sample size re-estimation." Instead, the total sample size is *determined* based on interim data.

In sample size determination at interim based on conditional power, we use the observed treatment difference to replace the expected mean difference in conditional power. Some, e.g., (Glimm 2012), point out that this may be suboptimal due to the variability in the observed treatment difference. It should be noted that other estimates can be used in estimating conditional power and the formula can be changed accordingly. In addition, we use a criterion of better efficacy for interim treatment selection. This is commonly used in the phase III setting where the doses under study are considered generally safe and a selection criterion needs to be spelled out clearly a priori for the data monitoring committee to conduct the dose selection.

Disclosure The support of this publication was provided by AbbVie. AbbVie participated in the review and approval of the content. Lanju Zhang and Lu Cui are employees of AbbVie, Inc. Yaoyao Xu is a former employee of AbbVie and is now employed by PAREXEL International.

Appendices

Appendix1: Conditional Power

From the definition of Z_i,

$$P\left(Z_i > C|Y_i^{(n)}\right) = P\left(w_1 \frac{Y_i^{(n)}}{\sigma\sqrt{2/n}} + w_2 \frac{Y_i^{(M_i-n)}}{\sigma\sqrt{2/(M_i-n)}} > C|Y_i^{(n)}\right)$$

$$= P\left(\frac{Y_i^{(M_i-n)} - \Delta_i}{\sigma\sqrt{2/(M_i-n)}} > \frac{1}{w_2}\left(C - \frac{w_1 Y_i^{(n)}}{\sigma\sqrt{2/n}}\right) - \frac{\Delta_i\sqrt{M_i-n}}{\sqrt{2}\sigma}|Y_i^{(n)}\right)$$

$$= 1 - \Phi\left(\frac{1}{w_2}\left(C - \frac{w_1 Y_i^{(n)}}{\sigma\sqrt{2/n}}\right) - \frac{\Delta_i\sqrt{M_i-n}}{\sqrt{2}\sigma}\right)$$

Appendix 2: Total Sample Size

We make the last formula in Appendix 1 at least $(1-\beta)$, so

$$\frac{1}{w_2}\left(C - \frac{w_1 Y_i^{(n)}}{\sigma\sqrt{2/n}}\right) - \frac{\Delta_i\sqrt{M_i-n}}{\sqrt{2}\sigma} \le -z_\beta,$$

where z_β is the upper quantile of standard normal distribution. So,

$$M_i \ge n + \frac{2\sigma^2}{w_2^2\Delta_i^2}\left(C + w_2 z_\beta - \frac{w_1 Y_i^{(n)}}{\sigma\sqrt{2/n}}\right)^2$$

Using observed effect size to replace Δ_i, i.e., $\hat{\Delta}_i = Y_i^{(n)}$, then total sample size should be,

$$M_i = n + \frac{2\sigma^2}{w_2^2}\left(\frac{C + w_2 z_\beta}{Y_i^{(n)}} - \frac{n w_1}{\sqrt{2n}\sigma}\right)^2$$

To apply a limit L, we use $M'_i = \min(M_i, \mathrm{L})$ in practice.

Appendix 3: Absolute Power

To find the absolute power, we need to integrate the conditional power in Appendix 1 with respect to the joint density of $Y_1^{(n)}$ and $Y_2^{(n)}$. Recall A is the event that occurs when treatment 1 is selected at the interim.

$$P(T_S > C) = P(Z_1 > C, A) + P(Z_2 > C, A')$$

$$= 1 - \iint_A \Phi\left(\frac{1}{w_2}\left(C - \frac{w_1 Y_1^{(n)}}{\sigma\sqrt{2/n}}\right) - \frac{\Delta_1\sqrt{M_1 - n}}{\sqrt{2}\sigma}\right) dF_n(y_1, y_2)$$

$$- \iint_{A\prime} \Phi\left(\frac{1}{w_2}\left(C - \frac{w_1 Y_2^{(n)}}{\sigma\sqrt{2/n}}\right) - \frac{\Delta_2\sqrt{M_2 - n}}{\sqrt{2}\sigma}\right) dF_n(y_1, y_2)$$

Treatment selection at interim can use many different criteria, so event A can be defined in many ways. Here we consider selecting the treatment with the larger effect at interim. In other words, A occurs or treatment 1 is selected when$Y_1^{(n)} > Y_2^{(n)}$. First, plugging in M_i,

$$P(T_S > C) = 1 - \iint_{y_1 > y_2} \Phi\left(\frac{1}{w_2}\left(C - \frac{w_1 Y_1^{(n)}}{\sigma\sqrt{2/n}}\right)\right.$$

$$\left. - \frac{\Delta_1}{w_2}\left|\frac{C + w_2 z_\beta}{y_1} - \frac{n w_1}{\sigma\sqrt{2n}}\right|\right) dF_n(y_1, y_2)$$

$$- \iint_{y_1 < y_2} \Phi\left(\frac{1}{w_2}\left(C - \frac{w_1 Y_2^{(n)}}{\sigma\sqrt{2/n}}\right)\right.$$

$$\left. - \frac{\Delta_2}{w_2}\left|\frac{C + w_2 z_\beta}{y_2} - \frac{n w_1}{\sigma\sqrt{2n}}\right|\right) dF_n(y_1, y_2)$$

where $F_n(y_1, y_2)$ is the joint distribution function of $Y_1^{(n)}$ and $Y_2^{(n)}$. Letting $v_i = \frac{y_i - \Delta_i}{\sqrt{2/n}\sigma}$,

$$= 1 - \iint_{v_1 > v_2 + \frac{n(\Delta_2 - \Delta_1)}{\sigma\sqrt{2n}}} \Phi\left(\frac{1}{w_2}\left(C - w_1\left(v_1 + \frac{n\Delta_1}{\sigma\sqrt{2n}}\right)\right)\right.$$

$$\left. - \frac{\Delta_1}{w_2}\left|\frac{C + w_2 z_\beta}{\Delta_1 + \sqrt{2/n}\sigma v_1} - \frac{n w_1}{\sigma\sqrt{2n}}\right|\right) dF\left(v_1, v_2\right)$$

$$- \iint_{v_1 < v_2 + \frac{n(\Delta_2 - \Delta_1)}{\sigma\sqrt{2n}}} \Phi\left(\frac{1}{w_2}\left(C - w_1\left(v_2 + \frac{n\Delta_2}{\sigma\sqrt{2n}}\right)\right)\right.$$

$$\left. - \frac{\Delta_2}{w_2}\left|\frac{C + w_2 z_\beta}{\Delta_2 + \sqrt{2/n}\sigma v_2} - \frac{n w_1}{\sigma\sqrt{2n}}\right|\right) dF\left(v_1, v_2\right)$$

$$= 1 - \iint_{v_1 > v_2 + \frac{n(\Delta_2 - \Delta_1)}{\sqrt{2n}\sigma}} \Phi\left(\frac{1}{w_2}\left(C - w_1\left(v_1 + \frac{n\Delta_1}{\sqrt{2n}\sigma}\right)\right)\right.$$

$$\left. - \frac{\Delta_1}{w_2}\left|\frac{C + w_2 z_\beta}{\Delta_1 + \sqrt{2/n}\sigma v_1} - \frac{n w_1}{\sigma\sqrt{2n}}\right|\right) f\left(v_2|v_1\right) \emptyset\left(v_1\right) dv_2 dv_1$$

$$- \iint_{v_1 < v_2 + \frac{n(\Delta_2 - \Delta_1)}{\sqrt{2n}\sigma}} \Phi\left(\frac{1}{w_2}\left(C - w_1\left(v_2 + \frac{n\Delta_2}{\sqrt{2n}\sigma}\right)\right)\right.$$

$$\left. - \frac{\Delta_2}{w_2}\left|\frac{C + w_2 z_\beta}{\Delta_2 + \sqrt{2/n}\sigma v_2} - \frac{n w_1}{\sigma\sqrt{2n}}\right|\right) f\left(v_1|v_2\right) \emptyset\left(v_2\right) dv_1 dv_2$$

$$= 1 - \int_{-\infty}^{\infty} \int_{-\infty}^{v_1 - \frac{n(\Delta_2 - \Delta_1)}{\sqrt{2n}\sigma}} \Phi\left(\frac{1}{w_2}\left(C - w_1\left(v_1 + \frac{n\Delta_1}{\sqrt{2n}\sigma}\right)\right)\right.$$

$$\left. - \frac{\Delta_1}{w_2}\left|\frac{C + w_2 z_\beta}{\Delta_1 + \sqrt{2/n}\sigma v_1} - \frac{n w_1}{\sigma\sqrt{2n}}\right|\right) f\left(v_2|v_1\right) \emptyset\left(v_1\right) dv_2 dv_1$$

$$- \int_{-\infty}^{\infty} \int_{-\infty}^{v_2 + \frac{n(\Delta_2 - \Delta_1)}{\sqrt{2n}\sigma}} \Phi\left(\frac{1}{w_2}\left(C - w_1\left(v_2 + \frac{n\Delta_2}{\sqrt{2n}\sigma}\right)\right)\right.$$

$$\left. - \frac{\Delta_2}{w_2}\left|\frac{C + w_2 z_\beta}{\Delta_2 + \sqrt{2/n}\sigma v_2} - \frac{n w_1}{\sigma\sqrt{2n}}\right|\right) f\left(v_1|v_2\right) \emptyset\left(v_2\right) dv_1 dv_2$$

Note $_2|v_1 \sim N(\rho v_1, (1 - \rho^2)), v_2|v_1 \sim N(\rho v_2, (1 - \rho^2)), \quad \rho = 1/2$

$$= 1 - \int_{-\infty}^{\infty} \Phi\left(\frac{1}{w_2}\left(C - w_1\left(v_1 + \frac{n\Delta_1}{\sqrt{2n}\sigma}\right)\right)\right.$$

$$\left. - \frac{\Delta_1}{w_2}\left|\frac{n\left(C + w_2 z_\beta\right)}{n\Delta_1 + \sqrt{2n}\sigma v_1} - \frac{n w_1}{\sqrt{2n}\sigma}\right|\right) \Phi\left(\frac{(1-\rho)\, v_1 - \frac{n(\Delta_2 - \Delta_1)}{\sqrt{2n}\sigma}}{\sqrt{1-\rho^2}}\right) \phi\left(v_1\right) dv_1$$

$$- \int_{-\infty}^{\infty} \Phi\left(\frac{1}{w_2}\left(C - w_1\left(v_2 + \frac{n\Delta_2}{\sqrt{2n}\sigma}\right)\right)\right.$$

$$\left. - \frac{\Delta_2}{w_2}\left|\frac{n\left(C + w_2 z_\beta\right)}{n\Delta_2 + \sqrt{2n}\sigma v_2} - \frac{n w_1}{\sqrt{2n}\sigma}\right|\right) \Phi\left(\frac{(1-\rho)\, v_2 + \frac{n(\Delta_2 - \Delta_1)}{\sqrt{2n}\sigma}}{\sqrt{1-\rho^2}}\right) \phi\left(v_2\right) dv_2$$

Since we need to use $M'_i = \min\left(M_i, L\right)$, the above fourmula should be modified to,

$$1 - \int_{-\infty}^{\infty} \Phi\left(\frac{1}{w_2}\left(C - w_1\left(v_1 + \frac{n\Delta_1}{\sqrt{2n}\sigma}\right)\right)\right.$$

$$\left. - \frac{\Delta_1}{w_2}\min\left(\left|\frac{n\left(C + w_2 z_\beta\right)}{n\Delta_1 + \sqrt{2n}\sigma v_1} - \frac{n w_1}{\sqrt{2n}\sigma}\right|, \quad \frac{w_2\sqrt{L-n}}{\sigma\sqrt{2}}\right)\right) \Phi$$

$$\times \left(\frac{(1-\rho)\, v_1 - \frac{n(\Delta_2 - \Delta_1)}{\sqrt{2n}\sigma}}{\sqrt{1-\rho^2}}\right) \phi\left(v_1\right) dv_1$$

$$- \int_{-\infty}^{\infty} \Phi\left(\frac{1}{w_2}\left(C - w_1\left(v_2 + \frac{n\Delta_2}{\sqrt{2n}\sigma}\right)\right)\right.$$

$$\left. - \frac{\Delta_2}{w_2}\min\left(\left|\frac{n\left(C + w_2 z_\beta\right)}{n\Delta_2 + \sqrt{2n}\sigma v_2} - \frac{n w_1}{\sqrt{2n}\sigma}\right|, \quad \frac{w_2\sqrt{L-n}}{\sigma\sqrt{2}}\right)\right) \Phi$$

$$\times \left(\frac{(1-\rho)\, v_2 + \frac{n(\Delta_2 - \Delta_1)}{\sqrt{2n}\sigma}}{\sqrt{1-\rho^2}}\right) \phi\left(v_2\right) dv_2$$

If total sample size is planned at the beginning of the study, say, M, we can replace M_1 and M_2 with M, and the above derivation still follows. In this case, if we choose $w_1^2 = \frac{n}{M}$, then the design accommodates dose selection with sample size adaption, similar to Stallard and Todd (2003). In other words, our design is generalized. Function "integrate" in R is often used to evaluate an integral like this. However, it doesn't work well for this truncated integrand. Instead, the function "adaptIntegrate" in package "cubature" should be used.

Appendix 4: Type I Error Rate

There are two other type I error configurations, in addition to that in (10.1). One is when $\Delta_1 = 0$, $\Delta_2 > 0$, treatment arm 1 is selected and rejected. The other is when $\Delta_1 > 0$, $\Delta_2 = 0$, treatment arm 2 is selected and rejected. We first show that these two cases are dominated by the case in (10.1). Let T_i, $i = 1, 2$, be the test statistics at the end of the trial with random final sample size.

Under $\Delta_1 = 0$, $\Delta_2 > 0$,the type I error rate is

$P_{\Delta_1=0,\Delta_2>0}\left(T_1 > C, Y_1^{(n)} > Y_2^{(n)}\right)$.

Under $\Delta_1 = 0$, $\Delta_2 = 0$ the type I error rate is

$P_{\Delta_1=0,\Delta_2=0}\left(T_1 > C, Y_1^{(n)} > Y_2^{(n)}\right) + P_{\Delta_1=0,\Delta_2=0}\left(T_2 > C, Y_1^{(n)} < Y_2^{(n)}\right)$

$\geq P_{\Delta_1=0,\Delta_2=0}\left(T_1 > C, Y_1^{(n)} > Y_2^{(n)}\right) \geq P_{\Delta_1=0,\Delta_2>0}\left(T_1 > C, Y_1^{(n)} > Y_2^{(n)}\right)$.

It is similar for the other case. So in the following we only consider the case of (10.1).

When $\Delta_1 = \Delta_2 = 0$, the absolute power formula gives the type I error rate.

$$P\left(T_S > C\right) = 1 - \int_{-\infty}^{\infty} \Phi\left(\frac{1}{w_2}\left(C - w_1 v_1\right)\right) \Phi\left(\frac{(1-\rho)\, v_1}{\sqrt{1-\rho^2}}\right) \o\left(v_1\right) dv_1$$

$$- \int_{-\infty}^{\infty} \Phi\left(\frac{1}{w_2}\left(C - w_1 v_2\right)\right) \Phi\left(\frac{(1-\rho)\, v_2}{\sqrt{1-\rho^2}}\right) \o\left(v_2\right) dv_2$$

$$= 1 - 2\int_{-\infty}^{\infty} \Phi\left(\frac{1}{w_2}\left(C - w_1 v_2\right)\right) \Phi\left(\frac{(1-\rho)\, v_2}{\sqrt{1-\rho^2}}\right) \o\left(v_2\right) dv_2$$

As noted in Appendix 3, this is the type I error rate for fixed sample size designs with treatment selection if $w_1^2 = \frac{n}{M}$. Next we show when $w_1 = 0$, the above result is equal to

$$1 - 2\int_{-\infty}^{\infty} \Phi(C)\Phi\left(\frac{(1-\rho)\, v}{\sqrt{1-\rho^2}}\right) \o(v)dv = 1 - \Phi(C).$$

To that end, we let

$$g(k) = \int_{-\infty}^{\infty} \Phi(kv)\o(v)dv$$

Then, $g'(k) = -\int_{-\infty}^{\infty} v\phi(kv)\phi(v)dv = 0$ since $\phi(.)$ is an even function. So $g(k)$ is a constant with respect to k. Taking $k = 0$, we got $g(0) = 0.5$.

Table 10.1 Critical values for different weight w_1

w_1	C	w_1	C	w_1	C
0.05	1.98	0.35	2.081	0.7	2.166
0.1	1.998	0.4	2.095	0.75	2.176
0.15	2.016	0.45	2.109	0.8	2.184
0.2	2.034	0.5	2.122	0.85	2.192
0.25	2.05	0.55	2.134	0.9	2.2
0.26	2.053	0.6	2.145	0.95	2.206
0.3	2.066	0.65	2.156	0.99	2.211

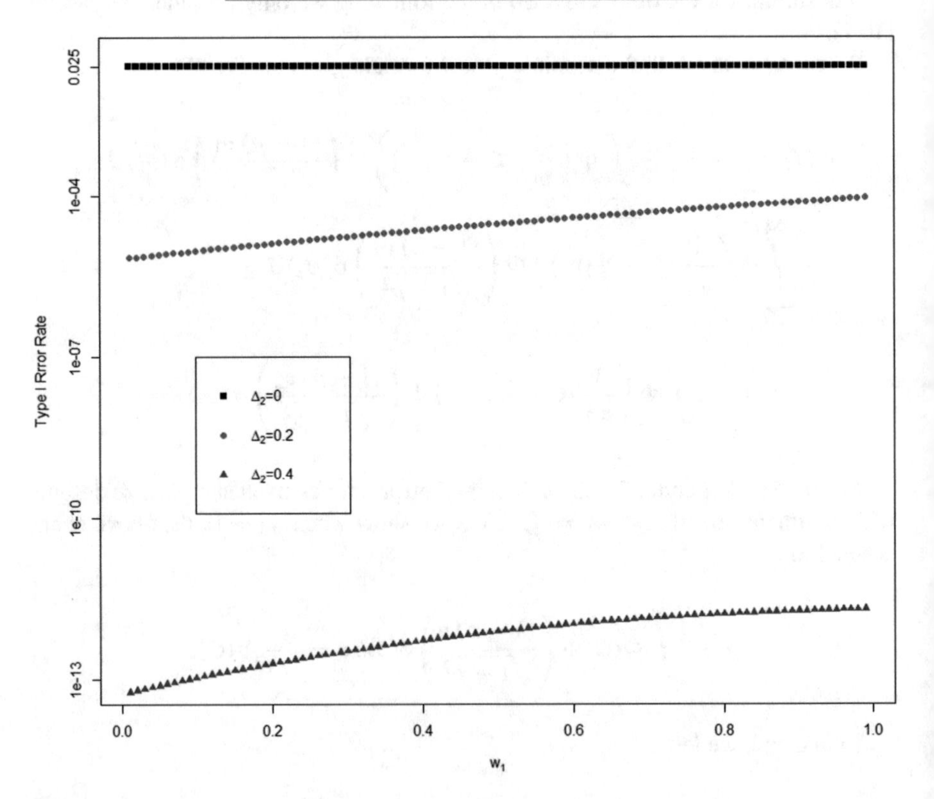

Fig. 10.1 Type I error rate with $\Delta_1 = 0$. When $\Delta_2 = 0$, the type I error rate is the probability to show the selected treatment to be significant; when $\Delta_2 > 0$, the type I error rate is the probability to select treatment 1 and conclude it is significant

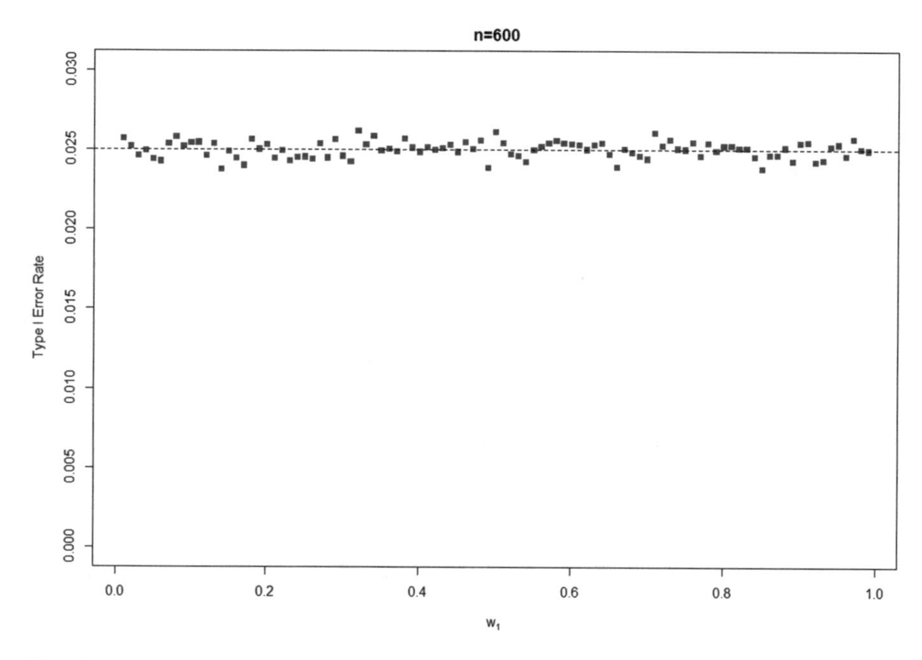

Fig. 10.2 Type I error rate simulation with 100,000 replications with $\Delta_1 = \Delta_2 = 0$ and critical values corresponding to w_1

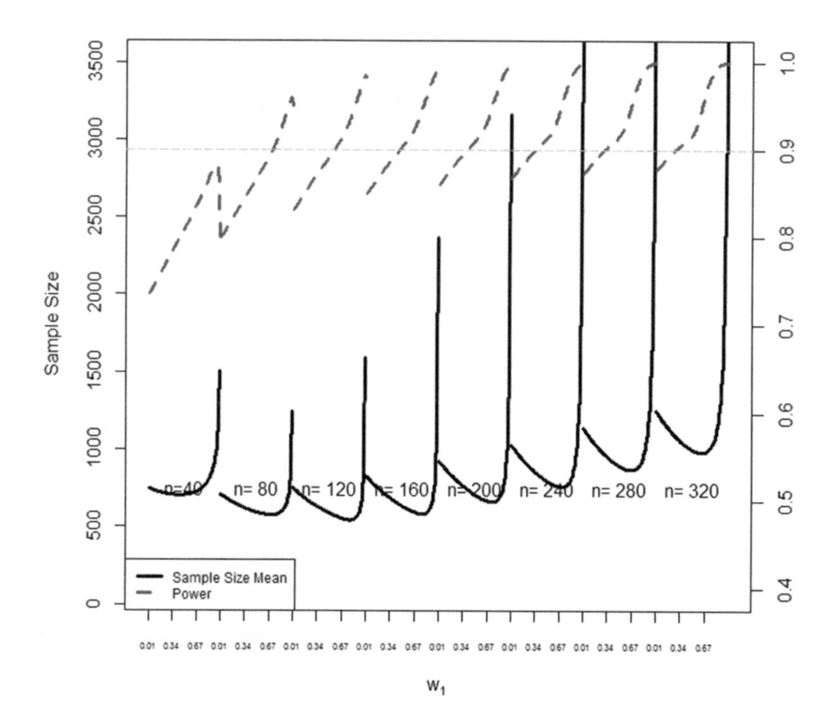

Fig. 10.3 Design optimization against interim sample size and weight with $\Delta_1 = 0.2$ and $\Delta_2 = 0.4$. For every value of n, w_1 varies from 0.01 to 0.99

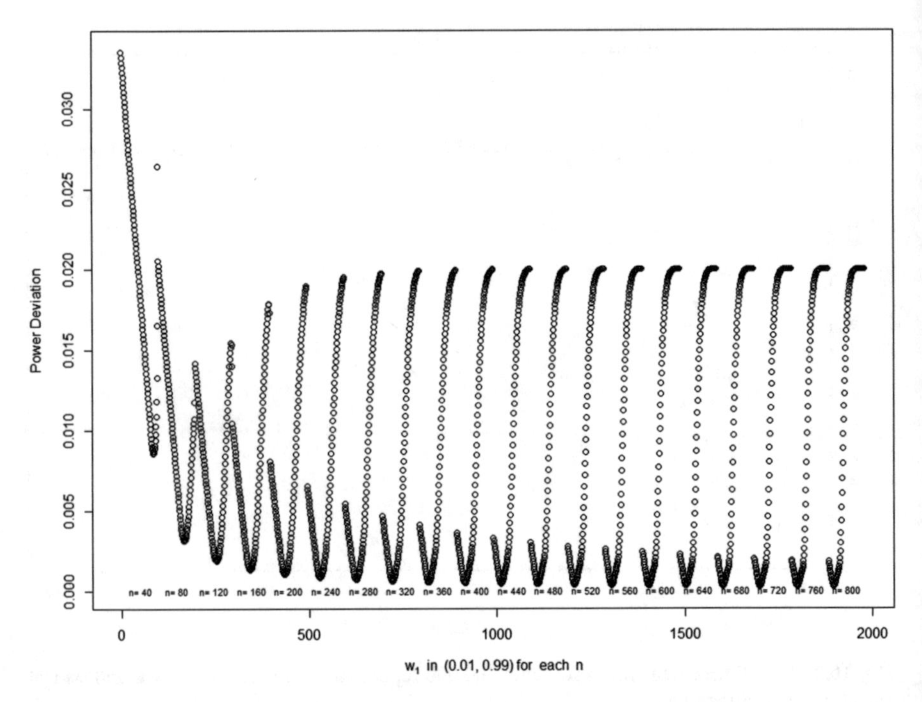

Fig. 10.4 Power deviation for all candidate designs

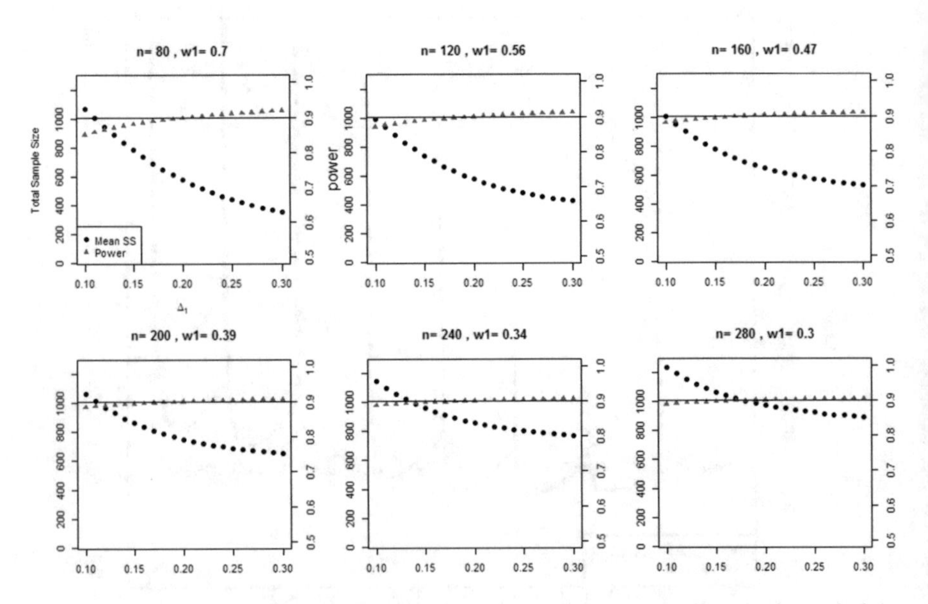

Fig. 10.5 Mean total sample size and power of selected designs over the effect size interval of $\boldsymbol{\Delta}_1$

References

Bauer, P., Köhne, K.: Evaluation of experiments with adaptive interim analyses. Biometrics. **50**, 1029–1041 (1994). correction in *Biometrics* 1996, 52:380

Bauer, P., Bretz, F., Dragalin, V., Konig, F., Wassmer, G.: Twenty-five years of confirmatory adaptive designs: opportunities and pitfalls. Stat. Med. **35**, 325–347 (2016)

Bretz, F., Koenig, F., Brannath, W., Glimm, E., Posch, M.: Adaptive designs for confirmatory clinical trials. Stat. Med. **28**, 1181–1217 (2009)

Chaturvedi, P.R., Antonijevic, Z., Mehta, C.: Practical considerations for a two-stage confirmatory adaptive clinical trial design and its implementation: ADVENT trial. In: He, W., Pinheiro, J., Kuznetsova, O.M. (eds.) Practical Considerations for Adaptive Trial Design and Implementation, pp. 383–411. Springer, New York (2014)

Cui, L., Hung, H.M., Wang, S.J.: Impact of changing sample size in a group sequential clinical trial. In: ASA Proceedings of Biopharmaceutical Section, pp. 52–57. American Statistical Association, Anaheim, CA (1997)

Cui, L., Hung, H.M., Wang, S.J.: Modification of sample size in group sequential clinical trial. Biometrics. **55**, 853–857 (1999)

Denne, J.S.: Sample size recalculation using conditional power. Stat. Med. **20**, 2645–2660 (2001)

FDA. Guidance for industry: adaptive design clinical trials for drugs and biologics (2010)

Gan, H.K., You, B., Pond, G.R., Chen, E.X.: Assumptions of expected benefits in randomized phase III trials evaluating systemic treatments for cancer. J. Natl. Cancer Inst. **104**, 1–9 (2012)

Glimm, E.: Comments on 'Adaptive increase in sample size when interim results are promising: a practical guide with examples' by C. R. Mehta and S. J. Pocock. Stat. Med. **31**, 98–99 (2012)

Lawrence, D., Bretz, F.: Approaches for optimal dose selection for adaptive design trials. In: He, W., Pinheiro, J., Kuznetsova, O.M. (eds.) Practical Considerations for Adaptive Trial Design and Implementation, pp. 125–137. Springer, New York (2014)

Lawrence, D., Bretz, F., Pocock, S.: IHANCE: an adaptive confirmatory study with dose selection at interim. In: Trifilieff, A. (ed.) Indacaterol-the First Once-Daily Long-Acting Beta2 Agonist for COPD, pp. 77–92. Springer, Basel (2014)

Lee, E.B., Fleischmann, R., Hall, S., Wilkinson, B., Bradley, J.D., Gruben, D., Koncz, T., Krishnaswami, S., Wallenstein, G.V., Zang, C., et al.: Tofacitinib versus methotrexate in rheumatoid arthritis. N. Engl. J. Med. **370**, 2377–2386 (2014)

Lehmacher, W., Wassmer, G.: Adaptive sample size calculations in group sequential trials. Biometrics. **55**, 1286–1290 (1999)

Liu, G.F., Zhu, G.R., Cui, L.: Evaluating the adaptive performance of flexible sample size designs with treatment difference in an interval. Stat. Med. **27**, 584–596 (2008)

Mehta, C., Pocock, S.J.: Adaptive increase in sample size when interim results are promising: a practical guide with example. Stat. Med. **30**, 3267–3284 (2011)

Muller, H.H., Schafer, H.: Adaptive group sequential designs for clinical trials: combining the advantages of adaptive and classical group sequential approaches. Biometrics. **57**, 886–891 (2001)

Posch, M., Maurer, W., Bretz, F.: Type I error rate control in adaptive designs for confirmatory clinical trials with treatment selection at interim. Pharm. Stat. **10**(2), 96–104 (2011)

Proschan, M.A., Hunsberger, S.A.: Designed extension of studies based on conditional power. Biometrics. **51**, 1315–1324 (1995)

Stallard, N., Todd, S.: Sequential designs for phase III clinical trials incorporating treatment selection. Stat. Med. **22**, 689–703 (2003)

Stallard, N., Todd, S.: Seamless phase II/III designs. Stat. Methods Med. Res. **20**, 623–634 (2011)

Thall, P.F., Simon, R., Ellenberg, S.S.: Two-stage selection and testing designs for comparative clinical trials. Biometrika. **75**, 303–310 (1988)

Zhang, L., Cui, L., Yang, B.: Optimal flexible sample size design with robust power. Stat. Med. **35**, 3385–3396 (2016)

Chapter 11
Critical Boundary Refinement in a Group Sequential Trial When the Primary Endpoint Data Accumulate Faster Than the Secondary Endpoint

Jiangtao Gou and Oliver Y. Chén

11.1 Introduction

In classical clinical trial studies, a clinical endpoint is defined as the time point at which a disease or symptom occurs. An individual reaching an endpoint during a clinical trial indicates either the conclusion of the trial, or there is strong evidence rendering the subject withdraws from the trial. To allow for early diagnosis, personalized treatment, and timely drug development, modern clinical trials are designed with customized endpoints. Consequently, the assessment time available to statistical analysis for each endpoint varies. For example, in oncology clinical trials, depending on the centering focus, the endpoints can be categorized into patient-centered endpoints and tumor-centered endpoints. An example of a patient-centered endpoint is the overall survival (OS), defined as the cumulative days a patient has lived, counting beginning from the date on which the disease is diagnosed or the date on which treatment is initiated; an example of a tumor-centered endpoint is progression-free survival (PFS), defined as cumulative days a patient has lived with cancer since the treatment and that the disease has not progressed (Fiteni et al. 2014). While OS is more reliable (since it covers a longer period) than PFS,

J. Gou (✉)
Department of Mathematics and Statistics, Hunter College of CUNY, New York, NY, USA

Department of Mathematics and Statistics, Villanova University, Villanova, PA, USA
e-mail: jgou@u.northwestern.edu

O. Y. Chén
Institute for Biomedical Engineering, University of Oxford, Oxford, UK
e-mail: yibing.chen@seh.ox.ac.uk

L. Zhang et al. (eds.), *Contemporary Biostatistics with Biopharmaceutical Applications*, ICSA Book Series in Statistics,
https://doi.org/10.1007/978-3-030-15310-6_11

the latter is usually used in practice as a surrogate for OS, when an accelerated evaluation is demanded, for example, during a drug test. However, we cannot at present ascribe such a replacement to any well-defined statistical theory, owning, in part, to the genuine differences between the two types of endpoints, and, in part, to the intellectual discovery of only modest correlation between PFS and OS (Amir et al. 2012; Michiels et al. 2017). By probing into the hierarchical basis of these two types of endpoints, statistical science can help us uncover the utility underlying each endpoint in addressing problems in clinical trials and improve statistical power. A beginning in this direction can be made by considering a hierarchical test embedded in a group sequentially design (Hung et al. 2007).

A group sequential design is a framework that allows statistical analysis during longitudinally ordered stages, defined as interim stages followed by a final stage (Jennison and Turnbull 2000). During each interim stage, a statistic (e.g. the estimated logarithm of the hazard ratio) is computed on data hitherto collected to determine whether or not to reject a null hypothesis (e.g. whether or not a treatment is more effective than the standard treatment), based upon a stopping criterion (called a critical boundary). Specifically, if the statistic exceeds the critical boundary, the null hypothesis is rejected, and the trial is subsequently terminated prior to the next interim stage. If a trial reaches the final stage, all data are utilized to test the null hypothesis.

Chief to a group sequential design is the critical boundary for early stopping. Pocock (1977) and O'Brien and Fleming (1979) individually proposed two now widely used critical boundaries for group sequential trials. Attributing to their contribution, these boundaries are commonly referred to as the Pocock (POC) boundary and the O'Brein-Fleming (OBF) boundary today, respectively. However, the POC and OBF boundaries require that the total number of decision times specified in advance. When this condition is not met, Lan and DeMets (1983) utilized a family of error spending functions to approximate the POC and the OBF boundaries. All of these approaches consider group sequential trials with a single primary endpoint. To address issues in group sequential trials involving multiple primary endpoints, Jennison and Turnbull (1993), Tang and Geller (1999), Maurer and Bretz (2013), Ye et al. (2013) and Xi and Tamhane (2015) provided various suggestions.

To raise any clinical finding related to an endpoint to the rank of science, one has to construct statistical hypotheses test for each endpoint. In a randomized trial consisting multiple endpoints, the endpoints often present a hierarchical structure. Statistical testing can be conducted serially for each ordered endpoint, or in parallel for all endpoints by applying the gatekeeping procedure (Dmitrienko and Tamhane 2007; Dmitrienko et al. 2009). A more flexible framework is the graph-theoretic-based procedure introduced by Bretz et al. (2009) and Burman et al. (2009), wherein nodes are used to represent hypothesis tests, coupled by directed and weighted edges indicating multiple test procedures. The above approaches were initially employed in single-stage designs with neither interim analysis nor trial extension. To extend these methods to multi-stage designs, Hung et al. (2007) first considered hierarchically testing multiple endpoints in a group sequential design.

The theoretical basis of group sequential designs involving multiple endpoints with complex hierarchical structure, one of the common practice in modern clinical trials, however, is not as-of-yet well-charted in statistical science. For instance, in an oncology trial, when the primary endpoint is PFS and the secondary endpoint is OS with the partially hierarchical design, can we improve upon the simple Bonferroni-based split between the primary and the secondary endpoint (which is the current practice), in a group sequential design? Prior work has built a reliable and useful repertoire that has offered us much insight, with which we build our theory. For example, Hung et al. (2007), Tamhane et al. (2010), Glimm et al. (2010), and Tamhane et al. (2018) considered the group sequential procedures for a primary and a secondary hypothesis with the same information fractions at interim analyses. In the light of their knowledge, in this article we attempt to address a few core issues in clinical trails when multiple objectives with hierarchical structures are present in group sequential designs.

11.2 Preliminaries

Consider a trial on a primary and a secondary endpoint hierarchically using a group sequential design with two stages. In the following, we use X to denote parameters and statistics that are related to the primary endpoint, and Y to denote parameters and statistics for the secondary endpoint. The number of interim looks at the secondary endpoint is permitted to be greater than the number of looks at the primary, if it takes longer to collect the secondary endpoint data than the primary endpoint data. We first consider a two-stage group sequential design that is applied to the primary endpoint, and a K-stage design that is used for the secondary endpoint ($K \geq 2$). For simplicity, we call it $[2|K]$-stage design. As a natural extension, we introduce the procedure with a K_X-stage design for the primary hypothesis and a K_Y-stage design for the secondary hypothesis. We denote this as a $[K_X|K_Y]$-stage design.

In a $[2|K]$-stage design, let $n_{1,X}$ and $n_{2,X}$ be the sample sizes for the two stages of the primary endpoint H_X, and $n_{1,Y}, n_{2,Y}, \ldots, n_{K,Y}$ for the K stages of the secondary endpoint H_Y. The total sample size is N, where $N = n_{1,X} + n_{2,X} = \sum_{i=1}^{K} n_{i,Y}$. The information time of the primary endpoint at the interim analysis is denoted as $t_X = n_{1,X}/N$. For the secondary endpoint, there are $K-1$ interim analyses, and the information times are $t_{i,Y} = \sum_{j=1}^{i} n_{j,Y}/N$, $i = 1, \cdots, K-1$. The information time or information fraction is the proportion of subjects or events already observed (Lan and DeMets 1989). The correlation between the two endpoints is denoted as ρ.

Let (X_1, X_2) and $(Y_1, Y_2, \ldots, Y_K)$ denote the standardized sample mean test statistics for the two endpoints at different stages, specified by a numeric subscript.

The normal theory applies asymptotically in this case. The correlations between the test statistics are shown as follows.

$$\operatorname{corr}(X_1, X_2) = \lambda, \quad \operatorname{corr}(Y_i, Y_j) = \gamma_i/\gamma_j \ (i < j),$$
$$\operatorname{corr}(X_1, Y_K) = \lambda\rho, \quad \operatorname{corr}(X_2, Y_K) = \rho,$$
$$\operatorname{corr}(X_1, Y_i) = \min\{\lambda/\gamma_i, \gamma_i/\lambda\} \cdot \rho, \quad \operatorname{corr}(X_2, Y_i) = \gamma_i\rho,$$

where $\lambda = \sqrt{t_X}, \gamma_i = \sqrt{t_{i,Y}}$ for $i = 1, \ldots, K-1$ and $\gamma_K = 1$.

Let $(\Delta_{1,X}, \Delta_{2,X})$ and $(\Delta_{1,Y}, \Delta_{2,Y}, \ldots, \Delta_{K,Y})$ denote the standardized treatment effects of the primary and the secondary endpoints at each stage. Noting that $\Delta_{1,X} = \lambda\Delta_{2,X}$ and $\Delta_{i,Y} = \gamma_i\Delta_{K,Y}$, we therefore simplify the notations by letting $\Delta_X = \Delta_{2,X}$ and $\Delta_Y = \Delta_{K,Y}$.

Denote H_X and H_Y as the primary and the secondary null hypotheses. Let (c_1, c_2) and $(d_1, d_2, \ldots, d_K)$ denote the primary boundary and the secondary boundary, respectively, in a group sequential procedure. Here, (c_1, c_2) correspond to (X_1, X_2) and $(\Delta_{1,X}, \Delta_{2,X})$; $(d_1, d_2, \ldots, d_K)$ are with respect to $(Y_1, Y_2, \ldots, Y_k)$ and $(\Delta_{1,Y}, \Delta_{2,Y}, \ldots, \Delta_{K,Y})$. Examples of common boundaries are discussed in Pocock (1977), O'Brien and Fleming (1979), and Lan and DeMets (1983).

In this article, we investigate three types of hierarchical testing scenarios: stage-wise hierarchical, overall hierarchical, and partially hierarchical scenarios. To conduct hypothesis testing with respect to each scenario, a scenario-specific decision rule needs to be defined a priori. Following Glimm et al. (2010), these decision rules are specified as below. Here, we define α_Y^S, α_Y^O, and α_Y^P, as the type I errors for a stagewise (S), an overall (O), and a partially (P) hierarchical rule, respectively, under the null hypothesis H_Y.

- Stagewise hierarchical rule $\mathcal{P}_S$. The primary hypothesis is tested sequentially. The secondary hypothesis will be automatically accepted if the primary hypothesis is not rejected. If the primary hypothesis is rejected, the secondary hypothesis will be tested only once at the same stage. The associated type I error is

$$\alpha_Y^S = \Pr(X_1 > c_1, Y_1 > d_1) + \Pr(X_1 \le c_1, X_2 > c_2, Y_2 > d_2).$$

- Overall hierarchical rule $\mathcal{P}_O$. Besides $\mathcal{P}_S$, the secondary hypothesis can be tested until its final stage if the primary hypothesis is rejected. The associated type I error is

$$\alpha_Y^O = \alpha_Y^S + \sum_{i=1}^{K-1} \Pr(X_1 > c_1, Y_1 \le d_1, \cdots, Y_i \le d_i, Y_{i+1} > d_{i+1})$$
$$+ \sum_{i=2}^{K-1} \Pr(X_1 \le c_1, X_2 > c_2, Y_2 \le d_2, \cdots, Y_i \le d_i, Y_{i+1} > d_{i+1}).$$

- Partially hierarchical rule $\mathcal{P}_P$. Besides $\mathcal{P}_O$, the secondary hypothesis can be tested from stage 2 to stage K if the primary hypothesis is failed to be rejected at its interim and final stage. The associated type I error is

$$\alpha_Y^P = \alpha_Y^O + \Pr(X_1 \le c_1, X_2 \le c_2, Y_2 > d_2) + \sum_{i=2}^{K-1} \Pr(X_1 \le c_1, X_2 \le c_2, Y_2 \le d_2, \cdots, Y_i \le d_i, Y_{i+1} > d_{i+1}).$$

Glimm et al. (2010) also listed another hierarchical rule called the coequal rule $\mathcal{P}_C$, where the primary and the secondary hypotheses are tested independently without any hierarchical structure. For a trial design using the coequal hierarchical rule, Bonferroni-type methods have been well developed, such as Maurer and Bretz (2013)'s method based on the graphical approach (Bretz et al. 2009, 2011), and Ye et al. (2013)'s method based on the Holm (1979) procedure. Other distribution-based or p-value-based tests can also be applied in trial designs using the coequal hierarchical rule, such as the Dunnett and Tamhane (1992) test, the Simes (1986) test, the generalized Simes test (Sarkar 2008; Gou and Tamhane 2014, 2018b), and their corresponding multiple testing procedures, such as Hommel (1988), Hochberg (1988), Rom (1990), and the hybrid Hochberg–Hommel procedure (Gou et al. 2014; Gou and Tamhane 2018a; Tamhane and Gou 2018). Since the endpoints under the coequal hierarchical rule are co-primary endpoints without a real hierarchical structure, we focus on the stagewise (S), the overall (O), and the partially (P) hierarchical rule in this article.

In a $[K_X|K_Y]$-stage design, we use terminologies and notations similar to those of a $[2|K]$-stage design. The sample sizes for H_X and H_Y in each stage are denoted as $n_{1,X}, \ldots, n_{K_X,X}$ and $n_{1,Y}, \ldots, n_{K_Y,Y}$ respectively, and the total sample size $N = \sum_{i=1}^{K_X} n_{i,X} = \sum_{i=1}^{K_Y} n_{i,Y}$. The cumulative sample sizes at stage i for H_X and H_Y are $N_{i,X} = \sum_{j=1}^{i} n_{j,X}$ and $N_{i,Y} = \sum_{j=1}^{i} n_{j,Y}$. The information times are calculated accordingly as $t_{i,X} = N_{i,X}/N$ and $t_{i,Y} = N_{i,Y}/N$, where $t_{K_X,X} = t_{K_Y,Y} = 1$. Let $\lambda_i = \sqrt{t_{i,X}}$, $\gamma_i = \sqrt{t_{i,Y}}$, and the correlation between X_{K_X} and Y_{K_Y} be ρ. The correlations between the standardized test statistics $(X_1, \cdots, X_{K_X})$ and $(Y_1, \cdots, X_{K_X})$ are

$$\operatorname{corr}(X_i, X_j) = \lambda_i/\lambda_j \ (i < j), \quad \operatorname{corr}(Y_i, Y_j) = \gamma_i/\gamma_j \ (i < j),$$
$$\operatorname{corr}(X_i, Y_j) = \min\{\lambda_i/\gamma_j, \gamma_j/\lambda_i\} \cdot \rho, \quad \operatorname{corr}(X_{K_X}, Y_{K_Y}) = \rho.$$

The standardized effects for H_X and H_Y at the final stage are denoted as Δ_X and Δ_Y, so the effects at interim stage i are $\lambda_i\Delta_X$ and $\gamma_i\Delta_Y$, respectively. The critical boundaries for standardized test statistics of H_X and H_Y are $(c_1, \cdots, c_{K_X})$ and $(d_1, \cdots, d_{K_Y})$. When $K_X = K_Y$, Tamhane et al. (2018) gave the expressions of type I error rates under H_Y for $\mathcal{P}_S$, $\mathcal{P}_O$, and $\mathcal{P}_P$. In a more general setting when $K_X \neq K_Y$, the corresponding type I error rates under H_Y are

$$\mathcal{P}_S : \alpha_Y^S = \sum_{i=1}^{K_X \wedge K_Y} \Pr\left(X_1 \le c_1, \ldots, X_{i-1} \le c_{i-1}, X_i > c_i, Y_i > d_i\right),$$

$$\mathcal{P}_O : \alpha_Y^O = \alpha_Y^S + \sum_{i=1}^{K_X \wedge \{K_Y - 1\}} \sum_{j=i+1}^{K_Y} \Pr\big(X_1 \le c_1, \cdots, X_{i-1} \le c_{i-1}, X_i > c_i,$$
$$Y_i \le d_i, \cdots, Y_{j-1} \le d_{j-1}, Y_j > d_j\big),$$

$$\mathcal{P}_P : \alpha_Y^P = \begin{cases} \alpha_Y^O + \Pr\left(X_1 \le c_1, \cdots, X_{K_X} \le c_{K_X}, Y_{K_Y} > d_{K_Y}\right), \text{ if} K_X \ge K_Y, \\ \alpha_Y^O + \sum_{i=K_X}^{K_Y} \Pr\left(X_1 \le c_1, \cdots, X_{K_X} \le c_{K_X}, Y_{K_X} \le d_{K_X}, \ldots,\right. \\ \left. Y_{i-1} \le d_{i-1}, Y_i > d_i\right), \text{ if} K_X < K_Y, \end{cases}$$

where $K_X \wedge K_Y = \min\{K_X, K_Y\}$.

Note that for a test on a primary and a secondary endpoint in a group sequential design, the control of familywise error rate (FWER) (Hochberg and Tamhane 1987; Tamhane et al. 2010; Zhang and Gou 2019a) requires that FWER $=$ $\Pr\left(\text{Reject at least one true } H \in \{H_X, H_Y\}\right) \le \alpha$. Following the closure principle (Marcus et al. 1976), the control of type I error under primary hypothesis H_X, the control under secondary hypothesis H_Y and the control under their intersection $H_X \cap H_Y$ are all at level α, leading to the control of the FWER at level α.

11.3 Stagewise Hierarchical Rule

The stagewise hierarchical rule $\mathcal{P}_S$ and the overall hierarchical rule $\mathcal{P}_O$ satisfy the gatekeeping condition, In other words, the secondary endpoint is tested only if the primary endpoint is significant (Dmitrienko and Tamhane 2007; Dmitrienko et al. 2009). Under this condition, the event $R_Y = \{\text{Reject } H_Y\}$ is a subset of the event $R_X = \{\text{Reject } H_X\}$. It follows that $\Pr\left(R_X \cup R_Y | H_X \cap H_Y\right) = \Pr\left(R_X | H_X\right)$. This indicates that once the primary endpoint is tested using an α-level boundary, then $\Pr\left(R_X \cup R_Y | H_X \cap H_Y\right) \le \alpha$ (Tamhane et al. 2010). Consequently, for testing procedures using the stagewise hierarchical rule $\mathcal{P}_S$ or the overall hierarchical rule $\mathcal{P}_O$, in order to control FWER at level α, the only requirement of type I error control for the secondary hypothesis is $\Pr\left(R_Y | H_Y\right) \le \alpha$, or more specifically, $\Pr\left(R_Y | \overline{H}_X \cap H_Y\right) \le \alpha$.

In a $[2|K]$-stage design, the primary hypothesis H_X can be tested flexibly using any α-level group sequential boundary (c_1, c_2). For example, the critical boundary (c_1, c_2) satisfies $\alpha_X = 1 - \Pr\left(X_1 \le c_1, X_2 \le c_2\right) \le \alpha$. The marginal significance level of the secondary hypothesis H_Y is defined as $\alpha_Y = 1 - \Pr\left(\cap_{i=1}^{K} \{Y_i \le d_i\}\right)$.

We consider using a more liberal secondary boundary $(d_1, \ldots, d_K)$ where α_Y can be greater than α with the control of FWER at level α.

Assume that the test statistics follow the multivariate normal distribution, which applies asymptotically to a wide range of test statistics. Namely,

$$\begin{pmatrix} X_1 \\ Y_1 \\ X_2 \\ Y_2 \end{pmatrix} \sim N\left(\begin{pmatrix} \lambda\Delta_X \\ \gamma_1\Delta_Y \\ \Delta_X \\ \gamma_2\Delta_Y \end{pmatrix}, \begin{pmatrix} 1 & \frac{\min\{\lambda,\gamma_1\}}{\max\{\lambda,\gamma_1\}}\rho & \lambda & \frac{\min\{\lambda,\gamma_2\}}{\max\{\lambda,\gamma_2\}}\rho \\ \frac{\min\{\lambda,\gamma_1\}}{\max\{\lambda,\gamma_1\}}\rho & 1 & \gamma_1\rho & \gamma_1/\gamma_2 \\ \lambda & \gamma_1\rho & 1 & \gamma_2\rho \\ \frac{\min\{\lambda,\gamma_2\}}{\max\{\lambda,\gamma_2\}}\rho & \gamma_1/\gamma_2 & \gamma_2\rho & 1 \end{pmatrix} \right). \tag{11.1}$$

In the following, Theorem 1 gives an upper bound of type I error of stagewise hierarchical rule $\mathcal{P}_S$. Unlike the results where the primary and the secondary endpoint have the same information fractions (Tamhane et al. 2010; Glimm et al. 2010; Tamhane et al. 2018) or the results with only one interim analysis for the secondary hypothesis H_Y where $\gamma_2 = 1$ (Gou and Xi 2019), the upper bound we provided for multiple interim stages with different information fractions is not sharp. In other words, the following theorem guarantees a more liberal secondary boundary unconditionally.

Theorem 1 (Upper Bound for Type I Error) *When using a stagewise hierarchical rule $\mathcal{P}_S$ under H_Y, the type I error α_Y^S is bounded from above by*

$$\alpha_Y^S < 1 - \Pr(Y_1 \le d_1, Y_2 \le d_2).$$

When $0 < \gamma_1 < \gamma_2 < 1$, this upper bound cannot be achieved.

Specifically, when the primary hypothesis data are obtained earlier than the secondary hypothesis data, at stage 1 we have $n_{1,X} > n_{1,Y}$. It follows that the information fraction of the primary hypothesis at stage 1 is greater than the corresponding information fraction of the secondary hypothesis. Starting from Theorem 1 along with the assumption that $t_X > t_{1,Y}$ and the correlation ρ between X_2 and Y_K is positive, we show in Theorem 2 below that the type I error rate for a stagewise hierarchical test under the secondary hypothesis H_Y, or α_Y^S, is uniformly monotonous.

Theorem 2 (Uniform Monotonicity of Type I Error) *Consider two group sequential designs using $\mathcal{P}_S$, one with the square roots of information fractions $(\lambda, \gamma_1', \gamma_2)$ and boundaries (c_1, c_2, d_1', d_2'), and the other with $(\lambda, \gamma_1'', \gamma_2)$ and (c_1, c_2, d_1'', d_2''). Denote the corresponding type I errors under H_Y by $\alpha_Y^{S'}$ and $\alpha_Y^{S''}$, respectively. Suppose that these two designs share the same boundary for the primary hypothesis (c_1, c_2), and the same information fraction $t_X = \lambda^2$ at the interim analysis of the primary hypothesis and the information fraction $t_{2,Y} = \gamma_2^2$ at the second stage of the secondary hypothesis. If $\gamma_1' \le \gamma_1'' \le \lambda$, $d_1' \ge d_1''$ and*

$d_2' \geq d_2''$, then for any $\rho \in [0, 1]$ and for any Δ_X,

$$\alpha_Y^{S'} \leq \alpha_Y^{S''}.$$

In order to apply Theorem 2 to the OBF-POC design, where an OBF boundary is used for the primary endpoint and a POC boundary is used for the secondary endpoint, we need the following result. The OBF-POC design in the stagewise hierarchical rule is recommended by Tamhane et al. (2010, 2018) and Zhang and Gou (2019b).

Lemma 1 *Consider two trials that use the Pocock test with two stages under the same significance level. In one trial, the interim analysis is performed at information time t', and the corresponding Pocock boundary is d'. In the other trial, the interim analysis is performed at t'' with Pocock boundary d''. If $t' < t''$, then $d' > d''$.*

An immediate consequence of Theorem 2 and Lemma 1 is that, when the information fraction of the secondary hypothesis at the interim analysis is small compared to the information fraction of the primary hypothesis, the statistical power of group sequential design using the stagewise hierarchical rule will benefit greatly from the secondary boundary refinement. Formally, this means that the OBF-POC design with unrefined boundaries becomes more conservative for testing the secondary hypothesis H_Y when the information time at the first stage $t_{1,Y}$ becomes smaller.

Figure 11.1 shows that the error rate α_Y^S under H_Y of an OBF-POC design, where the α-level boundaries (c_1, c_2) and (d_1, d_2) are used, say, $\alpha = 1 - \Pr(X_1 \leq c_1, X_2 \leq c_2) = 1 - \Pr(Y_1 \leq d_1, Y_2 \leq d_2)$. Figure 11.1 confirms the result in Theorem 1 that the error rate α_Y^S is strictly less than α. It also confirms that the uniform monotonicity of α_Y^S as a function of $t_{1,Y}$ in Theorem 2. The error rate α_Y^S of an OBF-OBF design, where both primary and secondary boundary are OBF, is also bounded by α, and is uniformly monotonic of $t_{1,Y}$, as shown in Fig. 11.2. The boundary values (d_1, d_2) can be refined to allow α_Y^S to achieve α.

The secondary boundary can be refined without knowing the correlation ρ between two hypotheses by assuming the least favorable situation where $\rho = 1$. If ρ is known or can be estimated (Tamhane et al. 2012a,b), we can further refine the boundary for the secondary hypothesis. Table 11.1 gives an example of the refined boundary (d_1', d_2') of the secondary hypothesis using OBF-POC and OBF-OBF designs, where $\rho = 1, 0.8, 0.5$. The error rate α_Y^S equals the level of significance α exactly with the boundary refinement of the secondary hypothesis.

Since $\lim_{\Delta_X \to +\infty} \alpha_Y^S(\rho, \Delta_X) = \Pr(Y_1 > d_1)$, for any ρ, λ, γ_1 and γ_2, the refined secondary boundary d_1 in an OBF-POC design is at least z_α, where z_α is the upper α critical point of the standard normal distribution. Note that the naïve strategy in Hung et al. (2007), where the secondary boundary $d_1 = d_2 = z_\alpha$, has been shown to be liberal when the information fractions for the primary and the secondary endpoint are the same. Gou and Xi (2019) first observed that the naïve strategy in Hung et al. (2007) actually control the FWER when the primary and the secondary

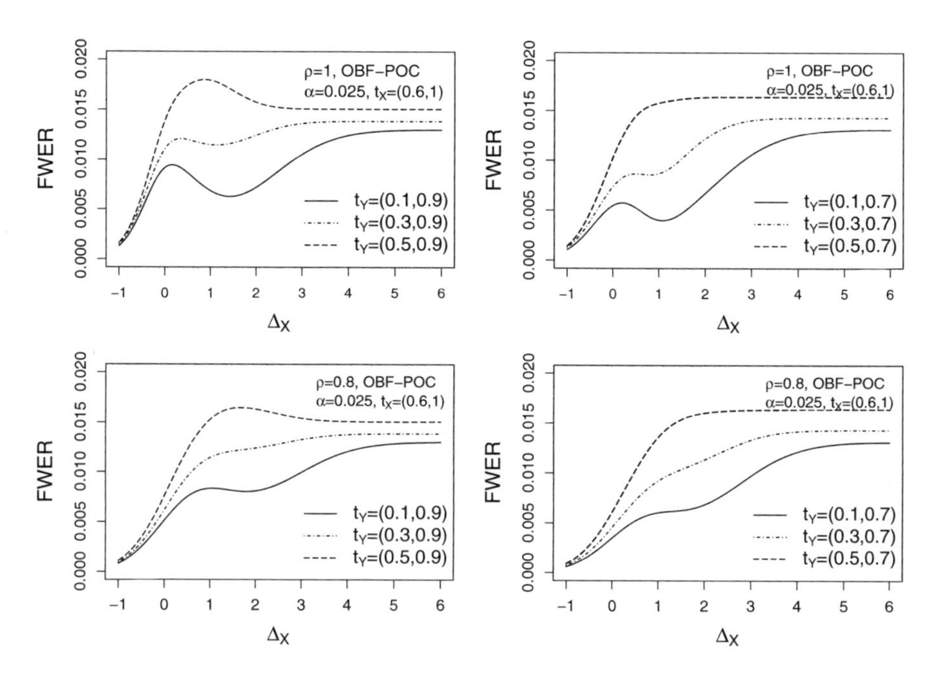

Fig. 11.1 FWER plot for O'Brien-Fleming primary and Pocock secondary boundary under $\mathcal{P}_S$ with $t_X = 0.6$, marginal level of significance $\alpha = 1 - \Pr(X_1 \leq c_1, X_2 \leq c_2) = 1 - \Pr(Y_1 \leq d_1, Y_2 \leq d_2) = 0.025$. Correlation $\rho = 1$ (top panels), $\rho = 0.8$ (bottom panels), $t_{2,Y} = 0.9$ (left panels), $t_{2,Y} = 0.7$ (right panels)

hypothesis have different information fractions, but without further discussion. A natural question here to ask is, when will the FWER inflation of the naïve strategy in Hung et al. (2007) not happen? Under an OBF-POC design, where an α-size OBF boundary (c_1, c_2) is chosen for the primary endpoint, and the boundary for the secondary endpoint is $d_1 = d_2 = z_\alpha$, Fig. 11.3 shows the admissible region of $(t_{1,Y}, t_{2,Y})$ for controlling the FWER of the naïve strategy in Hung et al. (2007) for different choices of the information fractions at the interim analysis of the primary hypothesis. The feasible region of $(t_{1,Y}, t_{2,Y})$ becomes larger when t_X increases. Generally speaking, when $(t_{1,Y}, t_{2,Y})$ are small enough compared with t_X, the naïve strategy controls the FWER. For example, in a phase III trial in Baselga et al. (2012), the primary endpoint is PFS with information fraction $\mathbf{t}_X = (0.6, 1)$, and the key secondary endpoint is OS with $\mathbf{t}_Y = (0.21, 0.44)$. If this trial follows the stagewise hierarchical strategy to control the FWER at level $\alpha = 0.025$ and uses an α-level OBF boundary for the PFS endpoint, then the boundary $d_1 = d_2 = z_\alpha = 1.960$ for the OS can be used since $t_{1,Y} = 0.21$ and $t_{2,Y} = 0.44$ fall into the admissible region when $t_X = 0.6$. This is shown in Fig. 11.3.

A simple empirical rule for properly using the naïve strategy in Hung et al. (2007) is followed: when $t_{1,Y}^2 \leq t_X$, a group sequential design with an 0.025-level OBF boundary for the primary hypothesis can directly apply $d_1 = d_2 = z_{0.025}$ as its boundary for the secondary hypothesis H_Y.

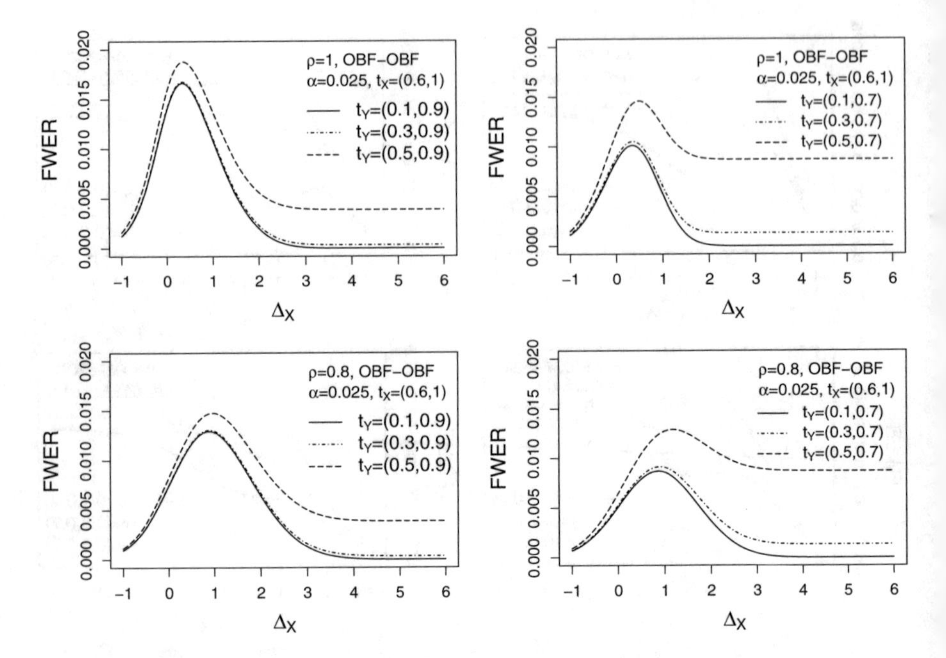

Fig. 11.2 FWER plot for O'Brien-Fleming primary and O'Brien-Fleming secondary boundary under $\mathcal{P}_S$ with $t_X = 0.6$, marginal level of significance $\alpha = 1 - \Pr(X_1 \le c_1, X_2 \le c_2) = 1 - \Pr(Y_1 \le d_1, Y_2 \le d_2) = 0.025$. Correlation $\rho = 1$ (top panels), $\rho = 0.8$ (bottom panels), $t_{2,Y} = 0.9$ (left panels), $t_{2,Y} = 0.7$ (right panels)

Table 11.1 Refined secondary boundaries for given correlation ρ under the stagewise hierarchical rule

OBF-POC	α-level boundary		Refined boundary		
ρ	d_1	d_2	d_1'	d_2'	Marginal error of H_Y
1	2.169	2.169	2.032	2.032	0.0345
0.8	2.169	2.169	1.996	1.996	0.0375
0.5	2.169	2.169	1.973	1.973	0.0394
OBF-OBF	α-level boundary		Refined boundary		
ρ	d_1	d_2	d_1'	d_2'	Marginal error of H_Y
1	2.664	1.985	2.511	1.872	0.0328
0.8	2.664	1.985	2.386	1.778	0.0408
0.5	2.664	1.985	2.308	1.721	0.0465

$t_X = 0.6$, $t_{1,Y} = 0.5$, $t_{2,Y} = 0.9$, the OBF boundary for the primary hypothesis is $c_1 = 2.572$, $c_2 = 1.992$ at $\alpha = 0.025$. The marginal error rate of H_Y is $1 - \Pr(Y_1 \le d_1', Y_2 \le d_2')$

In a $[K_X|K_Y]$-stage design following the stagewise hierarchical rule, similar conclusions on type I error rate can be achieved. The type I error rate α_Y^S is bounded from above by $1 - \Pr(Y_1 \le d_1, \ldots, Y_{K_X \wedge K_Y} \le d_{K_X \wedge K_Y})$, and this upper bound is

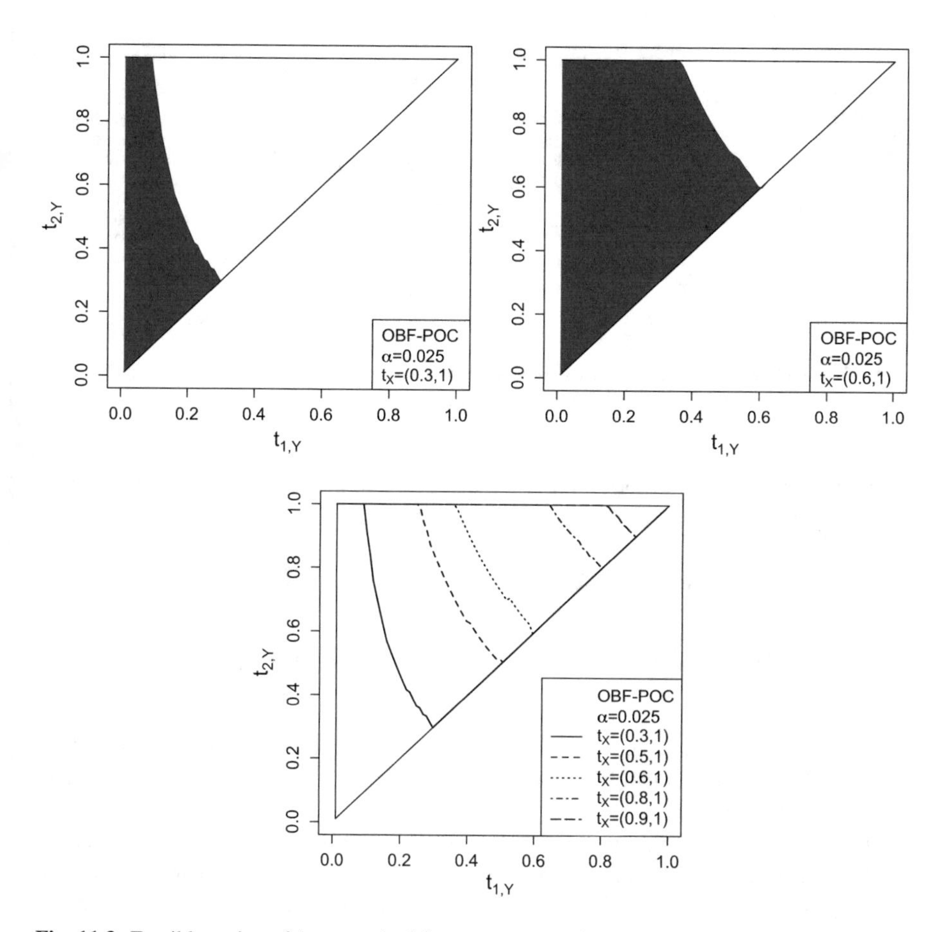

Fig. 11.3 Feasible region of $(t_{1,Y}, t_{2,Y})$ of the naïve strategy in Hung et al. (2007)

not sharp when $K_X \neq K_Y$. Under some conditions, the power gain for the secondary hypothesis H_Y by using the boundary refinement is significant when the information times of H_Y are less than the information times of the primary hypothesis H_X at interim stages.

11.4 Overall Hierarchical Rule

Compared with the stagewise hierarchical rule $\mathcal{P}_S$, a trial design using the overall hierarchical rule $\mathcal{P}_O$ allows testing the secondary hypothesis H_Y more than once if the primary hypothesis H_X is rejected. Following a similar argument in Tamhane et al. (2018), one cannot refine the secondary boundary unless there

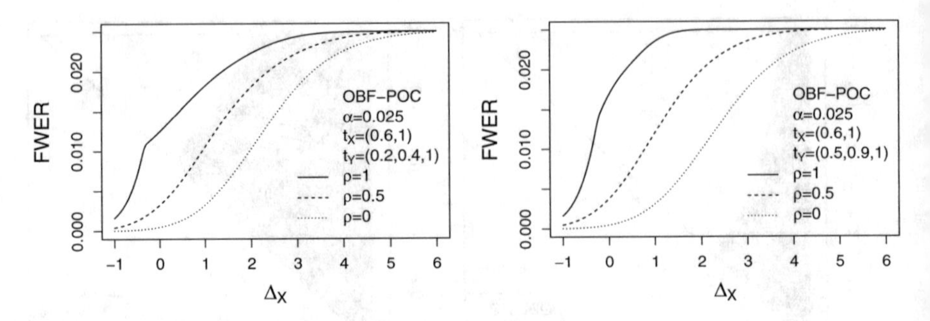

Fig. 11.4 FWER plot of an OBF-POC design using the overall hierarchical rule $\mathcal{P}_O$ with α-level boundary of the primary and the secondary hypothesis

is some prior information on Δ_X and ρ, since the difference between $1 - \Pr(Y_1 \leq d_1, \cdots, Y_K \leq d_K)$ and α_Y^O, which equals to

$$\Pr(X_1 \leq c_1, X_2 \leq c_2, Y_2 > d_2) + \Pr(X_1 \leq c_1, Y_1 > c_1, Y_2 \leq d_2, \cdots, Y_K \leq d_K)$$

$$+ \sum_{i=2}^{K-1} \Pr(X_1 \leq c_1, X_2 \leq c_2, Y_2 \leq d_2, \cdots, Y_i \leq d_i, Y_{i+1} > d_{i+1}),$$

in a $[2|K]$-stage design, goes to 0 when Δ_X goes to positive infinity. Similarly, a $[K_X|K_Y]$-stage design using the overall hierarchical rule cannot be refined without information on Δ_X and ρ.

Figure 11.4 shows the type I error under H_Y of an OBF-POC design with $\alpha = 0.025$. Refinement of the secondary boundary is possible only when an upper bound on Δ_X is known. If a reliable estimate of Δ_X is available, the refinement of the boundary of the secondary hypothesis will be relatively noticeable when the time fraction of the secondary hypothesis $\mathbf{t}_Y$ is small or when the correlation ρ is small.

11.5 Partially Hierarchical Rule

The partially hierarchical rule $\mathcal{P}_P$ allows continued testing of the secondary hypothesis when the primary hypothesis has been confirmed to be non-significant. Thus, besides controlling of type I error under H_Y, one needs to also control the type I error under $H_X \cap H_Y$. Since $\Pr(R_X|H_X) \leq \Pr(R_X \cup R_Y|H_X \cap H_Y)$, in general we cannot use an α-level significance for the primary endpoint in a design under the partially hierarchical rule.

A Bonferroni-based design splits the significance level α for H_X and H_Y whereby $\alpha = \alpha_X + \alpha_Y$, where $\alpha_X = 1 - \Pr(X_1 \leq c_1, X_2 \leq c_2)$ and $\alpha_Y = 1 - \Pr(Y_1 \leq d_1, \cdots, Y_K \leq d_K)$ in a $[2|K]$-stage design. This design controls the

FWER. When the correlation ρ is known or can be estimated, the boundary for the secondary hypothesis can be refined. The following theorem provides a refinement when ρ is known to be non-negative.

Theorem 3 (Improved Boundary in a [2|K]-Stage Design) *Consider a group sequential design using the partially hierarchical rule* $\mathcal{P}_P$, *where* $\alpha_X = \Pr(R_X|H_X) < \alpha$. *A necessary and sufficient condition for* $\Pr(R_X \cup R_Y|H_X \cap H_Y) \le \alpha$ *for any non-negative* ρ *is that* $\Pr\left(\cap_{i=2}^{K}\{Y_i \le d_i\}\right) \ge (1-\alpha)/(1-\alpha_X)$.

Based on Theorem 3, a simple design is followed when $\rho \ge 0$ is satisfied, where the boundary of the secondary hypothesis is refined.

1. $\alpha_X = 1 - \Pr(X_1 \le c_1, X_2 \le c_2) \le \alpha$,
2. $\alpha_Y = 1 - \Pr(Y_1 \le d_1, \cdots, Y_K \le d_K) \le \alpha$,
3. $1 - \Pr\left(\cap_{i=2}^{K}\{Y_i \le d_i\}\right) \le \frac{\alpha-\alpha_X}{1-\alpha_X}$.

Denote $\alpha_Y^{(-1)} = 1 - \Pr\left(\cap_{i=2}^{K}\{Y_i \le d_i\}\right)$, which is the type I error of $(K-1)$-stage group sequential design for H_Y. Comparing the original K-stage group sequential design for H_Y, this $(K-1)$-stage design skips the first stage. Therefore, the conditions can be rewritten as

$$\alpha_X \le \alpha, \quad \alpha_Y \le \alpha, \quad \alpha_Y^{(-1)} \le \frac{\alpha - \alpha_X}{1-\alpha_X}.$$

Theorem 3 can be easily generalized to a $[K_X|K_Y]$-stage design where the trial of the primary endpoint has more than two stages. The refined method maintains the FWER control across both endpoints.

Corollary 1 (Improved Boundary in a [$\mathbf{K_X}|K_Y$]-Stage Design) *Consider a group sequential design with* K_X *stages for the primary hypothesis and* K_Y *stages for the secondary hypothesis, using the partially hierarchical rule* $\mathcal{P}_P$, *where* $K_X \le K_Y$. *Let* $\alpha_Y^{(-(K_X-1))} = 1 - \Pr\left(\cap_{i=K_X}^{K_Y}\{Y_i \le d_i\}\right)$ *be the type I error of a* $(K_Y - K_X + 1)$*-stage group sequential design for* H_Y. *This procedure controls the FWER for arbitrary* $\rho \ge 0$ *if and only if:* $\alpha_X \le \alpha$, $\alpha_Y \le \alpha$, *and* $\alpha_Y^{(-(K_X-1))} \le (\alpha - \alpha_X)/(1-\alpha_X)$.

The Lan-DeMets error spending function is widely used in clinical trials to approximate OBF and POC boundary (Lan and DeMets 1983). Using the Lan-DeMets boundaries, Table 11.2 shows the refined boundary for the secondary hypothesis under various values of ρ compared with the boundary based on Bonferroni split. The refined boundary for $\rho = 0$ can be used for any $\rho \ge 0$, based on Theorem 3. Even without the knowledge of the sign of ρ, the refined boundary for $\rho = 0$ is still approximately valid for endpoints with any correlation ρ, as shown in Table 11.2.

Table 11.2 Refined Lan-DeMets boundaries for the secondary hypothesis for given correlation ρ under the partially hierarchical rule $\mathcal{P}_P$, $\alpha = 0.025$, $\alpha_X = 0.0125$, $\mathbf{t}_X = (0.6, 1)$, $\mathbf{t}_Y = (0.5, 0.9, 1)$, Lan-DeMets OBF boundary for the primary hypothesis $(c_1, c_2) = (3.021, 2.254)$

	Lan-DeMets POC for the secondary				Lan-DeMets OBF for the secondary			
ρ	d_1	d_2	d_3	α_Y	d_1	d_2	d_3	α_Y
1	2.157	2.242	2.373	0.0250	2.963	2.105	2.057	0.0250
0.5	2.184	2.270	2.402	0.0234	3.237	2.313	2.249	0.0153
0	2.230	2.319	2.452	0.0208	3.304	2.363	2.295	0.0135
−0.5	2.235	2.324	2.457	0.0205	3.310	2.368	2.299	0.0133
Bonferroni	2.420	2.530	2.656	0.0125	3.345	2.394	2.323	0.0125

11.6 Power Analysis

In order to evaluate the performance of boundary refinement for the secondary hypothesis, we compare the secondary power $\Pr\left(R_Y|\overline{H}_Y\right)$ under the partially hierarchical rule $\mathcal{P}_P$ between the OBF-POC design and OBF-OBF design. Here, we only consider the O'Brien-Fleming boundary for the primary endpoint, since it is more powerful than the POC boundary for the primary hypothesis (Tamhane et al. 2018). For the power analysis, the assumption of multivariate normal distribution is satisfied asymptotically, so we incorporate the distribution information into the analysis. In general, if the distribution information is unknown, the power analysis models based on the Dirac function (Finner et al. 2009) or the step function (Zhang and Gou 2016) can be considered.

Table 11.3 displays the power comparisons between two designs (OBF-POC and OBF-OBF) and between two boundaries (refined boundary for $\rho \geq 0$ and unrefined boundary based on Bonferroni split). We assume the significance level $\alpha = 0.025$, and the primary hypothesis is tested with a 0.0125-level Lan-DeMets OBF boundary $(c_1, c_2) = (3.021, 2.254)$, where the information fraction at the interim analysis is 0.6. For the secondary hypothesis, we include the Lan-DeMets OBF and the Lan-DeMets POC boundary. Two choices of information fractions of the secondary endpoint show the impact of a fast data accumulation ($\mathbf{t}_Y = (0.5, 0.9, 1)$) and slow accumulation ($\mathbf{t}_Y = (0.2, 0.4, 1)$) for the secondary hypothesis. We assume the true correlation between the primary and the secondary hypothesis is 0.5. Note that we do not need to know this correlation for boundary refinement. The standardized treatment effect for the primary hypothesis Δ_X is 3, and it ranges from 2 to 4 for the secondary hypothesis.

From Table 11.3, we observe that the OBF-OBF design is better than the OBF-POC design in a group sequential trial using the partially hierarchical rule $\mathcal{P}_P$. Note that for a trial using the stagewise hierarchical rule $\mathcal{P}_S$, Tamhane et al. (2010), Tamhane et al. (2018) and Gou and Xi (2019) have shown that the OBF-POC is the better choice. For the OBF-OBF design using $\mathcal{P}_P$, the power gain over the Bonferroni split method increases when the information fractions of the secondary hypothesis become smaller.

Table 11.3 Power (%) comparison between the refined the unrefined boundary under the partially hierarchical rule

$\mathbf{t}_X = (0.6, 1)$		Lan-DeMets OBF-POC		Lan-DeMets OBF-OBF	
$\mathbf{t}_Y$	Δ_Y	Refined	Unrefined	Refined	Unrefined
	2	39.1	31.7	40.7	39.5
(0.5, 0.9, 1)	3	75.4	68.6	77.5	76.6
	4	95.2	92.8	96.0	95.7
	2	38.5	34.1	44.8	40.4
(0.2, 0.4, 1)	3	75.5	71.7	80.7	77.6
	4	95.4	94.1	96.9	96.1

11.7 Example and Extension

In practice, it is common that the attained sample sizes and the planned sample sizes are different. Using the error spending function, we can update the boundaries at each stage by considering the exact information fractions. The refined boundary can be updated in a similar manner adaptively.

Consider a phase III placebo-controlled two-arm clinical trial evaluating the efficacy of a treatment in patients with lymphoma. The primary objective is to evaluate the efficacy with respect to the progression-free survival (PFS). The secondary objective is to evaluate the efficacy with respect to the overall survival (OS). Table 11.4 shows a 0.025-level test using the partially hierarchical rule with a Lan-DeMets error spending function OBF-OBF design. The trial design includes one interim analysis for the primary endpoint PFS, and two interim analyses for the secondary endpoint OS. At stage 0, all sample sizes are planned. The sample size per arm is planned to be 400. The planned cumulative sample size for the primary objective is 240 at stage 1, and 400 at stage 2. For the secondary objective, the planned cumulative sample size is 200 at stage 1320 at stage 2, and 400 at stage 3. The critical boundaries for the primary and the secondary hypothesis can be calculated. At stage 1, $n_{1,X}$ and $n_{1,Y}$ are obtained, and the planned sample sizes for other stages are modified accordingly. The observed sample sizes at stage 1 for the primary and the secondary endpoint are 264 and 168, and the planned cumulative sample sizes at stage 2 and 3 remain the same. The critical boundary (c_1, c_2) and (d_1, d_2, d_3) are recalculated, and c_1 and d_1 are compared with the test statistics to make decisions. We further observe $n_{2,X}$ and $n_{2,Y}$ at stage 2, update the information times by using the observed cumulative sample sizes, and calculate the boundary c_2, c_3 and (d_2, d_3) by fixing the value of c_1 and c_2 in stage 1. Finally, $n_{3,Y}$ is observed at stage 3, and the total sample size for OS is updated, and the boundary d_3 is recalculated based on updated information times. In this example, initially the planned sample size is $(n_{1,X}, n_{2,X}, n_{1,Y}, n_{2,Y}, n_{3,Y}) = (240, 160, 200, 120, 80)$. At the final stage, the attained sample size becomes $(n_{1,X}, n_{2,X}, n_{1,Y}, n_{2,Y}, n_{3,Y}) = (264, 168, 168, 132, 108)$.

Table 11.4 Boundary updates among stages in an OBF-OBF design using $\mathcal{P}_P$: a comparison between the unrefined boundary (d_1, d_2, d_3) and the refined boundary (d'_1, d'_2, d'_3)

Stage	$n_{1,X}$	$n_{2,X}$	$n_{1,Y}$	$n_{2,Y}$	$n_{3,Y}$	c_1	c_2	d_1	d_2	d_3	d'_1	d'_2	d'_3
0	240	160	200	120	80	3.0205	2.2543	3.3446	2.5694	2.2938	3.2314	2.4794	2.2148
1	264	136	168	152	80	2.8614	2.2625	3.6810	2.5629	2.2928	3.5651	2.4770	2.2180
2	264	168	168	132	100	2.8614	2.2672	3.6810	2.6625	2.2835	3.5651	2.6529	2.2754
3	264	168	168	132	108	2.8614	2.2672	3.6810	2.6625	2.3170	3.5651	2.6529	2.3136

Acknowledgements We thank Ajit C. Tamhane and Dong Xi for comments that greatly improved the manuscript. This work was partially supported by the Professional Staff Congress-City University of New York (PSC-CUNY) research grant, Cycle 48 (2017–2018). This research article extended the framework that was present at the 2017 ICSA Applied Statistics Symposium, Session 148, Recent Developments in Theory and Application of Multiple Comparison Methods, *A gatekeeping test on a primary and a secondary endpoint in a group sequential design*, by Dr. Ajit C. Tamhane. It was also present at the 2017 ICSA Applied Statistics Symposium, Session 121, Multiplicity in Clinical Trials, *A gatekeeping test in a group sequential design with multiple interim looks*, by Dr. Jiangtao Gou. The authors thank editor Dr. Lanju Zhang and an anonymous referee for suggestions that improved this paper.

Conflict of Interest

The authors have declared no conflict of interest.

Appendix

Proof of Theorem 1 Note that $1 - \Pr(Y_1 \le d_1, Y_2 \le d_2) - \alpha_Y^S = \Pr(X_1 > c_1, Y_1 \le d_1, Y_2 > d_2) + \Pr(X_1 \le c_1, X_2 \le c_2, Y_2 > d_2) + \Pr(X_1 \le c_1, Y_1 > d_1, Y_2 \le d_2)$. All three terms on the right hand side are strictly positive when $\rho < 1$. When $\rho = 1$, the probability $\Pr(X_1 > c_1, Y_1 \le d_1, Y_2 > d_2)$ and $\Pr(X_1 \le c_1, Y_1 > d_1, Y_2 \le d_2)$ can be 0 if $\lambda = \gamma_1$, and the probability $\Pr(X_1 \le c_1, X_2 \le c_2, Y_2 > d_2)$ can be 0 if $\lambda = \gamma_2$. Since $\gamma_1 < \gamma_2$, these three terms cannot be 0 at the same time. It follows that $1 - \Pr(Y_1 \le d_1, Y_2 \le d_2)$ is strictly greater than α_Y^S. □

Proof of Theorem 2 Under $\overline{H}_X \cap H_Y$, the standardized treatment effects at the final stage for the secondary endpoint is zero, namely, $\Delta_Y = 0$. For simplicity, we denote the non-centrality parameters for the primary endpoint by $\Delta = \Delta_X$ under $\overline{H}_1 \cap H_2$. The type I error rate with smaller information fraction at stage 1 of the secondary hypothesis is

$$\begin{aligned}\alpha_Y^{S'} &= \Pr\left(X_1 > c_1, Y'_1 > d_1\right) + \Pr\left(X_1 \le c_1, X_2 > c_2, Y'_2 > d_2\right) \\ &= \Pr\left(X_1 - \lambda\Delta > c_1 - \lambda\Delta, Y'_1 > d_1\right) \\ &\quad + \Pr\left(X_1 - \lambda\Delta \le c_1 - \lambda\Delta, X_2 - \Delta > c_2 - \Delta, Y'_2 > d_2\right)\end{aligned}$$

For the first term, note that $\mathrm{corr}\,(X_1, Y_1) = \gamma_1\rho/\lambda$. Since $\gamma_1' < \gamma_1''$ and $\rho \geq 0$, we have $\mathrm{corr}\left(X_1, Y_1'\right) \leq \mathrm{corr}\left(X_1, Y_1''\right)$. By Slepian's inequality (Plackett 1954; Slepian 1962) and $d_1' > d_1''$, it follows that

$$\Pr\left(X_1 - \lambda\Delta > c_1 - \lambda\Delta, Y_1' > d_1'\right) \leq \Pr\left(X_1 - \lambda\Delta > c_1 - \lambda\Delta, Y_1'' > d_1''\right).$$

For the second term, note that $\mathrm{corr}\,(X_1, X_2) = \lambda$, $\mathrm{corr}\,(X_2, Y_2) = \gamma_2\rho$, $\mathrm{corr}\,(X_1, Y_2) = \rho \cdot \min\{\lambda, \gamma_2\}/\max\{\lambda, \gamma_2\}$, which are the same for the two designs. Since $d_2' \geq d_2''$, we get

$$\Pr\left(X_1 - \lambda\Delta \leq c_1 - \lambda\Delta, X_2 - \Delta > c_2 - \Delta, Y_2 > d_2'\right)$$
$$\leq \Pr\left(X_1 - \lambda\Delta \leq c_1 - \lambda\Delta, X_2 - \Delta > c_2 - \Delta, Y_2 > d_2''\right).$$

Thus $\alpha_Y^{S'} \leq \alpha_Y^{S''}$, for any $0 \leq \rho \leq 1$. □

Proof of Lemma 1 Suppose that $\left(Y_1', Y_2'\right)$ and $\left(Y_1'', Y_2''\right)$ are the bivariate normal distributed test statistics under the null hypothesis. The correlation between Y_1' and Y_2' is $\sqrt{t'}$, and the correlation between Y_1'' and Y_2'' is $\sqrt{t''}$. Since two trials have the same significance level α, we have

$$\Pr\left(Y_1' \leq d', Y_2' \leq d'\right) = 1 - \alpha = \Pr\left(Y_1'' \leq d'', Y_2'' \leq d''\right).$$

Since $\sqrt{t'} < \sqrt{t''}$, by Slepian's inequality, we get

$$\Pr\left(Y_1' \leq d', Y_2' \leq d'\right) < \Pr\left(Y_1'' \leq d', Y_2'' \leq d'\right).$$

It follows that

$$\Pr\left(Y_1'' \leq d', Y_2'' \leq d'\right) > \Pr\left(Y_1'' \leq d'', Y_2'' \leq d''\right).$$

Clearly, we have

$$d' > d''.$$

□

Proof of Theorem 3 For a design using the partially hierarchical rule $\mathcal{P}_P$, the error rate
$\Pr\left(R_X \cup R_Y | H_X \cap H_Y\right)$ is greater than $\Pr\left(R_X | H_X\right)$. The difference is bounded by

$$\Pr\left(R_X \cup R_Y | H_X \cap H_Y\right) - \Pr\left(R_X | H_X\right)$$
$$= \Pr\left(X_1 \leq c_1, X_2 \leq c_2, Y_2 > d_2\right)$$

$$+\sum_{i=2}^{K-1} \Pr(X_1 \le c_1, X_2 \le c_2, Y_2 \le d_2, \cdots, Y_i \le d_i, Y_{i+1} > d_{i+1})$$

$$= \Pr(X_1 \le c_1, X_2 \le c_2) - \Pr(X_1 \le c_1, X_2 \le c_2, Y_2 \le d_2, \cdots, Y_K \le d_K)$$

$$\le \Pr(X_1 \le c_1, X_2 \le c_2) - \Pr(X_1 \le c_1, X_2 \le c_2) \Pr\left(\cap_{i=2}^K \{Y_i \le d_i\}\right),$$

where $\Pr(X_1 \le c_1, X_2 \le c_2, Y_2 \le d_2, \cdots, Y_K \le d_K) \ge \Pr(X_1 \le c_1, X_2 \le c_2)$ $\Pr\left(\cap_{i=2}^K \{Y_i \le d_i\}\right)$ holds for any non-negative ρ, and two sides are equal when $\rho = 0$. It follows that

$$\Pr(R_X \cup R_Y | H_X \cap H_Y) - \alpha_X \le (1 - \alpha_X)\left(1 - \Pr\left(\cap_{i=2}^K \{Y_i \le d_i\}\right)\right).$$

Also note that if

$$(1 - \alpha_X)\left(1 - \Pr\left(\cap_{i=2}^K \{Y_i \le d_i\}\right)\right) \le \alpha - \alpha_X$$

the error rate control under intersection hypothesis, which is $\Pr(R_X \cup R_Y | H_X \cap H_Y) \le \alpha$, is guaranteed. Thus, if

$$\Pr\left(\cap_{i=2}^K \{Y_i \le d_i\}\right) \ge \frac{1 - \alpha}{1 - \alpha_X},$$

then $\Pr(R_X \cup R_Y | H_X \cap H_Y) \le \alpha$ for any $\rho \ge 0$. □

References

Amir, E., Seruga, B., Kwong, R., Tannock, I. F., Ocaña, A.: Poor correlation between progression-free and overall survival in modern clinical trials: are composite endpoints the answer? Eur. J. Cancer **48**, 385–388 (2012)

Baselga, J., Campone, M., Piccart, M., Burris III, H. A., Rugo, H. S., Sahmoud, T., Noguchi, S., Gnant, M., Pritchard, K. I., Lebrun, F., et al.: Everolimus in postmenopausal hormone-receptor–positive advanced breast cancer. N. Engl. J. Med. **366**, 520–529 (2012)

Bretz, F., Maurer, W., Brannath, W., Posch, M.: A graphical approach to sequentially rejective multiple test procedures. Stat. Med. **28**, 586–604 (2009)

Bretz, F., Posch, M., Glimm, E., Klinglmueller, F., Maurer, W., Rohmeyer, K.: Graphical approaches for multiple comparison procedures using weighted bonferroni, simes, or parametric tests. Biom. J. **53**, 894–913 (2011)

Burman, C.-F., Sonesson, C., Guilbaud, O.: A recycling framework for the construction of bonferroni-based multiple tests. Stat. Med. **28**, 739–761 (2009)

Dmitrienko, A., Tamhane, A.C.: Gatekeeping procedures with clinical trial applications. Pharm. Stat. **6**, 171–180 (2007)

Dmitrienko, A., Tamhane, A.C., Bretz, F.: Multiple Testing Problems in Pharmaceutical Statistics. Taylor & Francis, Boca Raton (2009)

Dunnett, C.W., Tamhane, A.C.: A step-up multiple test procedure. J. Am. Stat. Assoc. **87**, 162–170 (1992)

Finner, H., Dickhaus, T., Roters, M.: On the false discovery rate and an asymptotically optimal rejection curve. Ann. Stat. **37**, 596–618 (2009)

Fiteni, F., Westeel, V., Pivot, X., Borg, C., Vernerey, D., and Bonnetain, F.: Endpoints in cancer clinical trials. J. Visc. Surg. **151**, 17–22 (2014)

Glimm, E., Maurer, W., Bretz, F.: Hierarchical testing of multiple endpoints in group-sequential trials. Stat. Med. **29**, 219–228 (2010)

Gou, J., Tamhane, A.C. : On generalized Simes critical constants. Biom. J. **56**, 1035–1054 (2014)

Gou, J., Tamhane, A.C.: A flexible choice of critical constants for the improved hybrid Hochberg–Hommel procedure. Sankhya B **80**, 85–97 (2018a)

Gou, J., Tamhane, A.C.: Hochberg procedure under negative dependence. Stat. Sin. **28**, 339–362 (2018b)

Gou, J., Tamhane, A.C., Xi, D., Rom, D.: A class of improved hybrid Hochberg–Hommel type step-up multiple test procedures. Biometrika **101**, 899–911 (2014)

Gou, J., Xi, D.: Hierarchical testing of a primary and a secondary endpoint in a group sequential design with different information times. Stat. Biopharm. Res. (2019). https://doi.org/10.1080/19466315.2018.1546613

Hochberg, Y.: A sharper Bonferroni procedure for multiple tests of significance. Biometrika **75**, 800–802 (1988)

Hochberg, Y., Tamhane, A.C.: Multiple Comparison Procedures. Wiley, New York (1987)

Holm, S.: A simple sequentially rejective multiple test procedure. Scand. J. Stat. **6**, 65–70 (1979)

Hommel, G.: A stagewise rejective multiple test procedure based on a modified Bonferroni test. Biometrika **75**, 383–386 (1988)

Hung, H.M.J., Wang, S.-J., O'Neill, R.: Statistical considerations for testing multiple endpoints in group sequential or adaptive clinical trials. J. Biopharm. Stat. **17**, 1201–1210 (2007)

Jennison, C., Turnbull, B.W.: Group sequential tests for bivariate response: interim analyses of clinical trials with both efficacy and safety endpoints. Biometrics **49**, 741–752 (1993)

Jennison, C., Turnbull, B.W.: Group Sequential Methods with Applications to Clinical Trials. Chapman and Hall/CRC, New York (2000)

Lan, K.K.G., DeMets, D.L.: Discrete sequential boundaries for clinical trials. Biometrika **70**, 659–663 (1983)

Lan, K.K.G., DeMets, D.L.: Group sequential procedures: calendar versus information time. Stat. Med. **8**, 1191–1198 (1989)

Marcus, R., Peritz, E., Gabriel, K.R.: On closed testing procedures with special reference to ordered analysis of variance. Biometrika **63**, 655–660 (1976)

Maurer, W., Bretz, F.: Multiple testing in group sequential trials using graphical approaches. Stat. Biopharm. Res. **5**, 311–320 (2013)

Michiels, S., Saad, E.D., Buyse, M.: Progression-free survival as a surrogate for overall survival in clinical trials of targeted therapy in advanced solid tumors. Drugs **77**, 713–719 (2017)

O'Brien, P.C., Fleming, T.R.: A multiple testing procedure for clinical trials. Biometrics **35**, 549–556 (1979)

Plackett, R.L.: A reduction formula for normal multivariate integrals. Biometrika **41**, 351–360 (1954)

Pocock, S.J.: Group sequential methods in the design and analysis of clinical trials. Biometrika **64**, 191–199 (1977)

Rom, D.M.: A sequentially rejective test procedure based on a modified Bonferroni inequality. Biometrika **77**, 663–665 (1990)

Sarkar, S.K.: Generalizing Simes' test and Hochberg's stepup procedure. Ann. Stat. **36**, 337–363 (2008)

Simes, R.J.: An improved Bonferroni procedure for multiple tests of significance. Biometrika **73**, 751–754 (1986)

Slepian, D.: The one-sided barrier problem for gaussian noise. Bell Syst. Tech. J. **41**, 463–501 (1962)

Tamhane, A.C., Gou, J.: Advances in p-value based multiple test procedures. J. Biopharm. Stat. **28**, 10–27 (2018)
Tamhane, A.C., Gou, J., Jennison, C., Mehta, C.R., Curto, T.: A gatekeeping procedure to test a primary and a secondary endpoint in a group sequential design with multiple interim looks. Biometrics **74**, 40–48 (2018)
Tamhane, A.C., Mehta, C.R., Liu, L.: Testing a primary and a secondary endpoint in a group sequential design. Biometrics **66**, 1174–1184 (2010)
Tamhane, A.C., Wu, Y., Mehta, C.R.: Adaptive extensions of a two-stage group sequential procedure for testing primary and secondary endpoints (I): unknown correlation between the endpoints. Stat. Med. **31**, 2027–2040 (2012a)
Tamhane, A.C., Wu, Y., Mehta, C.R.: Adaptive extensions of a two-stage group sequential procedure for testing primary and secondary endpoints (II): sample size re-estimation. Stat. Med. **31**, 2041–2054 (2012b)
Tang, D.-I., Geller, N.L.: Closed testing procedures for group sequential clinical trials with multiple endpoints. Biometrics **55**, 1188–1192 (1999)
Xi, D., Tamhane, A.C.: Allocating recycled significance levels in group sequential procedures for multiple endpoints. Biom. J. **57**, 90–107 (2015)
Ye, Y., Li, A., Liu, L., Yao, B.: A group sequential holm procedure with multiple primary endpoints. Stat. Med. **32**, 1112–1124 (2013)
Zhang, F., Gou, J.: A p-value model for theoretical power analysis and its applications in multiple testing procedures. BMC Med. Res. Methodol. **16**, 135 (2016)
Zhang, F., Gou, J.: Control of false positive rates in clusterwise fMRI inferences. J. Appl. Stat. (2019a). https://doi.org/10.1080/02664763.2019.1573883
Zhang, F., Gou, J.: Refined critical boundary with enhanced statistical power for non-directional two-sided tests in group sequential designs with multiple endpoints. (2019b) (submitted)

Chapter 12
Bayesian Dose Escalation Study Design with Consideration of Both Early and Late Onset Toxicity

Li Liu, Lei Gao, and Glen Laird

12.1 Introduction

In phase I oncology clinical trials, the goal is to identify safe doses that will be investigated in further clinical trials. The safety profile is normally measurable by rate of dose limiting toxicity (DLT), which is "traditionally defined by grade 3/4 non-hematological or grade 4 hematological toxicity at least possibly related to the treatment, occurring during the first cycle of treatment" with some adjustments (Paoletti et al. 2014). To conduct a dose finding study, an algorithm can be used to guide dose escalation and patient allocation until the maximum tolerated dose (MTD) with predefined DLT rate is observed. Common algorithms include rule based designs such as 3 + 3 designs, and model based designs such as continual reassessment method (CRM) designs (Berry et al. 2010; O'Quigley et al. 1990).

Given the changing landscape in oncology drug development, there may be several problems with the traditional designs for dose finding studies. The concept of DLT was initially introduced in the landscape of cytotoxic therapies, which could induce "irreversible lethal cellular damage following short-term exposure" (EMA 2017). However, these toxicity properties are not applicable to targeted therapy, for which toxic effects are likely mild and reversible. In particular, targeted therapies could cause late onset toxicities due to prolonged use (EMA 2017). Furthermore, it is operationally infeasible to wait for a long period to allow late onset toxicity to occur before making a dose escalation decision with newly enrolled patients. To account for this, many authors have considered using a time to event model

L. Liu (✉)
Sanofi, Bridgewater, NJ, USA
e-mail: li.liu@sanofi.com

L. Gao · G. Laird
Vertex, Boston, MA, USA

L. Zhang et al. (eds.), *Contemporary Biostatistics with Biopharmaceutical Applications*, ICSA Book Series in Statistics,
https://doi.org/10.1007/978-3-030-15310-6_12

to guide dose finding algorithms by censoring patients without late onset toxicity at the time of dose escalation decision (Braun 2006; Cheung and Chappell 2000; etc). Regarding the CRM, it has been observed that the CRM method tends to overshoot the MTD (Goodman et al. 1995), and thus measures have been introduced to escalate with over dose control (EWOC). For binary toxicity, EWOC has been implemented by Babb et al. (1998) and a Bayesian logistic regression model (BLRM) by Neuenschwander et al. (2008). For time to event toxicity, models are available by Mauguen et al. 2011 and Tighiouart et al. 2014.

It is possible that both early and late onset toxicities are relevant to a targeted therapy, and knowledge of these toxic effects is useful to optimize the safety of the therapy (Varricchi et al. 2017). However, according to our knowledge, there is no readily available approach to handle dose finding studies with both early and late onset toxicity. In this paper, we address this problem by proposing to define DLT with two components, one for more immediate toxicity in a binary model, and the other for late onset toxicity in a time to event model. Note that the types of the early toxicity and late onset toxicity are different. In addition, we use Bayesian EWOC methods on both endpoints to calculate the toxicity probability without waiting for the final late onset toxicity, while providing protection to the patients from overdosing. This approach is simple and can address the aforementioned problems simultaneously.

The paper is organized as follows: in Sect. 12.2, we introduce the notation and methodology of the proposed method. Then we conduct simulation in Sect. 12.3 to examine the operating characteristics. Discussion is provided in Sect. 12.4.

12.2 Methods

12.2.1 Escalation with Overdose Control (EWOC)

The dose finding design using escalation with overdose control is a Bayesian adaptive design (Babb et al. 1998). It is based on the Bayesian logistic model, and aims to minimize the number of patients receiving doses higher than MTD by controlling the probability of overdosing.

Let MTD γ be defined as the dose (x) at which a proportion θ (say 33%) of patients exhibit DLT, i.e.

$$P\,(DLT|x = y) = \theta,$$

Babb et al. considered a two-parameter logistic model for the dose-toxicity relationship:

$$P\,(Y = 1|Dose = x) = \frac{\exp\,(\beta_0 + \beta_1 x)}{1 + \exp\,(\beta_0 + \beta_1 x)}, \tag{12.1}$$

where it is assumed that $\beta_1 > 0$ so that the probability of DLT is a monotonically increasing function of dose.

Based on the Bayesian logistic model, the posterior probability of the DLT rate at each dose can be estimated. The dose for each subsequent patient will be selected so that the posterior probability that it exceeds the MTD is equal to 25% (a pre-specified feasibility bound as used in Babb et al). At the end of the trial, after including a fixed number of patients n, MTD can be defined based upon the posterior probability density function of the MTD.

Neuenschwander et al. (2008) also applied a Bayesian logistic regression model, but the dose escalation decision is based on the entire posterior distribution. Desirable and undesirable regions of toxicity are defined, such as under dosing, target dosing, overdosing and unacceptable overdosing. By limiting the intervals for overdosing and unacceptable overdosing, the probability of overdose can be controlled. This interval probability approach is used in our method. See Sect. 12.2.3 for details.

12.2.2 *Escalation with Overdose Control Using Proportional Hazards Model (EWOC-PH)*

To account for the late onset toxicity, Tighiouart et al. (2014) proposed the escalation with overdose control with time to event endpoint using the proportional hazards model (EWOC-PH).

Let MTD γ be defined as the dose at which a proportion θ of patients exhibit DLT during the observation window $[0, \tau]$, i.e.,

$$P\left(T \leq \tau | x = \gamma\right) = \theta$$

The value chosen for the target probability θ depends on the nature and clinical manageability of the DLT. It can be high when the DLT is a transient, correctable or non-fatal condition; and it can be low when the DLT is lethal or life-threatening. Also the length of the observation window could be a factor in the specification of θ.

The risk of DLT given dose is modeled using a Cox proportional hazards model by assuming that patients given different doses of an agent have proportional risks of DLT.

$$h\left(t|x\right) = h_0\left(t, u\right)\exp\left(\beta\left(x - X_{min}\right)\right), \qquad (12.2)$$

where X_{min} is the lowest dose administered in the study, $h_0(t, u)$ is the baseline hazard function corresponding to the risk of DLT for a patient given dose X_{min} and u is a vector of parameters associated with the parametric baseline hazard.

Let $D_n = \{(Y_i, x_i, \delta_i), i = 1, \ldots, n\}$ be the observed data, where $Y_i = \min(T_i, \tau)$, T_i is the time to DLT for patient i, x_i is the dose for patient i, and $\delta_i = I(T_i \leq \tau)$ After n patients have been enrolled in the trial, the likelihood function for the parameters is

$$\mathrm{L}(\beta, \mu | D_n) = \prod_{i=1}^{n} h(Y_i | x_i)^{\delta_i} \exp\left\{-\int_0^{Y_i} h(s|x_i)\, ds\right\}.$$

The model can be reparameterized in terms of $\gamma = \mathrm{MTD}$ and $\rho_0 =$ probability of a DLT for a patient given dose $x = X_{\min}$. The estimated γ and ρ_0 can then be given by:

$$\gamma = \frac{1}{\beta}\left\{\beta X_{min} + \log\left[\frac{\log(1-\theta)}{-H_0(\tau;\mu)}\right]\right\}$$

$$\rho_0 = 1 - \exp\{-H_0(\tau;\mu)\},$$

where $H_0(t;\mu) = \int_0^t h_0(u;\mu)\, du$ is the cumulative baseline hazard function.
Assuming that the baseline hazard function $h_0(t, \mu) = \mu$, we have

$$\mu = -\frac{1}{\tau}\log(1-\rho_0)$$

$$\beta = \frac{1}{\gamma - X_{min}} \log\left[\frac{\log(1-\theta)}{\log(1-\rho_0)}\right],$$

and the likelihood function can be written as

$$\mathrm{L}(\beta, \mu | D_n) = \prod_{i=1}^{n} \left[\mu \exp\left(\beta\left(x_i - X_{min}\right)\right)\right]^{\delta_i} \exp\{-\mu Y_i \exp\left(\beta\left(x_i - X_{min}\right)\right)\}.$$

The likelihood of the reparameterized model $L(\rho_0, \gamma | D_n)$ can be easily derived based on the above equations. Let $g(\rho_0, \gamma)$ be a prior distribution for ρ_0 and γ. The posterior distribution of the model parameters based on the Bayes rule is proportional to the product of the prior distribution and the likelihood. That is,

$$\pi(\rho_0, \gamma | D_n) \propto g(\rho_0, \gamma) \times L(\rho_0, \gamma | D_n).$$

The posterior distribution is approximated using a Markov chain Monte Carlo (MCMC) method, which is implemented in JAGS version 3.4.

Any existing information about γ (MTD) and ρ_0 ($P(DLT|x = X_{min})$) can be reflected in the choice of their prior distributions. The existing information could come from other clinical trials, published data, preclinical results, *etc*. It is re-parameterized to this convenient parameterization since any existing information is likely to be most related to lower doses. In the absence of prior information about the MTD (γ) and probability of DLT at X_{min} (ρ_0), independent vague priors can be selected for ρ_0 and γ. For example, $\rho_0 \sim \text{Unif}(0, \theta)$ and $\gamma \sim \text{Unif}(X_{min}, X_{max})$. Note these priors assume that the MTD must be between X_{min} and X_{max} and that the probability of DLT at X_{min} is no more than θ. Probability of DLT will be estimated at 100% at X_{max} regardless of what any data will later say. So we need to make sure X_{max} is higher than the highest dose that will ever be considered. If we are not sure whether the dose will be below X_{max}, we can choose a different prior that allows the support of the MTD to extend beyond X_{max} (Tighiouart et al. 2005).

The first patient will receive the minimum dose X_{min}. The subsequent k-th patient receives the dose x_k where x_k is the dose at which the current posterior probability of exceeding the MTD is equal to the feasibility bound α (say, 25%). This is the overdose protection property of EWOC.

When the k-th patient enters the trial at time t_k, we calculate the posterior probability. The time to event Y_i is the time to DLT, if the patient experienced DLT; or the time since patient i was given dose x_i (up to a maximum of τ) if this patient is still at risk by this time (censored). MTD can be defined based upon the posterior probability density function of the MTD, such as median of posterior probability density function of the MTD. If a patient is withdrawn from the study due to disease progression before time τ, it will be censored at the time of disease progression. Note that here we assume non-informative censoring.

12.2.3 *Proposed EWOC Method Incorporating the Two Components (EWOC-2C)*

We propose to incorporate both EWOC and EWOC-PH to model the short term toxicity and the long term toxicity, and use the posterior interval probability to provide the dose recommendations. The Bayesian model based on EWOC using the binary endpoint is used to estimate the probability of the short term toxicity, and the Bayesian model based on EWOC-PH is used to estimate the probability of the long term toxicity.

For the short term toxicity using the binary endpoint, a non-informative prior can be used (Neuenschwander et al. 2008). The bivariate normal prior distribution $\eta = (\mu_1, \mu_2, \sigma_1, \sigma_2, \rho)$ for (β_0, β_1) can be estimated by comparing the quantiles of the bivariate normal distribution and the quantiles for the probabilities of toxicity derived from the minimally informative Beta distributions at the lowest dose and the highest dose. The two beta distributions at the lowest/highest dose levels can be obtained based on the anticipated DLT distribution at the lowest/highest dose levels. Note that generally we can use the lowest/highest dose, but a different pair of a lower dose and a higher dose (such as the anticipated MTD) will also allow us to get the estimate.

For the long term toxicity using the time to event endpoint, independent vague priors can be selected for the probability of DLT at X_{min} ρ_0 and MTD γ. For example, $\rho_0 \sim$ Unif(0, θ) and $\gamma \sim$ Unif(X_{min}, X_{max}). There are also other options for priors (Tighiouart et al. 2005).

The posterior distributions of the probability of the short term toxicity, $\pi_\theta(d)$ can be classified into several categories (Neuenschwander et al. 2008). Here, we divide into three categories:

Under-target: $\pi_\theta(d) \in (0,0.16]$
Targeted toxicity: $\pi_\theta(d) \in (0.16,0.33]$
Above target: $\pi_\theta(d) \in (0.33,1.00]$

After each patient cohort, the posterior probability of the short term toxicity and the probability of the long term toxicity are estimated for each dose level. The posterior distributions of short term toxicity can be summarized by the three categories: under-target, target, above target, and the posterior probability of the above target long term toxicity can also be calculated. The dose may be selected if the posterior probability of short term toxicity (such as DLT) has more mass in the targeted interval 16–33% than any other dose and; the overdosing risk is controlled. That is, the risk of a DLT above 33% should not exceed a predefined threshold (for example, 25%); the risk of the long term toxicity above the predefined toxicity level (for example, 25%) should not exceed a predefined threshold (for example, 50%). The exact cutoffs for each toxicity interval and the thresholds can be adjusted based on the study objectives and the type of toxicities.

We propose to test the patients in cohorts of at least three patients for a set of pre-defined dose levels. This approach has been applied to a phase 1 study and was well received by the clinical colleagues as the setting is quite similar to the 3 + 3 design except that the decision rules are based on the Bayesian models with control of overdosing for both short term and long term toxicity without holding up the escalation process.

The recommended dose for the next cohort is decided when a minimum of three patients in the current cohort have been followed up for the short term toxicity observation period. At the time of dose recommendation, the first patient in the cohort may have been followed for longer; patients from previous cohorts also continue to be followed; and there is no need to wait for the long term toxicity time window to be completely observed to enroll new patients, which can potentially reduce the trial length significantly.

The Bayesian model based on EWOC-PH will incorporate the available information on the long term toxicity. The risk of the long term toxicity above the pre-specified toxicity level will be estimated. Dose levels with estimated probabilities of long term toxicity not greater than the pre-specified toxicity level (*i.e.* which meet the EWOC criteria) can be considered. It is acceptable not to choose the highest allowable dose. In practice, we can choose not to skip any predefined dose levels, and escalate to the next dose level even if a higher dose may also meet the criteria for both EWOC and EWOC-PH. The trial can be stopped if a dose has been tested in two cohorts with at least three patients in each cohort, and it is recommended again. Different stopping rules with more criteria can be used. Our simulations suggested that the proposed simple stopping rule performs well.

12.3 Simulation

Simulation studies were conducted to study the operating characteristics of the proposed methods with two components, and compared with the traditional 3 + 3 design. A number of scenarios with different dose toxicity relationships were considered, where the time window for the short term toxicity is 4 weeks, and the time window for the long term toxicity is 8 weeks. See Table 12.1. For each

Table 12.1 Different dose toxicity relationships for the short term and the long term toxicity in simulations

Dose level		1	2	3	4	5	6	7	8	9	10
Dose		5	10	20	30	40	60	80	100	120	150
Scenarios	Probability of toxicity										
S1	Short term	0.02	0.03	0.04	0.04	0.06	0.10	0.13	**0.18**	**0.25**	0.31
	Long term	0.05	0.06	0.07	0.08	0.09	0.11	0.13	**0.16**	**0.25**	0.39
S2	Short term	0.02	0.02	0.02	0.02	0.11	**0.20**	**0.30**	0.40	0.47	0.54
	Long term	0.05	0.06	0.07	0.08	0.09	**0.11**	**0.13**	0.16	0.25	0.39
S3	Short term	0.02	0.03	0.03	0.04	0.05	0.06	0.08	0.10	0.12	0.15
	Long term	0.02	0.03	0.04	0.05	0.06	0.07	0.08	0.10	0.12	0.15
S4	Short term	0.02	0.03	0.04	0.05	0.07	0.10	**0.17**	**0.19**	**0.27**	0.37
	Long term	0.05	0.06	0.06	0.07	0.08	0.10	**0.12**	**0.15**	**0.18**	0.25
S5	Short term	0.03	0.04	0.05	0.07	0.10	0.14	**0.19**	0.27	0.39	0.47
	Long term	0.05	0.06	0.07	0.09	0.11	0.17	**0.25**	0.36	0.50	0.73
S6	Short term	0.02	0.03	0.04	0.04	0.06	0.10	0.13	**0.18**	**0.25**	**0.31**
	Long term	0.05	0.06	0.06	0.07	0.08	0.10	0.12	**0.15**	**0.18**	**0.25**
S7	Short term	0.02	0.03	0.04	0.05	**0.26**	0.36	0.45	0.50	0.54	0.58
	Long term	0.05	0.06	0.07	0.08	**0.09**	0.11	0.13	0.16	0.25	0.39
S8	Short term	0.03	0.04	0.05	0.07	0.10	0.14	**0.19**	**0.27**	0.39	0.47
	Long term	0.05	0.06	0.06	0.07	0.08	0.10	**0.12**	**0.15**	0.18	0.25
S9	Short term	0.02	0.03	0.04	0.05	0.07	0.10	**0.17**	**0.23**	0.29	0.37
	Long term	0.03	0.04	0.05	0.06	0.07	0.11	**0.17**	**0.24**	0.36	0.59
S10	Short term	0.02	0.02	0.02	0.11	**0.20**	**0.25**	0.30	0.40	0.47	0.54
	Long term	0.06	0.06	0.09	0.11	**0.15**	**0.25**	0.40	0.60	0.81	0.98

The target doses are in bold

scenario, 500 trials were simulated. For the first two low dose levels, accelerated escalation is assumed with one patient per cohort. Starting from dose level 3, at least 3 patients are enrolled at each cohort. The probability of selection of a dose as MTD, the probability that a dose is tested, the probability of MTD selection by true category of toxicity and the average number of patients were summarized. The dose toxicity curve and the simulation results for scenarios 1 and 2 are presented graphically in Figs. 12.1, 12.2, 12.3, and 12.4, Tables 12.2, 12.3, 12.4, and 12.5. The key simulation results for all the scenarios were presented in Table 12.6.

Take scenario 1 as an example. The dose toxicity curve for the short term and long term toxicity are presented in Fig. 12.1. The target dose should be 100 or 120, which has probability of DLT between 0.16 and 0.33. As we can see, the probability of selecting one of the target doses based on EWOC-2C is 59% (44% for 120, and 15% for 100); while the probability of selecting one of the target doses based on 3 + 3 is 45% (25% for 120 and 20% for 100). The probability of selecting a dose higher than MTD is 7% based on EWOC-2C, while it is 24% based on the 3 + 3 design. Also the probability of testing a dose higher than MTD is 26%, while it is 46% based on 3 + 3 design. The mean number of sample size for EWOC-2C is slightly higher than the sample size for 3 + 3, but quite close. See Tables 12.2 and 12.3.

As we can see, compared to a traditional 3 + 3 design, the proposed EWOC design with two components has more favorable operating characteristics with higher chance of selecting a dose within the target toxicity interval, and lower or similar chance of testing a dose above the target. Considerable improvements can

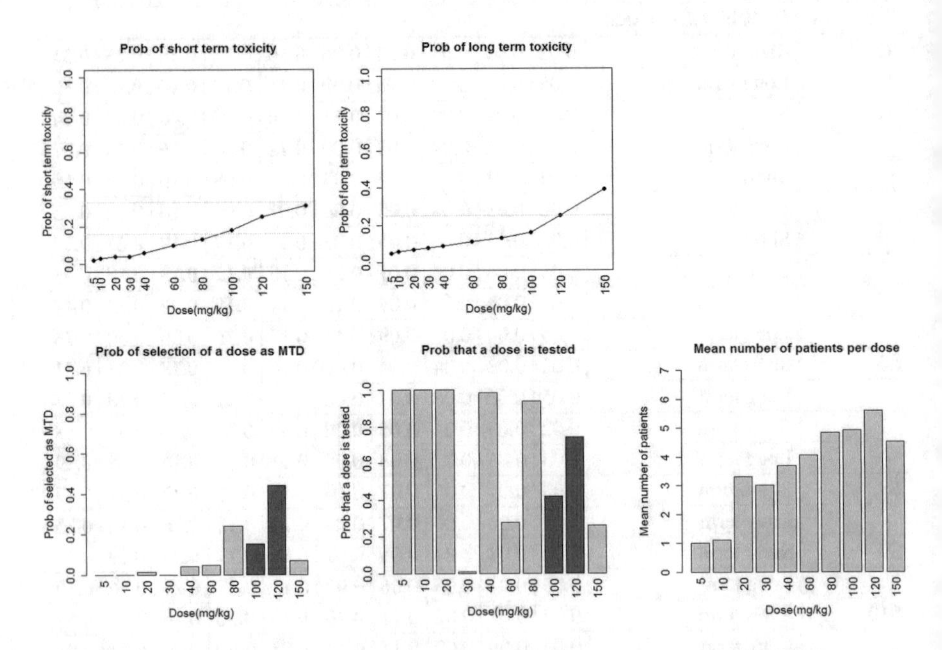

Fig. 12.1 Results based on the proposed methods with two components (EWOC-2C) for scenario 1

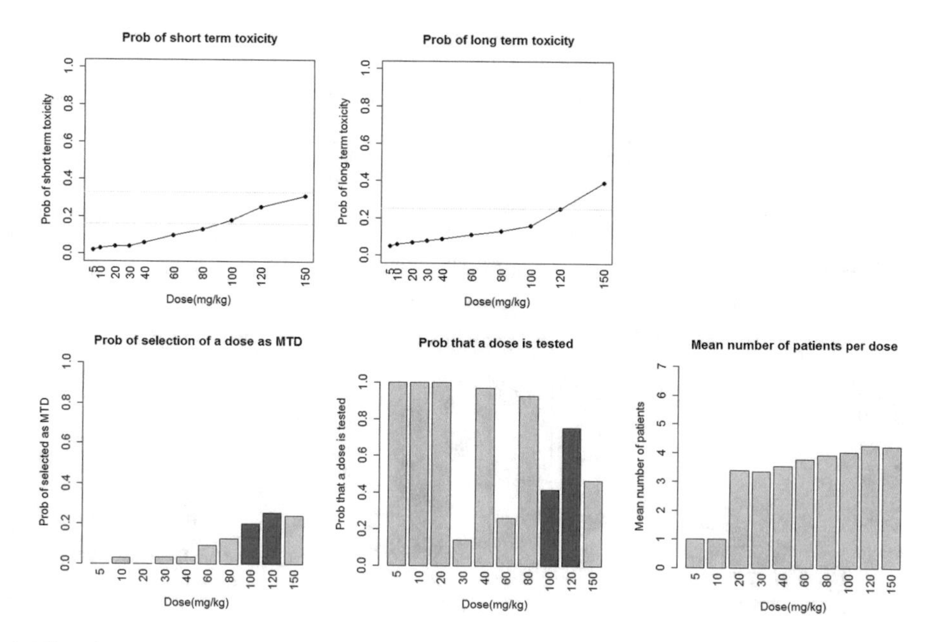

Fig. 12.2 Results based on traditional 3 + 3 design for scenario 1

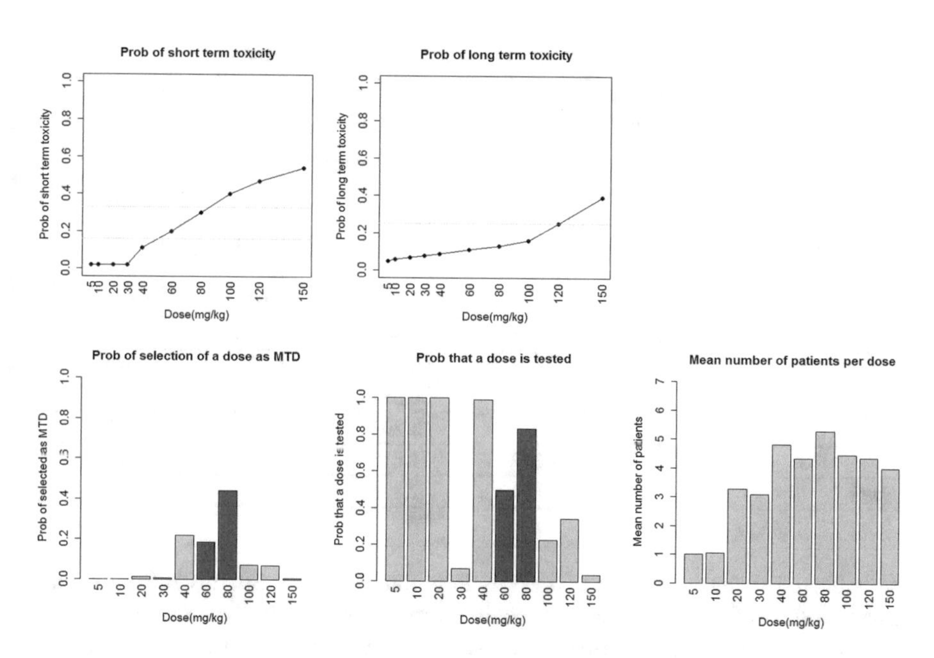

Fig. 12.3 Results based on the proposed methods with two components (EWOC-2C) for scenario 2

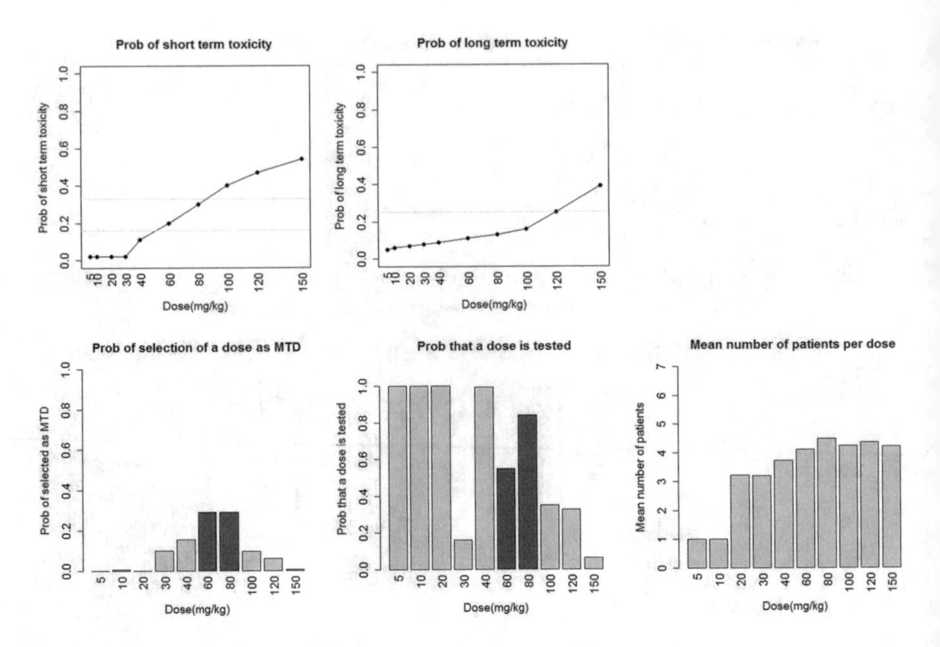

Fig. 12.4 Results based on traditional 3 + 3 design for scenario 2

Table 12.2 Summary of simulation results based on EWOC-2C for scenario 1

Dose	True short term Tox rate	True long term Tox rate	Mean # of short term Tox	Mean # of long term Tox	% Selected as MTD	Mean number of patients
5	0.02	0.05	0	0	0	1
10	0.03	0.06	0	0.006	0	1.102
20	0.04	0.07	0.116	0.242	0.014	3.306
30	0.04	0.08	0	0	0	3
40	0.06	0.09	0.191	0.345	0.042	3.675
60	0.10	0.11	0.381	0.345	0.046	4.058
80	0.13	0.13	0.623	0.571	0.238	4.842
100	0.18	0.16	0.810	0.652	0.152	4.929
120	0.25	0.25	1.317	1.293	0.440	5.589
150	0.31	0.39	1.392	1.615	0.068	4.523
Total			2.676	2.876	1	22.13

be achieved if the long term toxicity is important (S1, S5, S9, S10). Simulations using EWOC with binary endpoint for short term toxicity only were also performed, which showed that the proposed EWOC-2C has better performance if the long term toxicity is of concern. For example, in scenario 10, the above target probability based on EWOC with binary endpoint is 0.330, while it is 0.218 for EWOC-2C; the on target probability based on EWOC with binary endpoint is 0.608, while it is 0.708 for EWOC-2C.

Table 12.3 Summary of simulation results based on 3 + 3 for scenario 1

Dose	True short term Tox rate	True long term Tox rate	Mean # of short term Tox	Mean # of long term Tox	% Selected as MTD	Mean number of patients
5	0.02	0.05	0	0	0	1
10	0.03	0.06	0	0	0.028	1
20	0.04	0.07	0.17	0.236	0	3.384
30	0.04	0.08	0.114	0.243	0.032	3.343
40	0.06	0.09	0.218	0.309	0.032	3.531
60	0.10	0.11	0.454	0.338	0.092	3.762
80	0.13	0.13	0.504	0.530	0.124	3.899
100	0.18	0.16	0.745	0.558	0.200	4.010
120	0.25	0.25	1.040	1.066	0.254	4.241
150	0.31	0.39	1.259	1.168	0.238	4.203
Total			2.662	2.728	1	20.696

Table 12.4 Summary of simulation results based on EWOC-2C for scenario 2

Dose	True short term Tox rate	True long term Tox rate	Mean # of short term Tox	Mean # of long term Tox	% Selected as MTD	Mean number of patients
5	0.02	0.05	0	0	0	1
10	0.02	0.06	0	0	0	1.048
20	0.02	0.07	0.060	0.25	0.010	3.282
30	0.02	0.08	0	0.147	0.004	3.088
40	0.11	0.09	0.550	0.401	0.218	4.821
60	0.20	0.11	0.804	0.404	0.184	4.332
80	0.30	0.13	1.615	0.599	0.440	5.272
100	0.40	0.16	1.788	0.602	0.072	4.460
120	0.47	0.25	2.058	0.860	0.068	4.343
150	0.54	0.39	2.444	1.111	0.004	4
Total			3.552	1.830	1	19.520

Table 12.5 Summary of simulation results based on 3 + 3 for scenario 2

Dose	True short term Tox rate	True long term Tox rate	Mean # of short term Tox	Mean # of long term Tox	% Selected as MTD	Mean number of patients
5	0.02	0.05	0	0	0	1
10	0.02	0.06	0	0	0.006	1
20	0.02	0.07	0.078	0.240	0	3.210
30	0.02	0.08	0.062	0.162	0.098	3.188
40	0.11	0.09	0.386	0.318	0.156	3.724
60	0.20	0.11	0.796	0.336	0.290	4.106
80	0.30	0.13	1.356	0.504	0.290	4.468
100	0.40	0.16	1.726	0.480	0.096	4.217
120	0.47	0.25	2.104	0.963	0.060	4.344
150	0.54	0.39	2.333	1.400	0.004	4.200
Total			3.480	1.756	1	18.578

Table 12.6 Probability of selection as MTD by true category of toxicity for EWOC-2C, 3 + 3, and EWOC with binary endpoint in various simulation scenarios

Scenario	Method	Below Target	Target Toxicity	Above Target
S1	3 + 3	0.308	0.454	0.238
	EWOC	0.358	0.362	0.280
	EWOC-2C	0.340	0.592	0.068
S2	3 + 3	0.260	0.580	0.160
	EWOC	0.238	0.636	0.126
	EWOC-2C	0.232	0.624	0.144
S3	3 + 3	1	0	0
	EWOC	1	0	0
	EWOC-2C	1	0	0
S4	3 + 3	0.220	0.664	0.116
	EWOC	0.140	0.638	0.194
	EWOC-2C	0.148	0.762	0.090
S5	3 + 3	0.368	0.222	0.410
	EWOC	0.318	0.390	0.292
	EWOC-2C	0.378	0.434	0.188
S6	3 + 3	0.312	0.688	0
	EWOC	0.358	0.642	0
	EWOC-2C	0.294	0.706	0
S7	3 + 3	0.404	0.310	0.286
	EWOC	0.062	0.688	0.25
	EWOC-2C	0.122	0.660	0.218
S8	3 + 3	0.378	0.426	0.196
	EWOC	0.318	0.518	0.164
	EWOC-2C	0.248	0.524	0.228
S9	3 + 3	0.298	0.366	0.336
	EWOC	0.110	0.430	0.460
	EWOC-2C	0.182	0.588	0.230
S10	3 + 3	0.270	0.382	0.348
	EWOC	0.062	0.608	0.330
	EWOC-2C	0.074	0.708	0.218

12.4 Discussion

Simulation studies demonstrated the favorable operating characteristics of the proposed EWOC design with two components (EWOC-2C) compared to a traditional 3 + 3 design and EWOC using the binary endpoint. The short term toxicity is modeled using a binary endpoint, and can control the probability of overdosing. The late onset toxicity is modeled using EWOC-PH using a time to event endpoint, and can capture toxicity beyond the initial observation window while not requiring waiting beyond the initial window. Trial length can be potentially cut considerably versus holding dose escalation until full observation of long term toxicities, given the importance of the late onset adverse event. The patients are enrolled in a set of predefined dose levels with cohort size of at least three patients. When at least three patients are evaluable for the short term toxicities, the dose escalation recommendation is made based on the posterior probability of underdosing, target dosing and overdosing. This design has been implemented in a phase 1 dose escalation study with both short term DLT and long term AESI, which allowed the study team to consider both the short term DLT and long term AESI while maintaining the escalation timeline.

Whether to consider the correlation between the two endpoints is an important question. In light of the over-parameterizations problem in dose finding studies (Iasonos et al. 2016), it is recommended to use parsimonious models in dose finding studies with small sample size, as "increasing the dimension of parameter space, in the context of adaptive dose-finding studies, is usually counter productive". In particular, Cunanan and Koopmeiners (2014) explored copula models with joint distribution of toxicity and efficacy, and found that the model assuming independence performs as well as models with correlation. Therefore we model early and late onset toxicity separately without considering their correlation.

There are a few limitations with our work. First, efficacy is an important component for dose finding studies with targeted therapies (EMA 2017). Many authors have considered joint modeling of toxicity and efficacy in dose finding studies (see a survey in Sverdlov and Gao (2017)). Efficacy can be incorporated in our framework by introducing a dose-efficacy model. Second, one can consider to model time to first toxicity regardless if it is early (such as DLT) or late onset (such as adverse event of special interest) toxicity, and the univariate time to event dose finding algorithm (e.g. Cheung and Chappell 2000) can be applied. The benefit of such an approach is that it can avoid the competing risk problem: i.e. patient could drop off after early onset toxicity, and thus late onset toxicity becomes unobservable. As a future work, we will compare the two approaches to assess operating characteristics and estimation bias. Third, a time to event model can be used to model the short term toxicity. With this approach, we can enroll patients continuously without waiting patient to complete the DLT observation period (i.e. 4 weeks). As a future work, we can explore the statistical property of such an approach. Last, it is useful to consider a broad spectrum of data in dose finding studies, such as PK/PD, and/or biomarker guided dose finding studies (Kummar et al. 2006). To expand our work, it is possible to model the dose, exposure and response by considering the association among the PK/PD, biomarker, early and late onset toxicity. Due to small sample sizes of dose finding studies, one needs to weigh the benefit of the introduced information against the uncertainties in parameter estimation. Alternatively, one could consider non-parametric methods such as tree based learning (e.g. Ma et al. 2016).

Acknowledgement The authors appreciate the helpful discussions with Mourad Tighiouart and Pierre Colin. The authors thank referees and the associate editor for their excellent comments and suggestions.

References

Babb, J., Rogatko, A., Zacks, S.: Cancer phase I clinical trials: efficient dose escalation with overdose control. Stat. Med. **17**, 1103–1120 (1998)

Berry, S.M., Carlin, B.P., Lee, J.J., Müller, P.: Bayesian Adaptive Methods for Clinical Trials. CRC Press, Boca Raton (2010)

Braun, T.M.: Generalizing the TITE-CRM to adapt for early- and late-onset toxicities. Stat. Med. **25**, 2071–2083 (2006). https://doi.org/10.1002/sim.23372

Cheung, Y., Chappell, R.: Sequential designs for phase I clinical trials with late-onset toxicities. Biometrics. **56**, 1177–1182 (2000)

Cunanan, K., Koopmeiners, J.S.: Evaluating the performance of copula models in phase I-II clinical trials under model misspecification. BMC Med. Res. Methodol. **14**, 51 (2014)

EMA. Guideline on the evaluation of anticancer medicinal products in man. EMA/CHMP/205/95 Rev.5 (2017)

Goodman, S.N., Zahurak, M.L., Piantadosi, S.: Some practical improvements in the continual reassessment method for phase I studies. Stat. Med. **14**(11), 1149–1161 (1995)

Iasonos, A., Wages, N.A., Conaway, M.R., Cheung, K., Yuan, Y., O'Quigley, J.: Dimension of model parameter space and operating characteristics in adaptive dose-finding studies. Stat. Med. **35**, 3760–3775 (2016)

Kummar, S., Gutierrez, M., Doroshow, J.H., Murgo, A.J.: Drug development in oncology: classical cytotoxics and molecularly targeted agents. Br. J. Clin. Pharmacol. **62**(1), 15–26 (2006)

Ma, X., Zheng, W., Lu, Y.: Personalized effective dose selection in dose ranging studies. In: Lin, J., Wang, B., Hu, X., Chen, K., Liu, R. (eds.) Statistical Applications from Clinical Trials and Personalized Medicine to Finance and Business Analytics. Springer, Basel (2016)

Mauguen, A., Le Deley, M.C., Zohar, S.: Dose-finding approach for dose escalation with overdose control considering incomplete observations. Stat. Med. **30**(13), 1584–1594 (2011)

Neuenschwander, B., Branson, M., Gsponer, T.: Critical aspects of the Bayesian approach to phase I cancer trials. Stat. Med. **27**, 2420–2439 (2008)

O'Quigley, J., Pepe, M., Fisher, L.: Continual reassessment method: a practical design for phase 1 clinical trials in cancer. Biometrics. **46**, 33–48 (1990). https://doi.org/10.2307/2531628

Paoletti, X., Toumeau, C.L., Verweij, J., Siiu, L.L., Seymour, L., Postel-Vinay, S., Collette, L., Rizzo, E., Ivy, P., Olmos, D., Massard, C., Lacombe, D., Kaye, S.B., Soria, J.-C.: Defining dose-liming toxicity for phase 1 trials of molecularly targeted agents: results of a DLT-TARGETT international survey. Eur. J. Cancer. **50**(12), 2050–2056 (2014)

Sverdlov, O., Gao, L.: Phase I/II dose finding designs with efficacy and safety endpoints. In: O'Quigley, J., Iasonos, A., Bornkamp, B. (eds.) Handbook of Methods for Designing, Monitoring, and Analyzing Dose-Finding Trials. CRC Press, Boca Raton (2017)

Tighiouart, M., Liu, Y., Rogatko, A.: Escalation with overdose control using time to toxicity for cancer phase I clinical trials. PLoS One. **9**(3), e93070 (2014). https://doi.org/10.1371/journal.pone.0093070

Tighiouart, M., Rogatko, A., Babb, J.: Flexible Bayesian methods for cancer phase I clinical trials. Dose escalation with overdose control. Stat. Med. **24**, 2183–2196 (2005)

Varricchi, G., Galdiero, M.R., Marone, G., Criscuolo, G., Triassi, M., Bonaduce, D., Marone, G., Tocchetti, C.G.: Cardiotoxicity of immune checkpoint inhibitors. ESMO Open. **2**(4), e000247 (2017). https://doi.org/10.1136/esmoopen-2017-0002473

Chapter 13
A Bayesian Constancy-Enforced Non-Inferiority Design in Medical Device Trials with a Binary Endpoint

Ying Yang, Yunling Xu, Nelson T. Lu, and Ram C. Tiwari

13.1 Introduction

Noninferiority clinical trials have been commonly used for the evaluation of drugs, devices, biologics, and other medical treatments. Treatment with placebo control in a study may not be ethical when an effective treatment has already been established. Although some new treatments offer greater efficacy, others may promise greater safety or convenience, or maybe less expensive, while providing similar efficacy. Non-inferiority (NI) clinical trial aims to demonstrate that a new treatment is no worse than the active control treatment by an acceptably small amount (called a non-inferiority margin), with a given degree of confidence. The null hypothesis in a noninferiority study states that the primary effect for the experimental treatment is worse than that for the active control treatment by a prespecified margin, and rejection of the null hypothesis at a prespecified level of statistical significance is used to support a claim (that is, alternative hypothesis) that permits a conclusion of noninferiority. Although non-inferiority trials have been accepted for Premarket Approvals (PMAs) in the Center of Devices and Radiological Health, some challenging issues remain in a regulatory setting.

FDA's non-inferiority guidance (FDA 2016) says that "FDA regulations have recognized since 1985 the critical need to know, for an NI trial to be interpretable, that the active control had its expected effect in the trial". When the effectiveness endpoint is the primary endpoint, no subjects will be exposed to the placebo in a two-arm non-inferiority trial. The effect of the active control is not measured in the study but must be assumed. Therefore, one of the most challenging issues in a

Y. Yang (✉) · Y. Xu · N. T. Lu · R. C. Tiwari
Division of Biostatistics, Office of Surveillance and Biometrics, Center for Devices and Radiological Health, Food and Drug Administration, Silver Spring, MD, USA
e-mail: ying.yang@fda.hhs.gov

L. Zhang et al. (eds.), *Contemporary Biostatistics with Biopharmaceutical Applications*, ICSA Book Series in Statistics,
https://doi.org/10.1007/978-3-030-15310-6_13

regulatory setting is that the design of the new trial should preserve the condition of the trial in which the active control was shown to be effective, this is called the "constancy assumption". In other words, the constancy assumption requires that the measurable effect of the active control stays unchanged in the current active controlled trial and the historical trial. Any violation of the constancy assumption would put the conclusion of the non-inferiority trial in jeopardy. See, for example, Wang and Hung (2003), Hung et al. (2003), D'Agostino et al. (2003) and Fleming (2008) for a detailed discussion on the impact of constancy assumption on the non-inferiority study conclusion.

There are various factors that could cause violation of the constancy assumption, including rapid changes in medical practice and standard of care, differences in study population and trial conduct, etc. Some researchers have proposed a few approaches to address this problem. For example, violation of the constancy assumption could be minimized through a good trial conduct. To account for violations of the constancy assumption in non-inferiority clinical trials, Odem-Davis and Fleming (2013) develop a bias-adjusted noninferiority margin that accounts for both bias and uncertainty in the historical treatment effect of the active control. Liu et al. (2015) developed a robust range that allows investigators to estimate the degree to which the noninferiority margin is robust to bias in the historical estimate of the treatment effect of the active control. Nie and Soon (2010) proposed a covariate-adjustment generalized linear regression model approach to assess the new treatment effect when population difference causes constancy assumption violation. Koopmeiners and Hobbs (2018) proposed a Bayesian adaptive approach for detecting and accounting for violation of the constancy assumption in non-inferiority clinical trials. However, there are still cases where constancy assumption is still not satisfied even though the trial has been conducted according to the guideline for good clinical practice to minimize the inter-trial heterogeneity and/or analyses have been performed to adjust observed covariates that could cause violation of the constancy assumption. This violation of the constancy assumption may be caused by factors that could not be controlled or pinpointed out on the validity of the non-inferiority trial.

In this paper, we propose an approach to ensure the validity of a non-inferiority trial upfront to a desirable extent through devising a companion constancy test to enforce the non-inferiority test. A binary endpoint is considered for the proposed approach. The application of the proposed approach to other types of endpoints will be discussed elsewhere. The paper is organized as follows. In Sect. 13.2, we present the motivating example. It is followed by a detailed description of our proposed constancy-enforced non-inferiority design in Sect. 13.3. In Sect. 13.4, we discuss two Bayesian approaches to borrow historical data. In Sect. 13.5, we demonstrate its operating characteristics when this approach is applied to medical device studies presented in Sect. 13.2. Finally, in Sect. 13.6, we conclude with a discussion.

13.2 The Motivating Example

Consider a second-generation radio frequency (RF) ablation catheter for treating atrial fibrillation (AF). The first-generation RF ablation catheter was compared to the optimal medical management (OMM) in a randomized superiority trial, and the device was approved in a PMA. When a better treatment exists, the use of OMM as an active control raises ethical concerns and causes difficulties in patient recruiting. Therefore, it is impractical to use OMM as an active control for the second-generation RF ablation catheter study. In this situation, the sponsor proposes an alternative where a non-inferiority trial is considered to compare the second-generation RF to the first-generation RF ablation catheter. Since OMM is not part of the non-inferiority trial, the effectiveness of the second-generation RF ablation catheter with respect to OMM can only be addressed indirectly. The indirect way of determining effectiveness of the second-generation RF ablation catheter imposes challenges in the design, implementation, and analysis of non-inferiority trial for the sponsors and regulators (D'Agostino et al. 2003). In what follows, below we show how the non-inferiority margin is chosen for this study. In the historical trial, the observed chronic effectiveness success rate for OMM group was about 30%, and for the first-generation RF ablation catheter (the success rate) was about 60%. From an engineering perspective, the second-generation RF ablation catheter is expected, and should be, at least as good as the first generation. From a clinical perspective, the first-generation RF ablation catheter should not perform too inferior compared to its performance in the historical trial, and the chronic effectiveness success rate should be reasonably higher than 45% (also called performance goal (PG)). A performance goal is a numerical number considered sufficient by FDA for use as a comparison for an effectiveness endpoint (FDA 2013). Trying to show the second-generation RF ablation to be at least better than OMM for approval, a non-inferiority margin of 10% was chosen for the current study to evaluate the second-generation RF ablation catheter against the first-generation RF ablation catheter. Therefore, there is an explicit clinical requirement for constancy of the effectiveness of active control (i.e., the first-generation AF ablation catheter) to a specified extent. In the next section, we present a statistical approach addressing this requirement.

13.3 A Reinforced Non-Inferiority Design

Without loss of generality, we continue with the motivating example, and consider the non-inferiority study in which the second-generation of RF ablation catheter (referred as the investigational device) is compared to the first-generation of RF ablation catheter (referred as the active control device) regarding chronic effectiveness success. The non-inferiority (NI) hypotheses are commonly stated as:

$$H_0 : \pi_i - \pi_c \leq -\mathrm{M} \text{ vs.} H_a : \pi_i - \pi_c > -\mathrm{M}$$

where π_i and π_c represent the expected success rate for the investigational and the active control device, respectively. M (>0) is the pre-specified NI margin, which is 10% in the motivating example.

In the motivating example, there is a specific clinical requirement that the active control device preserves the treatment effect shown in the historical trial. In other words, to satisfy the constancy assumption, the success rate for the active control device in the current study needs to be at least 45% (i.e., performance goal).

To meet this clinical requirement, one way is to compare the observed success rate of the control device with the performance goal of 45%. Since the observed success rate is sample dependent, direct comparison is not statistically appropriate. A better way is to make this comparison an essential part of the statistical design and analysis so that the performance of the active control device and the overall study design can formally be evaluated. Specifically, we propose a companion hypotheses test to address the constancy issue. The companion hypotheses are then written as:

$$H_0{}^{com} : \pi_c < \text{PG vs.} H_a{}^{com} : \pi_c \geq \text{PG},$$

where, as mentioned before, π_c is the expected success rate for the active control device, and PG is the pre-specified performance goal for constancy/clinical requirement, which in the motivating example is equal to 45%.

Now, we have two sets of hypotheses: the main one is for the non-inferiority and the companion one is for the constancy assumption. We further specify that the investigational device is claimed to be non-inferior to the active control device if the null hypothesis for the non-inferiority and the null hypothesis for the constancy assumption are rejected. Since the success rule depends on winning on both H_a and $H_a{}^{com}$, there are no concerns regarding the multiplicity issue.

Following the usual inference, the rejection of the null hypotheses is based on whether (1) the lower bound of $100(1-\alpha)\%$ confidence interval of the difference $(\pi_i - \pi_c)$ is greater than $-M$, and (2) the lower bound of $100(1-\alpha)\%$ confidence interval of π_c is greater than PG.

13.4 Bayesian Approaches of Borrowing from Historical Data

While Lu et al. (2019) proposed a Frequentist method to tackle this problem, in this paper we approach it under Bayesian framework (FDA 2010). As the active control device data from the previous randomized trial is available at the design stage of this new non-inferiority trial, it is appealing in using (i.e., 'borrowing') this information when testing the main non-inferiority hypotheses. More trial resources can be devoted to the investigational device while retaining accurate estimates of the current control device parameters. This can result in more accurate point estimates, increased power, and reduce the false probability of claiming study success in a clinical trial, provided that the historical information is sufficiently similar to the current control data.

In this section, we incorporate the historical data for the active control device into the analyses using different approaches. The Bayesian decision criteria is defined through the following:

1. The investigational device is non-inferior to the active control if the posterior probability of the effect size, $(\pi_i - \pi_c)$, in the current non-inferiority trial exceeding the non-inferiority margin is larger than some pre-specified level λ_1, and
2. The posterior probability of π_c greater than the performance goal (PG) is larger than a pre-specified level λ_2.

Since the focus of the study is to test the non-inferiority, λ_2 is set smaller than λ_1 for the least burdensome for the study sponsor. Meanwhile, since the companion hypothesis test is devised to determine whether the active control treatment effect in the current non-inferiority study is the same as its effect presented in the historical study, no historical information should be borrowed when testing H_0^{COM}.

Let X_i and X_c denote the random variables corresponding to the investigational device and the active control in the current trial, respectively. Suppose these random variables are independent and that their distributions are $X_l \sim Bin(n_l, \pi_l)$ with π_l being unknown, where $l \in (i, c)$, $i =$ investigational, $c =$ active control.

13.4.1 Power Prior

Let X_{c_0} denote the random variable corresponding to the active control response in historical trials, and let its distribution be $X_{c_0} \sim Bin\left(n_{c_0}, \pi_c\right)$ with π_c unknown.

The "power prior" discussed by Ibrahim and Chen (2000) is a powerful tool, as an informative prior, for incorporation of historical data. The power prior distribution is constructed by raising the likelihood function of the historical data to a power a_0, where $0 \leq a_0 \leq 1$. a_0 is a parameter. $f_0(\pi_c)$ is the *initial prior* for π_c, before the historical data was observed. Let $L(\pi_c|X_c, n_c)$ and $L\left(\pi_c|X_{c_0}, n_{c_0}\right)$ denote the likelihood functions for the active control in the current trial and historical trial, respectively. The basic formulation of the power prior is written as

$$f\left(\pi_c|X_{c_0}, n_{c_0}, a_0\right) \propto L\left(\pi_c|X_{c_0}, n_{c_0}\right)^{a_0} f_0\left(\pi_c\right) \tag{13.1}$$

Using the power prior in (13.1), the corresponding posterior distribution of π_c in the current study is given by

$$f\left(\pi_c|X_c, n_c, X_{c_0}, n_{c_0}, a_0\right) \propto L\left(\pi_c|X_c, n_c\right) L\left(\pi_c|X_{c_0}, n_{c_0}\right)^{a_0} f_0\left(\pi_c\right) \tag{13.2}$$

It is noted from (13.2) that a_0 quantifies the heterogeneity between the current data and the historical data. It controls the influence of the historical data on

$L(\pi_c|X_c, n_c)$ and is also interpreted as the amount of the borrowing from the historical data. When $a_0 = 1$; that is when the historical and current data are fully exchangeable, there is full borrowing from the historical data, and (13.1) corresponds to the posterior distribution of π_c based on the historical data. When $a_0 = 0$, the prior distribution of π_c does not depend on the historical data, which is equivalent to a prior with no borrowing from historical data. Since we consider a binary outcome in this manuscript, a vague conjugate Beta prior is placed on π_c, i.e., $f_0(\pi_c) \sim Beta\left(\alpha_{c0}, \beta_{c0}\right)$ in the historical trial. After the historical data is observed, the power prior is updated to be

$$\pi_c \mid X_{c0}, \mathrm{n}_{c0} \sim Beta\left(\alpha_{c0} + a_0 \mathrm{X}_{c0}, a_0\left(\mathrm{n}_{c0} - \mathrm{X}_{c0}\right) + \beta_{c0}\right). \tag{13.3}$$

Therefore, the historical data was down weighted by a_0. We then use this posterior as an informative prior for the current active control.

For the current trial, we assume a non-informative Beta prior for π_i given by $f(\pi_i) \sim Beta(\alpha_i, \beta_i)$. After observing X_i responses on the investigational device and X_c responses on the active control device, combined with the power prior for π_c and a non-informative prior for π_i, the posterior distributions of π_i and π_c could be written as

$$\begin{aligned}
&\pi_i \mid X_i \sim Beta\left(\alpha_i + \mathrm{X}_i, \mathrm{n}_i + \beta_i - \mathrm{X}_i\right)\\
&\pi_c \mid X_c, \mathrm{X}_{c0} \sim Beta\left(\alpha_{c0} + a_0 \mathrm{X}_{c0} + \mathrm{X}_c, a_0\left(\mathrm{n}_{c0} - \mathrm{X}_{c0}\right) + \beta_{c0} + \mathrm{n}_c - \mathrm{X}_c\right)
\end{aligned} \tag{13.4}$$

In this paper, we consider the weight a_0 assigned to the historical data as fixed regardless of the current data. Some researchers have applied power prior in the non-inferiority setting. For example, Gamalo et al. (2013) considered a_0 as random and different a_0's for different historical studies to provide more flexibility in borrowing the historical information.

13.4.2 Hierarchical Model

Let $\pi_1, \pi_2, \ldots, \pi_H$ be the true success rates for the active control in the historical studies (here, the number of historical studies, H, could be 1). Define γ_c be the logit of true active control success rate in the current trial, i.e., $\gamma_c = \log it(\pi_c) = \log\left(\pi_c/(1 - \pi_c)\right)$. Define $\gamma_1, \gamma_2, \ldots, \gamma_H$ be the logit of true success rates in the H historical studies. Assume that $\gamma_c, \gamma_1, \gamma_2, \ldots, \gamma_H \sim N(\mu, \tau^2)$. Thus, μ represents the overall mean and τ represents between-study standard deviation. τ is the borrowing parameter, determining the degree of borrowing. Small τ implies that all γ values are similar and thus more borrowing would be appropriate. Large τ implies that true active control success rates are different in the different studies and thus minimal borrowing is recommended. Chen and Ibrahim (2006) described the asymptotic relationships between the power prior and hierarchical modeling for non-normal models such as generalized linear models. Under some specific conditions, the approximate power prior and the hierarchical model are identical. They also provided a formal methodology for eliciting a guide value for the power parameter a_0 via hierarchical models.

In general, μ and τ are unknown. We place priors on μ and τ, which consequently creates a hierarchical structure. Here, we consider priors $\mu \sim N\left(\mu_0, \tau_0^2\right)$ and $\tau^2 \sim IG(\alpha, \beta)$, an inverse gamma distribution with shape parameter α and scale parameter β. In this manuscript, we assign a $N(0, 100)$ prior for μ (non-informative on the logit scale) and $IG(1, 0.1)$ prior for τ^2 to allow a mild degree of prior information. There are other choices of priors of τ^2 as suggested in Gelman (2006); see also Hsu et al. (2019).

For the analysis, we assume $\gamma_i = \text{logit}(\pi_i) = \gamma_c + \theta$, where θ is the log odds treatment effect, on which a non-informative prior $\theta \sim N(0, 100)$ is assigned. All the conditional posterior distributions will be obtained via Markov Chain Monte Carlo (MCMC) techniques.

13.5 Simulation and Application

In this section, we assess the performance of the proposed approach using simulations.

13.5.1 Simulation Procedure

The steps for the simulations using power priors are outlined below.

1. Simulate a dataset from the binomial distribution with parameters n_c and π_c for the active control device and parameters n_i and $\pi_i = \pi_c - M$ for the investigational device independently. Under this scenario, the investigational device is inferior to the control device. Compute the number of successes X_c and X_i and the proportion of successes $\hat{\pi}_c$ and $\hat{\pi}_i$.
2. Assume a non-informative $Beta\left(\alpha_{c0}, \beta_{c0}\right)$ prior with $\alpha_{c0} = \beta_{c0} = 1$ for the historical control treatment success rate π_c and specify the power parameter a_0, then generate the posterior distribution of π_c with N (e.g., 10,000) MCMC iterations from (13.4).
3. Assume a non-informative $Beta(\alpha_i, \beta_i)$ prior for the investigational treatment success rate π_i and generate the posterior distribution of π_i by simulating N MCMC iterations from (13.4).
4. Compute the posterior differences $\pi_i - \pi_c$ for each of the N iterations.
5. Calculate the posterior probabilities $P(\pi_i - \pi_c \geq -M | X_i, X_c)$ by counting the number of iterations for which $\pi_i - \pi_c \geq -M$ then dividing the total by N.
6. As we mention in Sect. 13.4, no historical information is borrowed for the companion hypothesis H_a^{COM}. So the posterior probability $P(\pi_c \geq PG | X_c)$ is the proportion of $\pi_c \geq PG$ among N π_c's that are sampled from $Beta(1 + X_c, 1 + n_c - X_c)$. Here a $Beta(1, 1)$ prior is used for π_c.

7. Repeat steps 1–6, K (e.g., 1000) times, each time simulating a different dataset with the same parameters and obtaining the posterior probabilities $P(\pi_i - \pi_c \geq -M | X_i, X_c)$ and $P(\pi_c \geq PG | X_c)$.
8. Count the number of datasets for which the posterior probability $P(\pi_i - \pi_c \geq -M | X_i, X_c) > \lambda_1$ and $P(\pi_c \geq PG | X_c) > \lambda_2$, then divide the total by K to obtain the probability of falsely claiming study success when the investigational device is inferior to the control.

To evaluate the probability of correctly claiming study success when the investigational device is not inferior to the control device, data for the investigational device are simulated by setting $\pi_i = \pi_c$.

There is no close form for the posterior distribution for π_i and π_c when hierarchical model is used. The prior distributions for μ and τ described in Sect. 13.4.2 are used in the simulation. Draws from the posterior distributions can be obtained using MCMC sampling implemented in R2WinBUGS (2015).

13.5.2 Simulation Results

To evaluate the operating characteristics, the following scenarios are considered:

π_c = 0.35, 0.40, 0.45, 0.50, 0.55, 0.60, 0.65, $n_i = n_c = 300$. The power parameter $a_0 = 0, 0.35, 0.75, 1.00$ are considered. $N = 10{,}000$ MCMC samples were generated for each dataset. $K = 2000$ datasets were generated for each scenario. A non-informative *Beta*(1,1) prior was considered for the investigational device $Beta(\alpha_i, \beta_i)$ and the historical control $Beta\left(\alpha_{c0}, \beta_{c0}\right)$ for all scenarios.

Assuming the true success rates for the investigational and active control devices to be 0.6, a sample size of 300 per treatment group would demonstrate the non-inferiority of the investigational device by a non-inferiority margin (M) of 0.1 with 80% power at a 5% significance level based on pooled Z-test. In a historical study, 45 out of 75 (60%) subjects treated with the active control device achieved success. To ensure the effect of the active control in the non-inferiority trial to be consistent with the effect observed in the historical trial, the performance goal is set at 0.45.

For success criteria, λ_1 is set at 0.95 for the non-inferiority test so that it mimics the Type I error rate when the non-inferiority hypothesis was considered only. λ_2 is set at 0.85 for constancy requirement since it is a companion test and we would not set the criterion to be too strict.

Table 13.1 provides the posterior probability of falsely claiming study success when the investigational device is inferior to the control device, that is $\pi_i = \pi_c - M$. Notice that study success will not be claimed if the investigational device is inferior to the active control device or the constancy assumption is not met (i.e., $\pi_c < 0.45$). With no surprise, it is difficult to claim study success when the constancy assumption is not met regardless of amount of borrowing using power prior. Similar conclusion is observed with the hierarchical model.

When the constancy assumption is met, e.g., π_c = 0.50, 0.55, the posterior probability of falsely claiming study success is under controlled (i.e., <0.05) for both power prior and hierarchical model. However, when the true success rate for the active control arm in the current trial is higher than the observed success rate in the historical study (i.e., $\pi_c = 0.65$), the posterior probability of falsely claiming study success is higher than 0.05 and increases as more information was borrowed from the historical data. Depending on the practical/clinical need, λ_1 can be adjusted to achieve a desirable level of false study success. On the other hand, it is also noticed that the posterior probability of falsely claiming study success becomes smaller as power parameter a_0 becomes larger when the true success rate for the active control arm in the current trial is lower than the observed success rate in the historical study.

Table 13.1 Posterior probability of claiming study success when the investigational device is inferior to the control device ($\pi_i = \pi_c - M$)

π_c	Posterior prob. of $P(\pi_i - \pi_c \geq -M \mid X_i, X_c) > \lambda_1$ (Non-inferiority only)	Posterior prob. of $P(\pi_i - \pi_c \geq -M \mid X_i, X_c) > \lambda_1$ and $P(\pi_c \geq PG \mid X_c) > \lambda_2$				
		Power parameter a_0				Hierarchical model
		0	0.35	0.75	1	
0.35	0.0415	0	0	0	0	0
0.4	0.0460	0	0	0	0	0
0.45	0.0505	0	0	0	0	0
0.5	0.0510	0.0095	0.0055	0.0025	0.0005	0.0010
0.55	0.0505	0.0485	0.0320	0.0220	0.0185	0.0190
0.6	0.0515	0.0515	0.0475	0.0455	0.0460	0.0505
0.65	0.0510	0.0510	0.0585	0.0656	0.0740	0.0735

Table 13.2 summarizes the posterior probability of claiming study success when the investigational device and active control device perform equally in terms of the success rate, that is $\pi_i = \pi_c$. Note that when the non-inferiority hypothesis is only considered, there is at least 80% chance to claim non-inferiority even when the success rate for the active control device in the current trial is much lower than that in the historical study (i.e., $\pi_c = 0.35, 0.40$). In other words, there is high probability to claim non-inferiority when constancy assumption is not met. However, after incorporating constancy assumption hypotheses, that probability becomes extremely low as expected. Moreover, when the success rate for the control group in the current trial is consistent with the success rate in the historical study (i.e., $\pi_c = 0.60$), there is at least 80% chance to claim study success. It implies that incorporating constancy hypotheses test does not impair the chance of claiming the study success when the device truly works from the statistical perspective. Meanwhile the posterior probability of claiming study success increases as the power parameter a_0 increases. It is also noticed that there is still fairly high chance to claim study success when the success rate for the control group in the current trial is slightly lower than the success rate in the historical study but still meets the constancy requirement (for example, $\pi_c = 0.55$).

Table 13.2 Posterior probability of claiming study success when the investigational device is not inferior to the control device ($\pi_i = \pi_c$)

π_c	Posterior prob. of $P(\pi_i - \pi_c \geq -M \mid X_i, X_c) > \lambda_1$ (Non-inferiority only)	Posterior prob. of $P(\pi_i - \pi_c \geq -M \mid X_i, X_c) > \lambda_1$ and $P(\pi_c \geq PG \mid X_c) > \lambda_2$				
		Power parameter a_0				Hierarchical model
		0	0.35	0.75	1	
0.35	0.8275	0	0	0	0	0
0.4	0.8075	0.0010	0.0010	0.0010	0.0010	0
0.45	0.7940	0.0575	0.0430	0.0340	0.0310	0.0325
0.5	0.8005	0.5565	0.5185	0.4770	0.4485	0.4620
0.55	0.7975	0.7935	0.7805	0.7705	0.7650	0.7585
0.6	0.8140	0.8140	0.8310	0.8495	0.8585	0.8505
0.65	0.8230	0.8230	0.8705	0.9040	0.9235	0.9210

13.6 Discussion

In this article, we have developed a Bayesian methodology which is particularly suitable for designing and analyzing non-inferiority clinical trial to address violation of constancy assumption. For non-inferiority studies, constancy assumption requires that the active control device in the current study should have at least the effect that it was expected to have. If the assumption does not hold and the active control device does not have such an effect in the current non-inferiority trial, a conclusion that an ineffective device works can be erroneously made. In this paper, we proposed to incorporate a companion hypothesis test to ensure that the performance of the active control device in the non-inferiority study is as good as expected, with respect to historical trial. Consequently, the study becomes much harder to falsely claim study success when the constancy assumption is not met. On the other hand, it does not impose any difficulty to claim study success when the constancy assumption is met. Furthermore, if pre-planned, a superiority claim can be pursued after the non-inferiority claim is made as in the usual non-inferiority studies.

We also demonstrated how to incorporate historical data for the control device when the constancy assumption holds with a Bayesian approach. In this paper, we discussed two types of priors, namely, the power prior and the hierarchical prior to incorporate historical data. Similar operating characteristics were observed with the Bayesian approaches. After the companion test shows constancy, borrowing historical active control data usually would increase the power of the non-inferiority study. Please note that in cases where the current active control behaves quite different from its performance in the historical study, the probability of falsely claiming study success and the probability of correctly claiming study success can increase or decrease from the nominal level depending on the direction of the difference. To address this problem, a dynamic borrowing approach may be worthwhile to be explored, and it can serve as a future research topic.

As borrowing historical active control data usually increase the power, to obtain the same level of power, the sample size based on our proposed Bayesian approach may be lower than that derived based on a fixed design. In a future study, one may investigate the amount of sample size that can be saved by our proposed method comparing to the fixed design.

References

Chen, M.H., Ibrahim, J.G.: The relationship between the power prior and hierarchical models. Bayesian Anal. **1**(3), 551–574 (2006)

D'Agostino, R.B., Massaro, J.M., Sullivan, L.M.: Non-inferiority trials: design concepts and issues-the encounters of academic consultants in statistics. Stat. Med. **22**, 169–186 (2003)

FDA. Use of Bayesian statistics in medical device clinical trials. https://www.fda.gov/downloads/MedicalDevices/DeviceRegulationandGuidance/GuidanceDocuments/ucm071121.pdf (2010)

FDA. Design considerations for pivotal clinical investigations for medical devices. https://www.fda.gov/downloads/MedicalDevices/DeviceRegulationandGuidance/GuidanceDocuments/UCM373766.pdf (2013)

FDA. Non-inferiority clinical trials to establish effectiveness. https://www.fda.gov/downloads/Drugs/Guidances/UCM202140.pdf (2016)

Fleming, T.R.: Current issues in non-inferiority trials. Stat. Med. **27**, 317–322 (2008)

Gamalo, M.A., Tiwari, R.C., LaVange, L.M.: Bayesian approach to the design and analysis of non-inferiority trials for anti-infective products. Pharm. Stat. **13**, 25–40 (2013)

Gelman, A.: Prior distributions for variance parameters in hierarchical models. Bayesian Anal. **1**(3), 515–533 (2006)

Hsu, Y.Y., Zalkikar, J., Tiwari, R.C.: Hierarchical Bayes approach for subgroup analysis. Stat. Methods Med. Res. **28**(1), 275–288 (2019)

Hung, H.M.J., Wang, S.-J., Tsong, Y., Lawrence, J., O'Neil, R.T.: Some fundamental issues with non-inferiority testing in active controlled trials. Stat. Med. **22**, 213–225 (2003)

Ibrahim, J.G., Chen, M.-H.: Power prior distributions for regression models. Stat. Sci. **15**(1), 46–60 (2000)

Koopmeiners, J.S., Hobbs, B.P.: Detecting and accounting for violations of the constancy assumption in non-inferiority clinical trials. Stat. Methods Med. Res. **27**(5) 1547–1558 (2018)

Liu, Q., Li, Y., Odem-Davis, K.: On robustness of non-inferiority clinical trial designs against bias, variability, and nonconstancy. J. Biopharm. Stat. **25**, 206–225 (2015)

Lu, N.T., Xu, Y.L., Yang, Y.: Incorporating a companion test into the non-inferiority design of medical device trials. J. Biopharm. Stat. **29**(1), 143–150 (2019)

Nie, L., Soon, G.: A covariate-adjustment regression model approach to noninferiority margin definition. Stat. Med. **29**, 1107–1113 (2010)

Odem-Davis, K., Fleming, T.R.: Adjusting for unknown bias in noninferiority clinical trials. Stat. Biopharm. Res. **5**, 248–258 (2013)

R2WINBUGS. R Package. https://CRAN.R-project.org/package=R2WinBUGS (2015)

Wang, S.-J., Hung, H.M.J.: TACT method for non-inferiority testing in active controlled trials. Stat. Med. **22**, 227–238 (2003)

Chapter 14
Adaptive Randomization for Master Protocols in Precision Medicine

Jianchang Lin, Li An Lin, Veronica Bunn, and Rachael Liu

14.1 Introduction

In the era of precision medicine, especially in the oncology and hematology areas, there have been explosions in knowledge of the molecular profile of disease. With the genomic sequencing becoming more affordable, many tumors can now be classified from a molecular biology perspective, with different treatment options and tailored strategies for patients based on their tumor biomarker status. Under the drug development setting, new generation trials have emerged to target patient selection within any given tumor type based on specific underlying molecular and biologic characteristics: e.g. (1) 'Basket trials' usually are focused only on specific molecular aberrations, in several tumor types. (2) 'Umbrella (or Platform) trials' focus on drug development targeting several molecular subtypes in one tumor type. (3) 'Adaptive enrichment strategies' offer the potential to enrich for patients with a particular molecular feature that is predictive of benefit for the test treatment based on accumulating evidence from the trial. Among them, umbrella, basket and platform trials constitute a new generation of clinical trial design defined as master protocol, which allow for the study of multiple drugs, multiple diseases indications, or both within a single trial. These innovative approaches to clinical drug development have resulted in rapidly revolutionized methodologies, including adaptive randomization (Lin et al. 2016a, b, c), to conduct clinical trials in the setting of biomarkers and targeted therapies, whereas the traditional paradigm of treating

J. Lin (✉) · V. Bunn · R. Liu
Takeda Pharmaceuticals, Cambridge, MA, USA
e-mail: Jianchang.Lin@takeda.com; Veronica.Bunn@takeda.com; Yue.Liu@takeda.com

L. A. Lin (✉)
Merck & Co., Inc., Rahway, NJ, USA
e-mail: li.an.lin@merck.com

L. Zhang et al. (eds.), *Contemporary Biostatistics with Biopharmaceutical Applications*, ICSA Book Series in Statistics,
https://doi.org/10.1007/978-3-030-15310-6_14

a very large number of unselected patients is increasingly less efficient, lacks cost effectiveness and is ethically challenging.

In the past few years, there have been a variety of thought-provoking next generation master protocols conducted multi-institutionally in oncology: specific recognized examples include I-SPY2, BATTLE, NCI MATCH, LUNG-MAP, ALCHEMIST and FOCUS4 (Renfro et al. 2016). As a change from traditional clinical trial design paradigms, statisticians have partnered with clinicians to become fully integrated in these clinical trials and make critical contributions for advancing therapeutic development in this era of molecular medicine. Meanwhile, the new development of immunotherapeutic agents and implementation of next-generation sequencing (NGS) also brings many new and exciting opportunities in the design of biomarker driven trials. From a clinical trial operational perspective, there are some logistical challenges to implementing these innovative designs, e.g. central assay testing, drug supply, multiple institutional collaboration, real time data collection and integrations. However, these additional efforts are all worthwhile given the substantial improvement of efficient medicine development, and most importantly, the benefit of the patients.

In general, the goals of randomized clinical trials are to effectively treat patients and differentiate treatment effects efficiently. On one hand, a clinical trial tries to discriminate the effects of different treatments quickly, so that patients outside of the trial will sooner benefit from the more efficacious treatment. For this purpose, patients' allocation should be (nearly) balanced across the comparative arms. On the other hand, each trial participant should be given the most effective treatment, and patients themselves also hope that they would be assigned to the arm that performs better. This often leads to an unbalanced allocation through adaptive randomization by equipping a better arm with a higher allocation probability (Berry et al. 2010). Therefore, randomized clinical trials need to strike a balance between individual and collective ethics.

During the study planning stage, key components of the protocol such as primary endpoint, key secondary endpoints, clinically meaningful treatment effect difference, and treatment effect variability are pre-specified. Participating investigators and sponsors then collect all data in electronic data capture (EDC) system and perform statistical analyses. The success of the study depends on the accuracy of the original design assumptions or sample size calculation. Adaptive Designs are a way to address uncertainty about design parameters assumptions made during the study planning stage. Adaptive Designs allow a review of accumulating data or patient information during a trial to possibly modify trial characteristics and to promote multiple experimental objectives, while protecting the study from bias and preserving inferential validity of the results. The flexibility can translate into a more efficient drug development process by reducing the number of patients enrolled. This flexibility also increases the probability of success of the trial answering the question of scientific interest (finding a significant treatment effect if one exists or stopping the study as early as possible if no treatment effect exists).

Adaptive Designs have received a great deal of attention in the statistical, pharmaceutical, and regulatory fields. The US Food and Drug Administration (FDA) released a draft version of the "Guidance for Industry: Adaptive Design Clinical Trials for Drugs and Biologics" in 2010 (U.S. Food and Drug Administration 2010). The guidance defined an adaptive design as 'a study that includes a prospectively planned opportunity for modification of one or more specified aspects of the study design and hypotheses based on analysis of data (usually interim data) from subjects in the study.' The most common adaptive designs used in clinical trials include, but are not limited to, the following types: adaptive randomization design, seamless adaptive phase II/III design, adaptive dose-response design, biomarker adaptive design, adaptive treatment switching design, adaptive-hypothesis design, multiple arm adaptive design, group sequential design, sample size re-estimation design, et al (Kairalla et al. 2012).

14.2 Why Is Adaptive Randomization Important?

The design of any clinical trial starts with formulation of the study objectives. Most clinical trials are naturally multi-objective, and some of these objectives may compete. For example, one objective is to have sufficient power to test the primary study hypothesis, and consequently have sufficient sample size. However, cost considerations may preclude a large sample size, so the twin objectives of maximum power and minimum sample size directly compete. Other objectives may include minimizing exposure of patients to potentially toxic or ineffective treatments, which may compete with having sufficient numbers of patients on each treatment arm to conduct convincing treatment group comparisons. In the case of $K > 2$ treatments, where $(K - 1)$ experimental treatments are to be compared with the placebo group with respect to some primary outcome measure, the primary objective of the trial may be testing an overall hypothesis of homogeneity among the treatment effects, and a secondary objective may be performing all pairwise comparisons among the $(K - 1)$ experimental treatments versus placebo. Investigators may have an unequal interest in such comparisons. In addition to statistical aspects of a clinical trial design, there may be a strong desire to minimize exposure of patients to the less successful (or more harmful) treatment arms. Clearly, in these examples it is very difficult to find a single design criterion that would adequately describe all the objectives. Many of these objectives depend on model parameters that are unknown at the beginning of the trial. It is useful, and indeed sometimes imperative, to use accruing data during the trial to adaptively redesign the trial to achieve these objectives. These design considerations must be achieved without sacrificing the hallmark of the carefully conducted clinical trials—randomization—which protects the study from bias.

Once the study objectives are formally quantified and ranked in the order of their importance, the experimental design problem is to find a design that accommodates several selected design criteria. Frequently, the treatment allocations are unbalanced across treatment groups, and they depend on model parameters that are unknown a priori and must be calibrated through simulation. Adaptive randomization uses accruing information in the trial to update randomization probabilities to target the allocation criteria. Hu and Rosenberger (2006) classify adaptive randomization into four major types:

- *Restricted randomization*: a randomization procedure that uses past treatment assignments to select the probability of future treatment assignments, with the objective to balance numbers of subjects across treatment groups.
- *Covariate-adaptive randomization*: a randomization procedure that uses past treatment assignments and patient covariate values to select the probability of future treatment assignments, with the objective to balance treatment assignments within covariate profiles.
- *Response-adaptive randomization*: a randomization procedure that uses past treatment assignments and patient responses to select the probability of future treatment assignments, with the objective to maximize power or minimize expected treatment failures.
- *Covariate-adjusted response adaptive (CARA) randomization*: a combination of covariate-adaptive and response-adaptive randomization procedures.

A typical example of master protocol to screen three experimental treatments, A, B and C simultaneously is illustrated in Fig. 14.1. All patients recruited for the first stage of the trial are randomized to the treatment arms with equal probability. At each interim analysis, we update the Bayesian model used for setting the randomization probabilities. The proportion of patients that are randomized to better performing arms increases, and decreases to arms that are performing poorly.

14.3 Frequentist and Bayesian Approaches for Adaptive Randomization

The commonly used statistical approach to design and analyze clinical trials and other medical experiments is frequentist, while a Bayesian method provides an alternative approach. The Bayesian approach can be applied separately from frequentist methodology, as a supplement to it, or as a tool for designing efficient clinical trials that have good frequentist properties. The two approaches have rather different philosophies, although both use probability and deal with empirical evidence. Practitioners exposed to traditional, frequentist statistical methods appear to have been drawn to Bayesian approaches for three reasons (Ning and Huang 2010; Rosenberger et al. 2012; Thall and Wathen 2007; Yin et al. 2012; Yin 2013). One is that Bayesian approaches implemented with the majority of their informative content coming from the current available data, and not prior information, typically have good frequentist properties (e.g., low mean squared error (MSE) in repeated use). Second, these methods as now easily implemented in WINBUGS, OpenBUGS and other available MCMC software packages. These offer a convenient approach to hierarchical or random effect modeling, as regularly used in longitudinal data, frailty

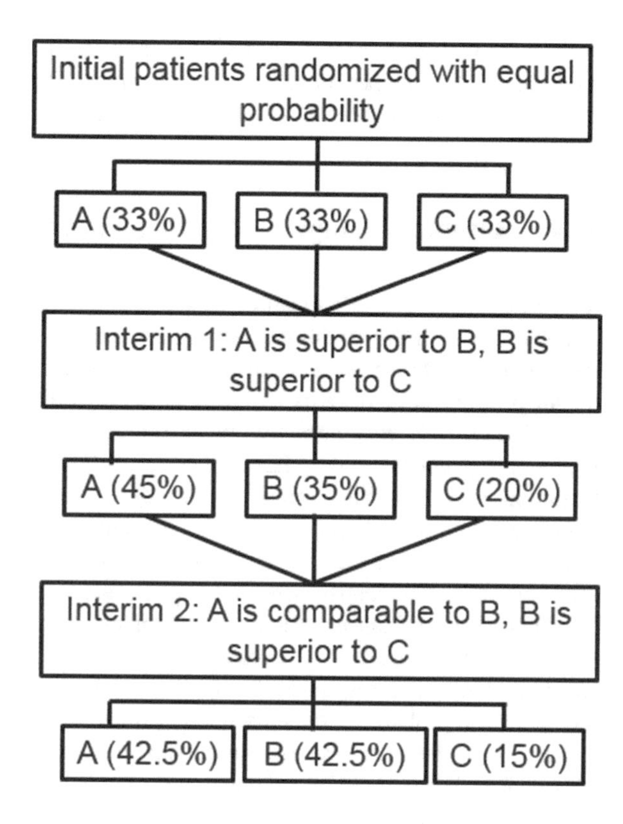

Fig. 14.1 Overview of adaptive randomization design

model, spatial data, time series data, and a wide variety of other settings featuring interdependent data. Third, practitioners are attracted to the increased levels of flexibility and adaptivity offered by the Bayesian approach which allows for early stopping for efficacy, toxicity, and futility, as well as facilitates a straightforward solution to a great many other advanced problems such as dosing selection, adaptive randomization, equivalence testing, and others.

Flexibility is the major difference between Bayesian and frequentist methods, in both design and analysis. In the Bayesian approach, experiments can be altered midcourse, disparate sources of information can be combined, and expert opinion can play a role in inferences. An important property of Bayesian design is that it can utilize prior information and Bayesian updating while still maintaining good frequentist properties (power and Type I error). Another major difference is that the Bayesian approach can be decision-oriented, with experimental designs tailored to maximize objective functions, such as company profits or overall public health benefit. Overall, designing a clinical trial is a decision problem, such as therapy selection, resource allocation, early stop etc., which involves costs and benefits consideration. In the Bayesian approach, these costs and benefits can be well assessed for all possible scenarios of future observations. However, frequentism fits naturally with the regulatory "gate-keeping" role, through its insistence on procedures that perform well in the long run regardless of the true state of nature. And indeed, frequentist operating characteristics (Type I and II error, power) are still very important to the FDA and other regulators.

14.4 Response-Adaptive Randomization

Response-adaptive randomization is one of the most important adaptive trial designs, in which the randomization ratio of patients assigned to the experimental treatment arm versus the control treatment arm changes overtime from 1:1 to a higher proportion of patients assigned to the arm that is performing better (Yuan et al. 2011). It is very attractive when ethical considerations or concerns make it potentially undesirable to have an equal number of patients assigned to each treatment arm. For the purpose of simplicity, suppose the trial objective is to compare treatments A and B. Patients are enrolled in sequential groups of size $\{N_j\}$, $j = 1, \ldots, J$, where N_j is the sample size of group j. When planning the trial, researchers typically have limited prior information regarding the superiority or effectiveness of the experimental treatment arms. Therefore, at the beginning stage of the trial, for the first j groups, e.g. j' = 1, patients are equally allocated to two treatments. The responses observed from these patients are utilized to update the allocation probability for subsequent groups of patients.

Let p_A be the response rate of treatment A and p_B be the response rate of treatment B. We set N to be the maximum sample size allowed for the trial and N_A (N_B) to be the maximum number of patient assigned to treatment A (B). We assign the first N_1 patients equally to two treatments (A, B) and observe the response Y_k (k = A, B). Assign p_k a noninformative prior of beta(α_k, β_k). If among n_k subjects treated in arm k, we observe y_k responses, then

$$Y_k \sim \text{binomial}\,(n_k, p_k) \tag{14.1}$$

and the posterior distribution of p_k is

$$p_k \mid \text{data} \sim \text{beta}\,(\alpha_k + x_k, \beta_k + n_k - x_k) \tag{14.2}$$

During the trial, we could continuously update the Bayesian posterior distribution of p_k, and allocate the next N_j patients to the kth treatment arm according to the posterior probability that treatment k is superior to all other treatment arms

$$\pi_k = \Pr\,(p_k = \max\,\{p_l, 1 \le l \le K\}\,|\text{data}) \tag{14.3}$$

One of the advantages of a Bayesian approach to inference is the increased flexibility to include sequential stopping compared to the more restrictive requirements of a classical group sequential approach in terms of number of interim analysis, stopping rules, etc. Noninformative stopping rules are irrelevant for Bayesian inference. In other words, posterior inference remains unchanged regardless of why the trial was stopped. Several designs make use of this feature of Bayesian inference to introduce early stopping for futility and/or for efficacy.

- *Futility*: if Pr ($p_k < p.min$|data) > θ_u, where p.min denotes the clinical minimum response rate, that is, there is strong evidence that treatment k is inferior to the clinical minimum response rate, we drop treatment arm k.
- *Superiority*: if Pr ($p_k > p.target$|data) > θ_l, where p.target denotes the target response rate, that is, there is strong evidence that treatment k is superior to prespecified response rate, we terminate the trial early and claim the treatment *k is promise*.

At the end of the trial, if Pr ($p_k > p.min$|data) > θ_t, then treatment k is selected as the superior treatment. Otherwise, the trial is inconclusive. To achieve desirable operating characteristics (type 1 error and power), we use simulations to calibrate the pre-specified cut-off points θ_u, θ_l, and θ_t.

We conducted simulations to show the procedure for design parameters calibration. The patient allocation probability is determined by algorithm (14.3). The minimum allocation probability is 10% to ensure a reasonable probability of randomizing patients to each arm. The minimum clinical response rate ($p.min$) is 0.2 and the target response rate ($p.target$) is 0.4. In this trial, we set maximum sample sizes of 90 and maximum sample size of 30 per treatment arm. We equally assigned the first 15 patients to three treatments (A, B, or C) and started using the adaptive randomization at the 16th patient. The sequential group size is set as 10, so that the early stopping rule and allocation probability updating will act after 10 new patient's responses cumulated. Although the design allows continuous monitoring after every patient's response outcome becomes available, from the operational and computational point of view, it's more convenient to monitor the trial for early termination with a cohort size of 10. A total of 5,000 independent simulations were performed for each configuration.

In the first stage, we set $\theta_u = \theta_l = 1$, so that the trial would not be terminated early, to determine the threshold values of θ_t. we performed a series of simulation studies with different values of θ_t and compared the corresponding type 1 error rates and powers. Table 14.1 shows the simulation results. Similarly, we can obtain a set of values of θ_t that reached the desired power. The value of θ_t with type 1 error (defined as the selection probability of Arm A) close to 5% and a desired power (defined as the selection probability of Arms B or C) will be selected for the next stage selection.

Table 14.1 Type 1 error rates and power, without early termination

	θ_t			
Arm (response rate)	0.9	0.91	0.92	0.93
A (0.2)	0.07	0.065	0.056	0.049
B (0.4)	0.842	0.838	0.832	0.825
C (0.6)	0.998	0.994	0.992	0.989

Table 14.2 Type 1 error rates and power, with early termination

		θ_l			
θ_u	Arm	0.92	0.93	0.94	0.95
0.85	A	0.073	0.071	0.064	0.066
	B	0.828	0.819	0.84	0.817
	C	0.991	0.996	0.989	0.992
0.86	A	0.082	0.071	0.065	0.057
	B	0.821	0.824	0.814	0.839
	C	0.993	0.99	0.989	0.986
0.87	A	0.08	0.07	0.067	0.057
	B	0.822	0.819	0.838	0.845
	C	0.996	0.994	0.994	0.993
0.88	A	0.078	0.07	0.069	0.053
	B	0.843	0.847	0.82	0.801
	C	0.996	0.993	0.995	0.996
0.89	A	0.079	0.072	0.069	0.048
	B	0.852	0.832	0.845	0.819
	C	0.991	0.994	0.997	0.994
0.9	A	0.069	0.063	0.062	0.048
	B	0.831	0.83	0.821	0.826
	C	0.997	0.989	0.992	0.994

In the second stage, fixing $\theta_t = 0.92$, we followed the similar procedure to calibrate (θ_u, θ_l), which determine the early termination of a trial due to equivalence or superiority respectively. Note that θ_l has to be greater or equal to θ_t because the decision criteria must be tighter during the trial than at the end of trial. Our goal is still to maintain a treatment-wise type 1 error rate of 5% or lower and to achieve desired power when the trial can terminate early (Table 14.2).

Alternatively, we can set $\theta_t = \theta_l$ which means that we will not relax the decision criteria at the end of the trial. Extensive simulation for various scenarios should be carried out to ensure controlled type 1 error and satisfied power for all possible situations in real trial (Table 14.3).

Suppose the trial require 0.1 type 1 error and at least 0.85 power for treatment B and 0.99 power for treatment C, we chose the design parameters as $\theta_t = \theta_l = 0.89$ and $\theta_u = 0.9$. The operation characteristics is list in Table 14.4.

14.5 Response-Adaptive Randomization for Survival Outcomes

The response-adaptive randomization design with binary outcomes is commonly used in clinical trial where "success" is defined as the desired (or undesired) event occurring within (or beyond) a clinically relevant time. Given that patients

Table 14.3 Type 1 error rates and power with $\theta_t = \theta_l$

		θ_u					
$\theta_t = \theta_l$	Arm	0.8	0.82	0.84	0.86	0.88	0.9
0.85	A	0.142	0.127	0.127	0.126	0.115	0.105
	B	0.871	0.869	0.866	0.864	0.857	0.857
	C	0.992	0.994	0.991	0.995	0.993	0.994
0.87	A	0.121	0.121	0.117	0.108	0.1	0.085
	B	0.876	0.877	0.87	0.87	0.869	0.852
	C	0.996	0.994	0.992	0.994	0.996	0.986
0.89	A	0.109	0.102	0.093	0.091	0.071	0.075
	B	0.855	0.861	0.849	0.861	0.847	0.857
	C	0.992	0.994	0.995	0.988	0.996	0.995
0.91	A	0.097	0.082	0.08	0.078	0.075	0.077
	B	0.83	0.848	0.848	0.84	0.849	0.825
	C	0.993	0.994	0.996	0.987	0.989	0.988
0.93	A	0.095	0.074	0.071	0.071	0.064	0.06
	B	0.797	0.809	0.835	0.833	0.817	0.799
	C	0.994	0.99	0.991	0.996	0.99	0.992
0.95	A	0.065	0.042	0.039	0.039	0.036	0.025
	B	0.784	0.792	0.775	0.764	0.79	0.778
	C	0.988	0.995	0.989	0.989	0.986	0.994

Table 14.4 Operation characteristics with $\theta_t = \theta_l = 0.89$ and $\theta_u = 0.9$

Arm	Response rate	Pr (selected early)	Pr (stopped early)	# patients (2.5%, 97.5%)
A	0.2	0.012	0.386	24.15 (6, 35)
B	0.4	0.496	0.077	27.72 (6, 37)
C	0.6	0.827	0.005	16.45 (7, 32)

enter a trial sequentially, only a fraction of patients will have sufficient follow-up during interim analysis. This results in a loss of information as it is unclear how patients without sufficient follow-up should be handled. Adaptive designs for survival trials have been proposed for this type of trial. However, current practice generally assumes the event times follow a pre-specified parametric distribution. In this section, we adopt a nonparametric model of survival outcome which is robust to model event time distribution, and then apply it to response-adaptive design. The operating characteristics of the proposed design along with parametric design are compared by simulation studies, including their robustness properties with respect to model misspecifications.

Patients are enrolled in sequential groups of size $\{N_j\}$, $j = 1, \ldots, J$, where N_j is the sample size of the sequential group j. Typically, before conducting the trial, researchers have little prior information regarding the superiority of the treatment arms. Therefore, initially, for the first j' groups, e.g. j' = 1, patients are allocated to K treatment arms with an equal probability 1/K. As patients accrue, the number of current patients increases. Let T_i be the event time for patient i, τ be the clinically relevant time and $\theta = \Pr(T > \tau)$ be the probability of interest. For example, a trial is conducted to assess the progression-free survival probability at 9 months. Let N(s) denote the current number of patients who have been accrued and treated at a given time s during the trial. Without censoring, θ can be modeled by binomial model where the likelihood function evaluated at time s is

$$\mathrm{L}\left(\mathrm{data}|\theta\right) = \prod_{i=1}^{N(s)} \theta^{I(T_i > \tau)} (1 - \theta)^{I(T_i \le \tau)} \tag{14.4}$$

However, censoring is unavoidable in clinical practice. As patients enter into the trial sequentially, the follow-up time for certain patients may be less than τ when we evaluate θ at any calendar time s. Other reasons for censoring may include patient dropout, failure to measure the outcome of interest, etc. If we ignore the censored patients, substantial information will be lost. Cheung and Chappell (2000) introduced a simple model for dose-finding trial. Later, Cheung and Thall (2002) adopted this model to continuous monitoring for phase II clinical trials. With censoring, the likelihood function (14.4) can be rewritten as

$$\mathrm{L}\left(\mathrm{data}|\theta\right) = \prod_{i=1}^{N(s)} \Pr\left\{T_i \le \min\left(x_i, \tau\right)\right\}^{Y(x_i)} \Pr\left\{T_i > \min\left(x_i, \tau\right)\right\}^{1-Y(x_i)} \tag{14.5}$$

where $x_i = \min(c_i, t_i)$ is the observed event time, c_i is the censoring time, and $Y(x_i) = I\{T_i \le \min(x_i, \tau)\}$ is the censoring indicator function.

Furthermore, the parameter θ will be plugged into the likelihood function through probability transformation. Let $t = \min(x_i, \tau)$,

$$\begin{aligned} \Pr(T_i \le t) &= \Pr(T_i \le t, T_i \le \tau) + \Pr(T_i \le t, T_i > \tau) \\ &= \Pr(T_i \le t | T_i \le \tau) \Pr(T_i \le \tau) + \Pr(T_i \le t | T_i > \tau) \Pr(T_i > \tau) \\ &= w(t)\,(1 - \theta) \end{aligned} \tag{14.6}$$

where $w(t) = \Pr(T_i \le t |\, T_i \le \tau)$, is a weight function

Finally, we can obtain a working likelihood with unbiased estimation of $w(t)$.

$$\mathrm{L}\left(\mathrm{data}|\theta\right) = \prod_{i=1}^{N(t)} \widetilde{w}\left(x_i\right)(1 - \theta)^{Y(x_i)} \left\{1 - \widetilde{w}\left(x_i\right)(1 - \theta)\right\}^{1-Y(x_i)} \tag{14.7}$$

Theorem *if $\widetilde{w}\,(x_i)$ converges almost surely to $w(x_i)$ for all i as $N(s) \to \infty$, then $\hat{\theta} = \mathrm{argmax} L\,(data|\theta)$ is strongly consistent for true survival probability θ.*

Cheung and Chappell (2000) assumed the nuisance parameter $\widetilde{w}\,(x_i)$ as a linear function $\widetilde{w}\,(x_i) = x_i/\tau$. Ji and Bekele (2009) show that these estimated weights are based on strong assumption of linearity and independence, and may lead to biased results when the assumptions are violated. We propose to estimate $\widetilde{w}(x_i)$ with Kaplan–Meier (KM) estimation of $\widetilde{S}(x_i)$, where

$$\widetilde{\mathrm{w}}\left(\mathrm{x_i}\right)=\frac{1-\widetilde{\mathrm{S}}\left(\mathrm{x_i}\right)}{1-\widetilde{\mathrm{S}}\left(\tau\right)}$$

It's easy to show that $\widetilde{w}(x_i)$ is an unbiased estimation of $w(x_i)$.

Assign a noninformative prior of beta (α, β), we can obtain posterior distribution of θ. However, the posterior distribution is not available in closed form and standard integral approximations can perform poorly. Without knowing the exact posterior distribution, we can easily draw random MCMC samples and obtain posterior estimation using standard MCMC software packages.

Under model (14.7), the survival probability evaluated at time τ is used as a conventional measure of treatment efficacy. However, such a survival probability at time τ ignores the entire path of survival curve. One of the interests in a clinical trial is the estimation of the difference between survival probability for the treatment groups at several points in time. As shown in Fig. 14.2, the survival curve under treatment B declines faster than that under treatment A, although both treatments have the same survival probability at time τ. In a renal cancer trial, this indicates that patients under treatment B would experience disease progression much faster than those under treatment A. Because delayed disease progression typically leads to a better quality of life, treatment A would be preferred in this situation (Ning and Huang 2010). Another example is showed in Fig. 14.3. The survival curves are almost identical between two treatments before time 20. If we compare the survival probability between two treatments at the time before 20, the treatment effect is inconclusive. To provide a comprehensive measure of efficacy by accounting for the shape of the survival curve, we propose to evaluate survival probability at several points in time. Let θ_{kj} be the survival probability at time τ_{j} for treatment k where j=1, ..., J. The treatment allocation probability for treatment k is defined as,

$$\pi_{\mathrm{k}}=\sum_{\mathrm{j}=1}^{\mathrm{J}}\mathrm{w_j}\,\mathrm{Pr}\left(\theta_{\mathrm{kj}}=\max\left\{\theta_{\mathrm{lj}},\,1\leq 1\leq \mathrm{J}\right\}|\mathrm{data}\right)$$

where $\mathrm{w_j}$ is the prespecified weight. Currently, we use equal weight with $\mathrm{w_j}=1/\mathrm{J}$.

During the trial, we continuously monitor posterior probability of π_k. When the efficacy of π_k is lower than the prespecified lower limit $\mathrm{p_l}$, then the treatment arm k will be terminated early due to futility. When π_k is higher than $\mathrm{p_u}$, the treatment arm k will be selected as promising treatment. At the end of the trial, if π_k is higher than $\mathrm{p_t}$, then treatment k is selected as the superior treatment. Otherwise, the trial is inconclusive. In practice, the values of $\mathrm{p_l}$, $\mathrm{p_u}$, and $\mathrm{p_t}$ are chosen by simulation studies to achieve desirable operating characteristics for the trial.

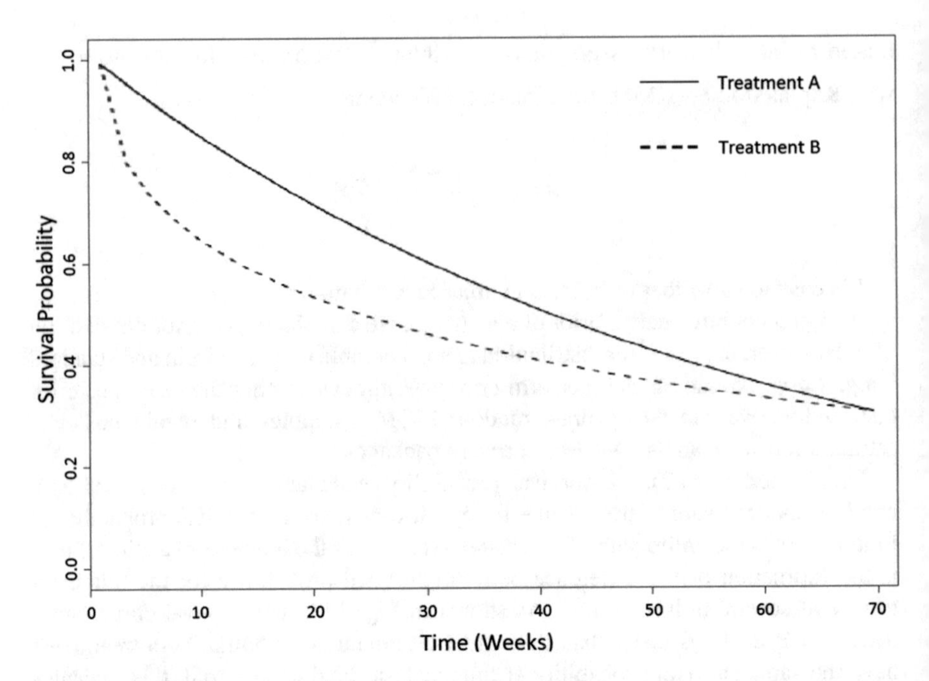

Fig. 14.2 Survival curves of the time to disease progression, where the two survival curves have the same survival probability at the follow-up time $\tau = 70$ weeks, but different areas under the survival curves until τ

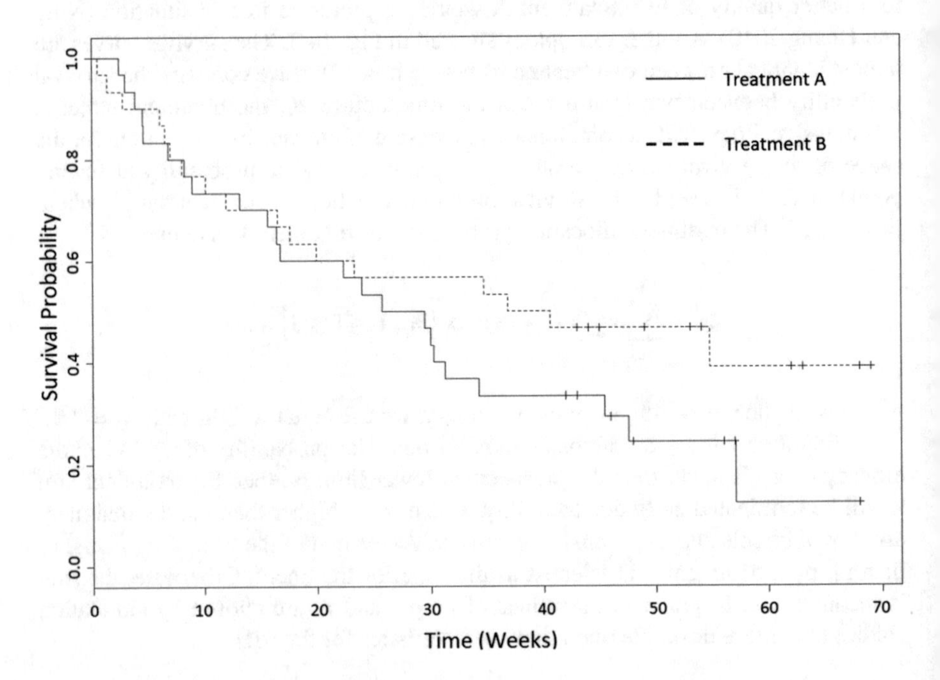

Fig. 14.3 Survival curves of the time to disease progression, where the two survival curves have the similar survival probability before week 20, but gradually show difference as time increase

We simulate a single arm trial where the event times follow a Weibull distribution with $\alpha = 2$ and $\lambda = 50$, where α is the shape parameter and λ is the scale parameter of the distribution. Patients enter the trial sequentially with accrual rate of one per week. At week 50, we stop enrolling the patients and continue to follow the trial for additional 30 weeks. The parameter of interest is $\theta = \Pr(T > 40)$.

The purpose of this simulation study is to compare the performance of estimation with different methods and to show whether the estimation at different trial monitoring time is consistent. Four estimation methods will be evaluated, including the proposed method, the true parametric method (estimate $S(x_i)$ by Weibull distribution), the misspecified parametric method (estimate $S(x_i)$ by exponential distribution), and the original method ($\widetilde{w}(x_i) = x_i/\tau$). Trial monitoring starts at week 40 and continues until the end of study. Figure 14.4 shows the estimated θ at different monitoring times. The results show that both the true parametric method and proposed method provide unbiased estimation over monitoring time while the original method and misspecified parametric method give large bias. It should be noted that the original method gives small bias at the end of trial because the number of censored observations (e.g. due to treatment ongoing) decreased as follow-up time increased. In Fig. 14.5, we present the coverage probability along the monitoring times. The figure shows that the proposed method and true parametric method provide constant coverage probability over the monitoring time which is close to the nominal value of 95%. In contrast, the original method and misspecified parametric method both give low coverage probability.

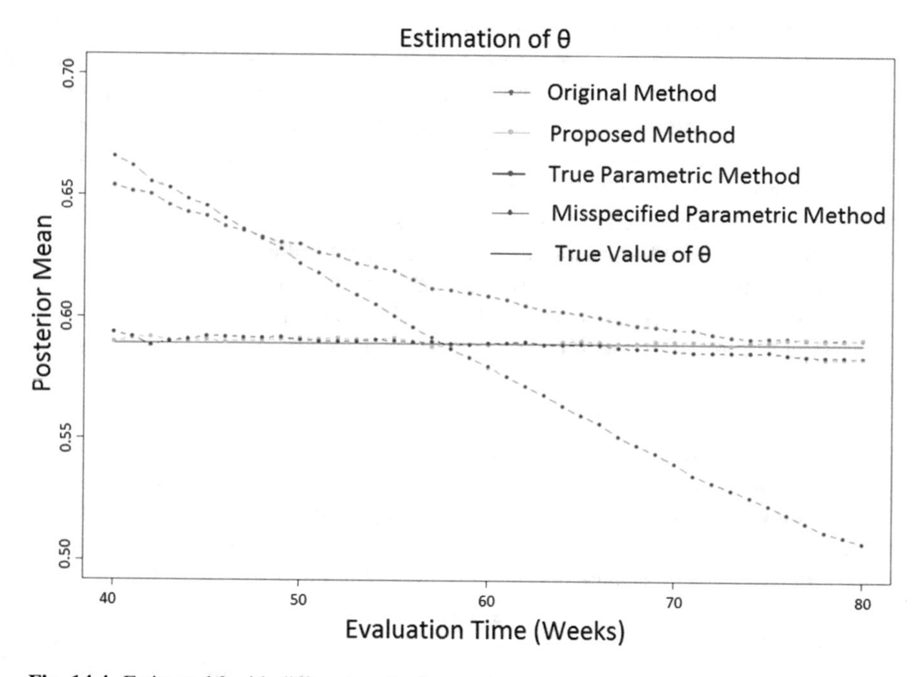

Fig. 14.4 Estimated θ with different methods

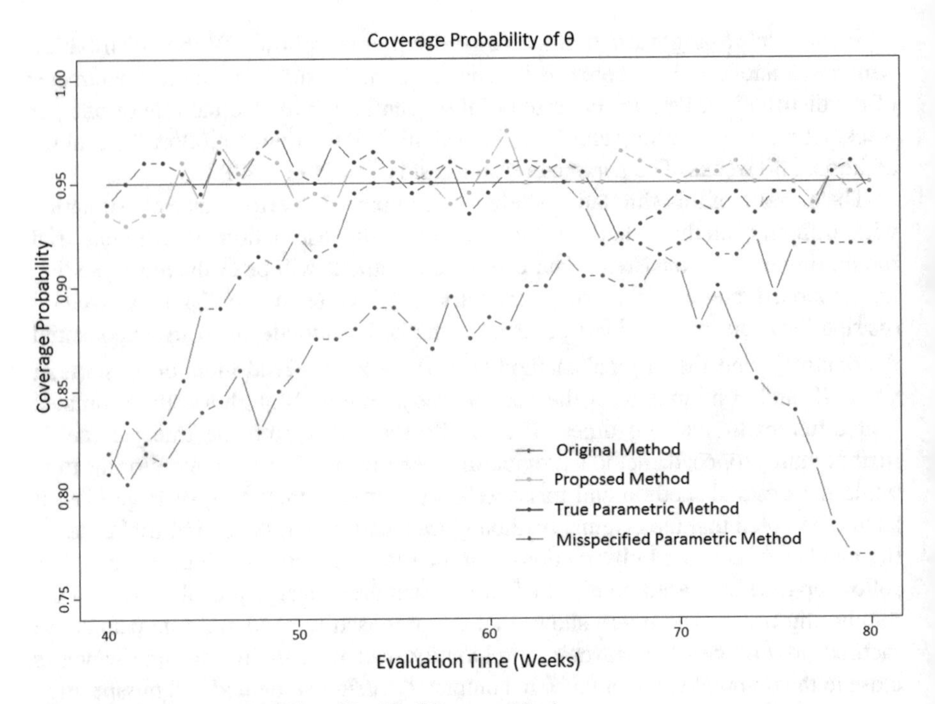

Fig. 14.5 Coverage probability of θ with different methods

We conducted a second set of simulations to evaluate the performance of the proposed adaptive randomization design under various clinical scenarios (1000 simulations per scenario). For the simulations, we set the accrual rate to two patients per week. The maximum number of patients is 120. After the initial 60 weeks of enrollment time, there is an additional follow-up period of 40 weeks. The event times are simulated from a Weibull distribution with $\alpha = 1$ in scenario I and $\alpha = 0.5$ in scenario II. We assigned the first 30 patients equally to two arms (A or B) and started using the adaptive randomization at the 31st patient. The randomization probability was evaluated every 5 weeks. The proposed design will be compared with different estimation methods for the weight function $w(t)$: proposed method, parametric method (estimate $S(x_i)$ by exponential distribution), and original method $(\tilde{w}(x_i) = x_i/\tau)$.

Table 14.5 shows the simulation results from scenario I, without early termination ($p_u = 1, p_l = 0$). For each method, we list the average number of patients (with percentage of total patients in the trial) assigned to each treatment arm, and the chance of a treatment being selected as promising. When comparing the parametric method, the proposed method provides comparable operational characteristic where both designs assign more patients to more promising treatment (69% for proposed design and 70.3% for parametric design) and both designs provide the sample level of power (0.978 for proposed design and 0.979 for parametric design). The original method achieves lower power than both the proposed method and parametric method.

Table 14.5 Simulation result for scenario I

		Proposed method ($p_t = 0.955$)		Exponential method ($p_t = 0.99$)		Original method ($p_t = 0.995$)	
Arm	λ	# of patients	Pr (select)	# of patients	Pr (select)	# of patients	Pr (select)
A	40	26.3 (31%)	0.005	24.72 (29.6%)	0.003	36.5 (36.1%)	0.003
B	100	58.6 (69%)	0.978	61.36 (70.3%)	0.979	64.5 (63.9%)	0.749
		84.9		90.08		101.0	

Table 14.6 shows simulation results for scenario II, without early termination ($p_u = 1$, $p_l = 0$). In the presence of event time distribution misspecification, the parametric method provides lower power than the proposed method (0.836 vs 0.647). In addition, the proposed method assigns more patients to the more promising treatment. Once again, the original method has lower power than the other two methods.

Table 14.6 Simulation result for scenario II

		Proposed method ($p_t = 0.965$)		Exponential method ($p_t = 0.995$)		Original method ($p_t = 0.995$)	
Arm	λ	# of patients	Pr (select)	# of patients	Pr (select)	# of patients	Pr (select)
A	50	27.8 (28.5%)	0.005	32.48 (32.2%)	0.0003	35.6 (33.8%)	0.001
B	200	69.6 (71.5%)	0.836	68.34 (67.8%)	0.647	69.8 (66.2%)	0.51
		97.4		100.82		105.4	

14.6 Case Studies

14.6.1 Investigation of Serial Studies to Predict Therapeutic Response with Imaging and Molecular Analysis 2 (I-SPY 2)

I-SPY 2 is an adaptive phase II clinical trial that pairs oncologic therapies with biomarkers for women with advanced breast cancer. The goal is to identify improved treatment regimens for patient's subsets based on molecular characteristics (biomarkers) of their disease (Barker et al. 2009).

The trial (Fig. 14.6) is initialized with two standard-of-care arms, and five treatment arms. Each treatment is tested on a minimum of 20 patients, and a maximum of 120 patients. Patient's biomarkers are determined at enrollment, and patients are randomized to treatment arms based on their biomarker signature. Bayesian methods of adaptive randomization are used to achieve a higher probability of efficacy. Thus, treatments that perform well within a biomarker subgroup will have an increased probability of being assigned to patients with that biomarker.

Treatments will be dropped for futility if they show a low Bayesian predictive probability of being more effective than the standard of care with any biomarker. Treatment regimens that show a high Bayesian predictive probability of being more effective than the standard of care will stop for efficacy at interim time-points. These treatments will advance (with their corresponding biomarkers) to phase III trials. Depending on the patient accrual rate, new drugs can be added to the trial as other drugs are discontinued for either futility or efficacy.

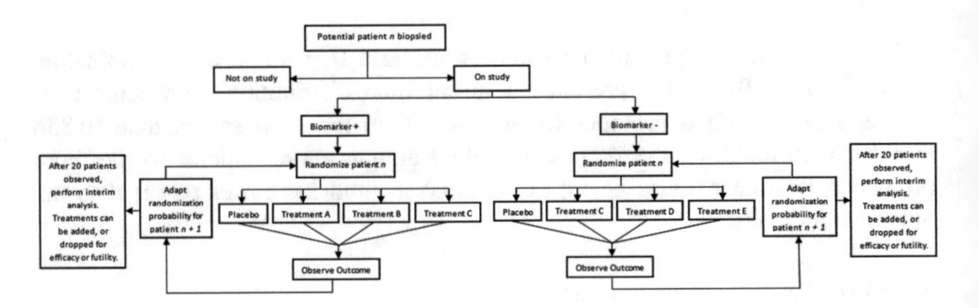

Fig. 14.6 I-SPY 2 trial

As of March 2017, 12 experimental treatment arms have been explored. Five agents, after showing promise within their biomarker groups, advanced to further studies and others are in queue for entry. A new I-SPY 3 master protocol is under planning to provide further evidence of effectiveness for agents successfully graduating from I-SPY2.

14.6.2 Gastric Cancer Umbrella Design for an Investigational Agent

This is an open-label, multicenter, phase 1b study of an investigational agent in combination with regimen A, regimen B, paclitaxel, or docetaxel in adult patients with locally advanced and metastatic gastric or gastroesophageal adenocarcinoma (Fig. 14.7). The study consists of a dose escalation phase (Part 1) and a dose expansion phase (Part 2). In Part 2, this study uses equal and adaptive randomization.

Any patient who enters Part 2 of the study is screened to determine whether their tumor tissue is positive for EBV (approximately 9% of patients with gastric cancer). An estimated 28 patients who are EBV-positive are assigned to treatment with regimen A in combination with the investigational agent (Cohort A). Patients who are EBV-negative initially are randomized equally to 1 of the other treatment cohorts, 5 patients per group: investigational agent + egimen B (Cohort B), investigational agent + paclitaxel (Cohort C), or investigational agent + docetaxel (Cohort D). These patients' data are assessed using a proportional weighted clinical utility function (allocating specific weights for complete response [CR], partial response [PR], stable disease [SD], and progressive disease [PD]). New patients are then randomized to treatment according to an adaptive randomization algorithm, which incorporates a weighted clinical utility function. The resulting probability is continually updated per accumulating data on the associations between the response rate and Bayesian stopping rules.

Adaptive randomization increases the opportunity for each patient to receive the most effective experimental treatment possible based on posterior probabilities. Up to an additional 25 patients may be enrolled in each treatment regimen. Based on simulation results, the sample size for Part 2 (umbrella portion) of the study may be between 61 and 90 patients.

Overall response rate is used as the efficacy benchmark. Target effect size of 25% (0.25) and an undesirable effect size of 10% (0.1) are chosen based on clinical judgment. Early stopping rules are prespecified if there is a clear signal of efficacy or lack of efficacy. The stopping rules are as follows:

1. achieve maximum sample size of each arm (30 patients);
2. stop an arm if posterior probability Pr (response rate [RR] > 0.25/Data) >80% and Pr (RR > 0.10)/Data) >90%;
3. suspend accrual to an arm if Pr (RR $\leq$ 0.10/Data) >80%.

The treatment arm(s) is/are chosen in relation to the efficacy bar prespecified (target and undesirable); therefore, it is possible to select multiple treatment arms per this study design.

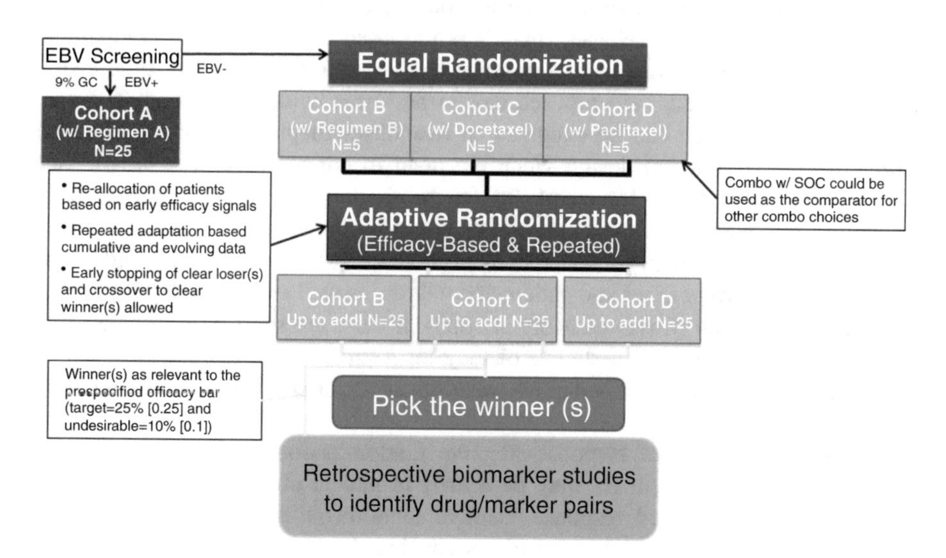

Fig. 14.7 Gastric cancer umbrella design

14.7 Discussion

With the closer collaborations between government, academia and industry, as well as the need to increase the probability of success of drug development across varied therapeutic areas, there are significant growing uses of innovative adaptive designs in master protocols, including the umbrella or platform trials, to screen multiple drugs simultaneously (Woodcock and LaVange 2017). Though different master protocols come with different sizes and settings, they share many common features, e.g. additional planning from the beginning of trial design, coordination between different stakeholders and increasingly sophisticated infrastructures for the research effects. Adaptive randomization is becoming a critical component and statistical methodology under these settings. While response-adaptive randomization procedures are not appropriate in clinical trials with a limited recruitment period and/or outcomes that occur after a long follow up, there is no reason why response-adaptive randomization cannot be used in clinical trials with moderately delayed response. Sequential estimates and allocation probabilities can be updated as data become available. For ease of implementation, updates can also be made after groups of patients have responded, rather that individually. From a practical perspective, there is no logistical difficulty in incorporating delayed responses into the response-adaptive randomization procedure, provided some responses become available during the recruitment and randomization period.

We have developed a Bayesian response-adaptive randomization design for survival trial. A nonparametric survival model is applied to estimate the survival probability at a clinical relevant time. The proposed design provides comparable operational characteristics as true parametric design. When the event time distribution is misspecified, the proposed design performs better than parametric one. The proposed design can be extended to Response-Adaptive Covariate-Adjusted Randomization (RACA) design when we need to control important prognostics among treatment arms (Lin et al. 2016a, b, c). Another potential approach of updating treatment allocation probability could be based on the restricted mean survival time. The benefits of adaptive randomization for survival trial depend on the distributions of event times and patient accrual rate as well as on the adaptive design under consideration (Case and Morgan 2003). If there are short-term response that are quickly available and predictive of long-term survival, we can use those short-term response to "speed up" adaptive randomization for survival trial (Huang et al. 2009).

A major criticism of response-adaptive randomization is that, despite stringent eligibility criteria, there may be a drift in patient characteristics over time. Using covariate-adjusted response-adaptive randomization can be a solution to this problem if the underlying covariates causing the heterogeneity are known in advance. This may not cause issues with large sample sizes since the randomization automatically balances prognostic factors among treatment groups asymptotically. For clinical trials with small or moderate sample sizes, the impact from the imbalance of the prognostic factors can be substantial when using response-adaptive randomization designs, and thus causes difficulties to the interpretation after randomization. Thus, it is encouraged to have a randomization procedure that could also actively balance the covariate across treatment arms. Consequently, such design can help balance patient characteristics between different treatment arms, and thereby control the inflated type I error rates that occur in response-adaptive randomization (Lin et al. 2016a, b, c; Lin and Bunn 2017).

References

Barker, A., Sigman, C.C., Kelloff, G.J., Hylton, N.M., Berry, D.A., Esserman, L.J.: I-SPY 2: an adaptive breast cancer trial design in the setting of neoadjuvant chemotherapy. Clin. Pharmacol. Ther. **86**(1), 97–100 (2009)

Berry, S.M., Carlin, B.P., Lee, J.J., Muller, P.: Bayesian Adaptive Methods for Clinical Trials. CRC Press, Boca Raton (2010)

Case, L.D., Morgan, T.M.: Design of phase II cancer trials evaluating survival probabilities. BMC Med. Res. Methodol. **3**(1), 6 (2003)

Cheung, Y.K., Chappell, R.: Sequential designs for phase I clinical trials with late-onset toxicities. Biometrics. **56**(4), 1177–1182 (2000)

Cheung, Y.K., Thall, P.F.: Monitoring the rates of composite events with censored data in phase II clinical trials. Biometrics. **58**(1), 89–97 (2002)

Hu, F., Rosenberger, W.F.: Introduction. In: The Theory of Response-Adaptive Randomization in Clinical Trials. Wiley, Hoboken, NJ (2006)

Huang, X., Ning, J., Li, Y., Estey, E., Issa, J.P., Berry, D.A.: Using short-term response information to facilitate adaptive randomization for survival clinical trials. Stat. Med. **28**(12), 1680–1689 (2009)

Ji, Y., Bekele, B.N.: Adaptive randomization for multi-arm comparative clinical trials based on joint efficacy/toxicity outcomes. Biometrics. **65**(3), 876–884 (2009)

Kairalla, J.A., Coffey, C.S., Thomann, M.A., Muller, K.E.: Adaptive trial designs: A review of barriers and opportunities. Trials. **13**(1), 145 (2012)

Lin, J., Bunn, V.: Comparison of multi-arm multi-stage design and adaptive randomization in platform clinical trials. Contemp. Clin. Trials. **54**, 48–59 (2017)

Lin, J., Lin, L., Sankoh, S.: A general overview of adaptive randomization design for clinical trials. J. Biom. Biostat. **7**, 294 (2016a)

Lin, J., Lin, L., Sankoh, S.: A Bayesian response-adaptive covariate-adjusted randomization design for clinical trials. J. Biom. Biostat. **7**, 287 (2016b)

Lin, J., Lin, L., Sankoh, S.: A phase II trial design with Bayesian adaptive covariate-adjusted randomization. In: Statistical Applications from Clinical Trials and Personalized Medicine to Finance and Business Analytics, ICSA Book Series in Statistics, pp. 61–73. Springer, Cham (2016c)

Ning, J., Huang, X.: Response-adaptive randomization for clinical trials with adjustment for covariate imbalance. Stat. Med. **29**(17), 1761–1768 (2010)

Renfro, L.A., Mallick, H., An, M.-W., Sargent, D.J., Mandrekar, S.J.: Clinical trial designs incorporating predictive biomarkers. Cancer Treat. Rev. **43**, 74–82 (2016). https://doi.org/10.1016/j.ctrv.2015.12.008

Rosenberger, W.F., Sverdlov, O., Hu, F.: Adaptive randomization for clinical trials. J. Biopharm. Stat. **22**(4), 719–736 (2012)

Thall, P.F., Wathen, J.K.: Practical Bayesian adaptive randomisation in clinical trials. Eur. J. Cancer. **43**(5), 859–866 (2007)

U.S. Food and Drug Administration. Guidance for industry: Adaptive design clinical trials for drugs and biologics. Draft guidance (2010)

Woodcock, J., LaVange, L.: Master protocols to study multiple therapies, multiple diseases, or both. N. Engl. J. Med. **377**, 62–70 (2017)

Yin, G.: Clinical Trial Design: Bayesian and Frequentist Adaptive Methods, vol. 876. Wiley, Hoboken (2013)

Yin, G., Chen, N., Jack Lee, J.: Phase II trial design with Bayesian adaptive randomization and predictive probability. J. R. Stat. Soc.: Ser. C: Appl. Stat. **61**(2), 219–235 (2012)

Yuan, Y., Huang, X., Liu, S.: A Bayesian response-adaptive covariate-balanced randomization design with application to a leukemia clinical trial. Stat. Med. **30**(11), 1218–1229 (2011)

Chapter 15
Some Statistical Considerations in Design and Analysis for Nonrandomized Comparative Studies Using Existing Data as Controls for Medical Device Premarket Evaluation

Nelson Lu, Yunling Xu, and Lilly Q. Yue

15.1 Introduction

It is not uncommon that non-randomized studies are utilized to provide primary evidence in medical device pre-market evaluation. In the safety and effectiveness evaluation of medical products, although well-controlled and well-conducted randomized clinical trials (RCT) are considered to be the designs to produce the highest level of evidence and thus are viewed as a gold standard, there may be some constraints and limitations associated with them. Sometimes, conducting a RCT may not be feasible due to ethical or practical reasons. Greater expense and span of time are generally required in running a RCT. Therefore, when seeking for a comparative claim, sometimes a non-randomized study is adopted as it is considered a less burdensome approach. This is acceptable if such a study provides sufficient, valid scientific evidence for a pre-market application (21CFR860.7 2012).

A typical case in medical device arena is to compare the results of a one-arm clinical study such as an investigational device exemption (IDE) study to the results of a control extracted from other data sources. Traditionally in medical device applications, the control group are generally formed from the data of an earlier IDE study. Started recently, there have been more and more studies in which the control subjects are selected from a patient or device registry of sufficient quality. This trend may be a result of the greater public and regulatory interest in utilizing the real-world data (RWD). For example, U.S. Food and Drug Administration (FDA) Center for Device Evaluation and Radiological Health (CDRH) issued the guidance "Use of real-world evidence to support regulatory decision making for medical devices" (US FDA 2017) in year 2017. As a control, the data from the registry could be

N. Lu (✉) · Y. Xu · L. Q. Yue
CDRH, U.S. Food and Drug Administration, Silver Spring, MD, USA
e-mail: Nelson.Lu@fda.hhs.gov

L. Zhang et al. (eds.), *Contemporary Biostatistics with Biopharmaceutical Applications*, ICSA Book Series in Statistics,
https://doi.org/10.1007/978-3-030-15310-6_15

prospectively or retrospectively collected, or concurrent or non-concurrent, with respect to the (IDE) clinical study.

When conducting a non-randomized study, additional sources of bias can be introduced as opposed to a RCT. Special attentions are needed to ensure the validity of study design and interpretability of study results. It is paramount to mitigate such potential bias with proper design and analysis. In Sect. 15.2, a brief discussion on the estimands associated with a non-randomized study is presented. Issues regarding the design utilizing the propensity score methodology are described in Sect. 15.3. Some analysis issues are addressed in Sect. 15.4. Concluding remarks are included in Sect. 15.5.

15.2 Estimands

The importance of selecting an appropriate estimand in conducting a clinical study has recently received great recognition and emphasis. It is one of the focal topics in the currently developing addendum to the International Conference on Harmonization E9 guideline. As pointed out in the final concept paper of ICH E9(R1) (2014), the incorrect choice of estimand and unclear definitions for estimands may lead to "problems in relation to design, conduct and analysis and introduce potential for inconsistencies in inference and decision making".

Upon the formulation of the scientific questions of interest or the study objective, the endpoint is determined, and a proper estimand of the treatment effect based on this question needs to be selected. Common factors to define and describe an estimand include outcome measure, treatment received, analysis population, time period of interest, and treatment adherence status (final concept paper, page 2). Study design and analysis then are planned accordingly. One key message presented in Mehrotra et al. (2016) is that "the importance of clearly articulating, in order, the trial objectives, endpoint, estimand, design, and analysis." They further point out that "confusion in regulatory submissions has arisen, in part, due to this order being essentially reversed in practice, with the estimand being implicitly defined as a consequence of the trial design and statistical analysis methodology."

Although the principles presented in the aforementioned development mainly target the RCT, many of them still apply to non-randomized comparative clinical studies. In the context of such studies, for the purpose of pre-market medical device evaluation, the study objective is to determine how the clinical outcomes of the investigational device compare with what would have been occurred to the same subjects under the control treatment. The appropriate estimand should be formulated right after the determination of study objective and study endpoint(s) and ideally before the study design phase.

One special consideration in determining the estimand for a non-randomized comparative study is due to patient populations from two treatment groups coming from different data sources. Two commonly adopted estimands are the average treatment effect (ATE) and the average treatment effect on the treated (ATT).

Under the potential outcomes framework (Rubin 1974), two potential outcomes for a subject are denoted as $Y(T = 0)$ and $Y(T = 1)$, where $T = 1$ represents that the subject is treated with investigational device and $T = 0$ with control. ATE, which can be expressed as $E[Y(1) - Y(0)]$, represents the average effect, at the population level, of moving an entire population from control to treatment (investigational device). ATT, which can be expressed as $E[Y(1) - Y(0) \mid T = 1]$, represents the average effect of treatment on those subjects who ultimately received the treatment (Austin 2011a).

When treatment effects vary among patients with different baseline characteristics, ATT and ATE do not necessarily coincide, since the distribution of patient population for subjects treated with investigational device may not be the same as that of the overall population in practice. For example, between the options of medical management and implanting a certain device requiring an invasive operation, older patients may more likely choose the medical management over the device. Therefore, the age of patient population for the users of this device tends to be younger than the overall intended population.

In many medical device pre-market applications, the study objective is to compare an investigational device to a similar device (control), which may be a competitor's marketed device or the previous generation device. A common question of interest for these cases is: "What is the treatment effect on outcomes if all patients (eligible for receiving both devices) are only offered the investigational device?" (Yue et al. 2016) The ATE may be used in this type of situation. From our experience, oftentimes the patient populations of the two treatment groups, as manifested from their associated data sources, are relatively comparable, if there is no great time lag between the collection of the two data sources.

When the study objective is to assess the performance of the investigational device compared to that of a control such as surgical operation, medical management, or a dissimilar type of device, etc., the patient populations may be different between the two treatment groups. The question of interest is often like "What is the treatment effect on outcomes in patients who receive the investigational device?" (Yue et al. 2016) In this situation, the ATT may be better suited for answering the question. The type of estimands usually determines how subjects are selected into the analysis set and the analysis methods. This is further discussed in Sects. 15.3.3 and 15.4.1.

15.3 Design

15.3.1 Bias

One important consideration when designing a study is to mitigate potential bias. Per Merriam-Webster dictionary, bias is a systematic error introduced into sampling or testing by selecting or encouraging one outcome or answer over others.

Bias may be introduced at any phase of a study, including study design, subject selection, study conduct, data collection, data analysis procedures, interpretation, and reporting/publication. Bias may cause erroneous results and thus sabotage the validity of the conclusion drawn from a study. Therefore, it is crucial to identify flaws that may cause the bias and take actions to eliminate these flaws so that the bias can be reduced.

Comparing to RCTs, non-randomized studies utilizing external control are prone to many more factors that may cause bias. As the external data often coming from another clinical study or real world data that reflect real world practice, some factors such as timeframe, space, study conduct, measurements, evaluation of the outcome, etc., may not be similar to those of the current IDE study. Treatment differences in study outcomes may be confounded with aspects such as distinct protocols (or lack thereof), medical practice, health policy, facilities, physician skills, rigor of data monitoring, data quality, etc., between studies. In addition, the distributions of patient baseline characteristics between two treatment groups are likely to be dissimilar. If data are analyzed without recognizing these factors, the estimated treatment effect may be greatly biased.

Clinical judgment is essential in determining whether there would be any significant bias in the estimated treatment effect based on the differences in the study level factors. Bias related to the differences in patient characteristics may be addressed using the statistical methodology discussed next.

15.3.2 Propensity Score Methodology

For the bias that may be caused by the differences in patient characteristics, it is possible to mitigate such bias via appropriate statistical methods. Propensity score methodology (Rosenbaum and Rubin 1983) has been widely applied in the premarket medical device applications (Yue et al. 2014, 2016).

Propensity score is the probability of receiving the treatment given the observed baseline covariates. It can be viewed as a one-dimensional summary of observed covariates. While the true propensity scores for subjects are never known, they usually are estimated by a logistic regression model in which the treatment is the response variable and the observed baseline covariates are the predictors. After the estimated propensity scores are obtained, the matching design is performed in which subjects from both treatment groups are matched based on the closeness of their estimated propensity scores. One nice theoretical property is that the baseline covariate distributions between two treatment groups are expected to be comparable based on the resulting propensity score design. The data analysis will then be performed based on the propensity score design. The tutorial on the propensity score methodology can be found in literature such as D'Agostino (1998) and Austin (2011b).

The analysis population is defined after the design is finalized. Oftentimes not all subjects initially used in building the propensity score estimation model are retained in the analysis population. For example, subjects may be discarded if their propensity score are far away from the propensity scores in the other group. In the pre-market medical device applications, it is important to note that all subjects in the investigational device group are recommended to be included in the analysis set. Throwing away any subjects in the investigational device group would inevitably alter the intended population and thus cause difficulty in defining the intended population in the labeling (Li et al. 2016; Yue et al. 2014, 2016).

Two main types of the propensity score design used in pre-market medical device applications are one-to-one (or 1:1) matching and stratification. The following gives some brief discussions of the two.

15.3.2.1 1:1 Matching

In the 1:1 matching technique, matched pairs (1 subject in the control group and 1 in the investigational device group) of subjects are formed if they share a similar value of the estimated propensity score. The matching can be done without replacement and with replacement (Rosenbaum 2002). When matching with the replacement is performed, the associated analysis method in variance estimation is more complicated as it needs to account for the dependence structure, and it may be more challenging in assessing whether the covariates reach the reasonable balance. Therefore, to have better interpretability in the pre-market medical device evaluation applications, it is suggested that the matching be performed without replacement. That is, each subject is included in at most one matched pair.

To obtain better matches, researchers often apply the criteria on the distance measure in the estimated propensity scores of the subjects in the matched pair. A pair is formed only when the distance measure of the estimated propensity scores of the two subjects is within a specified caliper. Under this scheme, a subject is thrown away if there are no available subjects (in another group) whose propensity score fall within the caliper of his/her estimated propensity score. Such a scheme may be redundant in the pre-market medical device applications due to the constraint that all subjects in the investigational device group are better to remain in the analysis set.

The 1:1 matching design is not encouraged as, based on our experience, it does not usually work well. The main reason is that the size of the pool for the control is usually relatively limited. As a result, some subjects in the investigational device group may be forced to match up with control subjects with faraway estimated propensity scores. However, with the possibility of huge data sets from the real world setting with reasonable quality, there may be more cases utilizing the one-to-one matching, or even many-to-one matching, design in the future.

15.3.2.2 Stratification

Stratification is the most commonly adopted approach in the pre-market medical device applications. In performing this design, all subjects from both groups are ranked based on their estimated propensity scores, and then grouped into several strata with roughly equal size accordingly. A common approach is to place subjects into five groups using the quintiles of the estimated propensity score. By doing so, approximately 90% of the bias due to measured confounders may be eliminated, when the linear treatment effect is estimated (Rosenbaum and Rubin 1984).

15.3.3 Two-Stage Design

As a RCT is designed before the outcomes are available, naturally the design and analysis are separated. This may not be the case for a non-randomized study without careful planning and execution. If the propensity score design and analysis are performed simultaneously, data dredging exercise may take place, and the integrity and the objectivity of the study may be suffered. To avoid such a situation, the design needs to be carried out without the access to the outcome data. This practice enhances the consistency, transparency, predictability, and effectiveness of regulatory decision making (Rubin 2008; Yue 2012).

Two-stage design process, proposed and discussed in (Li et al. 2016; Yue et al. 2014, 2016) can be implemented to fulfill the principle of the separation of design and analysis under the regulatory requirements. During the design process, the outcome data need to be entirely blinded.

The two-stage design is briefly discussed in the following.

15.3.3.1 First Stage

First stage design occurs before the start of the investigational study, similar to any typical pivotal clinical studies aimed for the medical device pre-market approval. The data source for the control group is identified.

One of the main tasks is to plan the sample size of the investigational device study. If the control data are collected concurrently (most likely from an ongoing registry), the sample size for the control group may be unknown at this stage. For the purpose of the sample size estimation, the propensity score design and statistical analysis methods may need to be specified. When the pool of the control group is expected to be large and thus a one-to-one matching is planned, the sample size estimation is straightforward. However, if the stratification technique is planned, the sample size estimation may be trickier as the comparability of the distributions of the patients' baseline characteristics is unknown in this stage. As poorer comparability would result in a larger sample size in order to achieve the same level of power, our recommendation is to take a more conservative approach by assuming that the treatment and control groups are less comparable.

Covariates to be used in building the propensity score model and to be checked for balance to validate the propensity score design are specified at this stage based on the clinical judgment. These covariates need to be collected in the investigational study and the external data source. It is recommended that all covariates related to the clinical outcomes and the treatment assignments be included in the propensity score model in order to better fulfill the ignorable treatment assignment assumption required in the propensity score methodology (Hill et al. 2004; Rubin and Thomas 1996). As reducing bias may be more important than obtaining better statistical efficiency from the regulatory perspective, it is recommended to include as many covariates as possible (Yue et al. 2014).

Some actions can be taken to ensure that the study is designed prospectively. For example, the propensity score design is better to be implemented by an independent statistician who is blinded to the outcome data. This is to reduce the possibility of data dredging practice. It is recommended that such a person is identified at this stage. Masking schemes such as building a firewall to the clinical outcome is suggested to be in place.

15.3.3.2 Second Stage

The second stage design can be started right after data on covariates for all subjects is available. The independent statistician identified in the first-stage design works on the propensity score estimation and design.

The primary goal in this stage is to obtain a propensity score design in which the covariate distributions between the groups is balanced. After a propensity score design is built based on the estimated propensity scores that are derived from a propensity score model, the covariate balance may need to be evaluated. If the covariate distribution is not balanced, the propensity scores may be re-estimated. Therefore, it may involve an iterative procedure between the propensity score estimation and matching/grouping design until the covariate balance is satisfactorily reached. Note that the evaluation of the covariate balance is based on the particular propensity score design and planned analysis (Stuart 2010). Examples regarding the covariate balance diagnostic based on the stratification design can be found in Yue et al. (2014, 2016) and Li et al. (2016). For covariate balance diagnostic based on one-to-one matching designs, the reader is referred to Austin (2009).

In the process to finalize the design, the subjects included in the design and analysis set need to be identified. While all subjects treated with investigational device need to be included as mentioned above, sometimes in practice some control subjects are left out to obtain better covariate balance. Whether it is reasonable to throw away control subjects may depend on the estimand being used. For ATE, if the remaining subjects, after discarding part of control subject, do not well represent the population of the control, the combined subjects may consequently not well represent the overall population. As a result, the estimated ATE based on this sample may not be the intended estimand. On the other hand, ATT can always be legitimately estimated as long as the patient set enrolled into the investigational device study well represent the population.

Once the propensity score design is agreed upon among the stakeholders, the analysis population is defined, and the statistical analysis methods are finalized in this stage. The study power can be re-calculated. If the sample size was underestimated in the first stage design due to a false expectation of good comparability of patient characteristics, it is possible that the study is found underpowered at this stage. This would put the investigator in an awkward situation.

If the patient distribution of the control subjects is deemed to be so different that they do not provide good matches to the subjects in the investigational device group, "additional" control subjects may need to be obtained from other data sources. If no proper control data are available, the comparative claim cannot be assessed. In this situation, the results of the investigational device group may be compared with a performance goal, if such a plan has been proposed in the first stage. Note that this approach is valid only if the outcome data are not accessed in the entire design stage.

Example 1

Results from a one-arm study with subjects using a new device were proposed to be compared with control group (medical management) data formed from a registry, and the treatment effect was to be estimated using the estimand ATT. The primary endpoint is the 12-month event rate, and it was proposed to be evaluated using Kaplan-Meier method. Based on the sponsor's original proposal regarding the selection of the control subjects from the registry a subject is selected if (1) s/he meets the same inclusion/exclusion criteria of the investigational study, and (2) a primary endpoint event occurs to the subject before 12 months or the subject completes the 12-month evaluation. Note that, a lost-to-follow-up (LTF) registry patient is not a candidate for the control group.

The proposal of excluding LTF subjects from the registry data appears to be problematic as it may introduce bias. The control subjects should be selected solely based on their baseline characteristics, not the clinical outcome or follow-up information. Registry subjects with comparable baseline characteristics to the subjects in the treatment group should be included even if no complete clinical outcome is available. The missing clinical outcome data for such subjects should be addressed by censoring or some other missing data analysis methods, but not by excluding them from the data analysis.

Example 2

The performance of a new device was evaluated by comparing the clinical outcomes with a control group, which was selected from a registry database. The primary endpoint is 30-day adverse event rate. Based on the protocol submitted in the IDE, a single arm study of 250 subjects treated with the investigational device was proposed.

A total of 500 control subjects from the registry were selected based on the same inclusion/exclusion criteria used in the IDE study. After the single-arm IDE study was finished, the 250 subjects were matched with 500 control subjects using the 1:1 matching method (without replacement) based on the logit of estimated propensity scores. The matching was performed with different caliper sizes. The sponsor proposed to use a caliper size of a 0.4 of the standard deviation of the logit of the estimated propensity scores. It resulted in 200 matched pairs. By doing so, a total of 50 treated subjects and 300 control subjects were discarded based on this proposal. It appears that the control group does not provide good matches.

At least two concerns were raised. First, there were no plans to establish a firewall to mask outcome, and no independent statistician was identified. As it was unknown whether the matching was performed with outcome data in sight, the objectivity of the study was in doubt. Second, it is problematic to discard subjects in the treatment group. Thereby it would be difficult to identify the intended population. On a related note, it was unclear what the estimand was to be estimated based on the 200 matched pairs.

Example 3

A sponsor intended to conduct a one-arm investigational study to assess the non-inferiority regarding the primary endpoint of a cardiovascular device to a control where the control group was extracted from a national registry. The primary endpoint is the treatment success rate at 12 months.

In the first design stage, the sponsor anticipated that 400 control subjects would be available. The sponsor expected that the success rate for both treatments would be 80%. With the 10% non-inferiority margin using a one-sided α of 0.025 of the Wald test, the sponsor proposed 300 subjects for the IDE study. This was derived by treating the study design as if it was a randomized controlled trial with a power of 90% in attempt to compensate the potential imbalance in sample size distribution among strata.

After the enrollment of the study was completed and all baseline covariates were collected, an independent statistician who had no access to the outcome data performed the propensity score modeling and design. The final agreed study design was based on the stratification method. The sample size distribution of the design is displayed in Table 15.1. It can be observed that sample sizes are relatively unbalanced across strata. As a result, the re-calculated power based on this distribution became 74%, which was less than the desired level of 80%. This relatively awkward situation may be prevented if the sample size was calculated more conservatively in the first design stage.

Table 15.1 Sample size distribution among strata (quintiles) for Example 3

	Quintile (k)					
	1	2	3	4	5	Total
$N_{t(k)}$	20	38	55	78	109	300
$N_{c(k)}$	120	102	85	62	31	400
Total	140	140	140	140	140	700

15.4 Analysis

15.4.1 Analysis Based on the Type of the Design

15.4.1.1 1:1 Matching

When each subject in the treatment group is matched with K subjects from the control group in the propensity score design, the data can be analyzed in two different ways: One way is to account for the feature of the matched-pair data, and the other is to treat the matched samples as if they were collected from a RCT.

Some researchers such as Imbens (2004) and Austin (2011a) and practitioners believe that the correlation should be accounted for in the matched-pair data. Subjects within a pair have similar propensity scores, and their observed baseline covariates come from the same multivariate distribution. In the presence of confounding, baseline covariates are related to outcomes, and thus matched subjects are more likely to have similar outcomes. Others have a different perspective. They point out that the theory of propensity scores only indicates that covariate distributions of the subjects with the same propensity scores between two treatment groups are similar. Subjects with the same propensity score may have very different values in the baseline covariates. In addition, there may be no reasons that the outcomes of matched subjects are correlated in any way (Schafer and Kang 2008). Additional discussions can be found in Stuart (2008, 2010).

Whichever approach an applicant intends to use needs to be clearly specified in the design stage. For either approach, note that, as each of the subjects in the investigational device arm is matched with one control subject and some of the control subjects are likely thrown out, only ATT may be estimated.

15.4.1.2 Stratification

With the stratification design, the treatment effect is estimated within each stratum and then all estimated effects are aggregated across strata (Rosenbaum and Rubin 1984). To obtain the estimate of the ATT, the estimates of strata are weighted by the number of subjects in the investigational device group within the associated stratum. To obtain the estimate of the ATE, the estimates of strata are weighted equally. ATT can be always estimated since all subjects in the investigational device group are retained in the analysis set. However, if some subjects in the control group are thrown away, the ATE may not be able to be estimated as the overall patient population may be altered by discarding subjects.

Example 4
An applicant proposed to conduct a one-arm study to study the investigational device that was intended to treat patients with de novo and non-stented restenotic lesions in superficial femoral and proximal popliteal arteries. The primary endpoint was the primary patency within 12 months. To support the claim that the

investigational device was non-inferior to a similar device that had already been approved on the market, the applicant proposed to design a non-randomized comparative study. The control subjects were to be selected from a national registry with satisfactory data quality.

The primary patency rate is expected to vary greatly among patients with different characteristics. For example, it is believed that the primary patency rate would decrease with the increased length of the lesion to be treated.

The applicant proposed to apply the propensity score methodology to mitigate the bias introduced by the potential differences in patient baseline characteristics. Subjects were to be grouped into quintiles based on their estimated propensity scores. A Z test was proposed to test the non-inferiority hypothesis. In doing so, the stratum-specific estimates of difference in 12-month primary patency rates were proposed to be weighted by

$$w_i = \left(\frac{1}{N_{c(k)}} + \frac{1}{N_{t(k)}}\right)^{-1} / \sum_i \left(\frac{1}{N_{c(k)}} + \frac{1}{N_{t(k)}}\right)^{-1},$$

where k is kth quintile ($K = 5$); $N_{t(k)}$ and $N_{c(k)}$ are the associated sample size in the treatment and control arm in the kth quintile, respectively.

The likely rationale for this proposal was that the estimate with such weights has better statistical efficiency. To see this, first note that the variance of the estimate of difference in 12-month primary patency rates in strata k is proportional to

$$\frac{1}{N_{c(k)}} + \frac{1}{N_{t(k)}}.$$

It can be observed that, since $N_{c(k)} + N_{t(k)}$ are roughly the same for all k, the less balanced the sample size between two groups within a stratum, the higher the variance is. With the proposed weights w_i, the impact of the strata with less balanced sample size distribution is reduced.

The proposed estimator is a valid estimator for the average treatment effect of the target population only if the treatment effects across the propensity score strata are the same. As there was no evidence or plausible explanation to suggest that the treatment effects between two devices would stay constant across strata, it was unclear what estimand was to be estimated based on the proposed weights.

15.4.2 Missing Data

Because there may be differences in the treatment intervention and data collection mechanisms between the two groups, missing rates and missing patterns may be different. For example, higher missing rate may be observed in real-world data sources as opposed to the closely-monitored, well conducted clinical studies. This may need to be considered when handling the missing data in the analysis.

Multiple imputation is a popular method adopted to handle missing data in medical device pre-market evaluation. Due to the nature described in the previous paragraph, the missing values may be imputed based on separate model built from data of each treatment group. In a propensity score stratification design, as missing data may reside in different strata, issue regarding how to account for the strata in the imputation need some considerations.

15.4.3 Subgroup Analysis

15.4.3.1 1:1 Matching

When the analyses are performed by treating the samples as if the samples were collected from a RCT, considerations and methods involved in conducting the subgroup analyses based on RCT design also apply.

However, if the data are analyzed to account for the nature of the matched-pair data, it may not be possible to conduct a subgroup analysis in align with the primary analysis. Subjects within a pair may belong to different subgroups for some matched pairs.

15.4.3.2 Stratification

In this type of design, each stratum may be viewed as a quasi-RCT. Therefore, the subgroup analyses may be conducted within each stratum. Sometimes a statistical test is desired to evaluate the similarity of outcomes among subgroups. For example, a test such as Breslow-Day test may be used for evaluating binary endpoint.

15.5 Conclusions

The observational, nonrandomized comparative studies have been utilized in medical device premarket evaluation. This type of studies is gaining momentum recently as there has been great interest in employing real world data of acceptable quality in the premarket applications.

Proper practice can enhance the validity of the nonrandomized study. It is suggested that the type of the estimands be identified as soon as possible, ideally in the early phase of the study design. The propensity score methodology has widely been used to design and analyze such studies. Appropriate implementation is essential in obtaining a valid conclusion drawn from the study. In particular, the two-stage design, which ensures the separation of design and analysis, can be adopted under the regulatory framework to mitigate bias. Careful considerations are needed in analyzing the data to better assess the evidence. We envision that, through proper and careful study design and analysis, the integrity of the study can be maintained, and the interpretability of study results can be enhanced.

References

Austin, P.: Balance diagnostics for comparing the distribution of baseline covariates between treatment groups in propensity-score matched samples. Stat. Med. **28**, 3083–3107 (2009)

Austin, P.: An introduction to propensity score methods for reducing the effects of confounding in observational studies. Multivar. Behav. Res. **46**(3), 399–424 (2011a)

Austin, P.: A tutorial and case study in propensity score analysis: an application to estimating the effect of in-hospital smoking cessation counseling on mortality. Multivar. Behav. Res. **46**(1), 119–151 (2011b)

Code of Federal Regulations 21CFR860.7(c)(2).: U.S. Government Publishing Office. https://www.ecfr.gov/cgi-bin/text-idx?SID=b7516459b287456a33a3a31c67da82b0&mc=true&node=se21.8.860_17&rgn=div8 (2012). Accessed 30 Nov 2017

D'Agostino, R.: Propensity score methods for bias reduction in the comparison of a treatment to a non-randomized control group. Stat. Med. **17**, 2265–2281 (1998)

Hill, J.L., Reiter, J.P., Zanutto, E.L.: A comparison of experimental and observational data analyses. In: Gelman, A., Meng, X.-L. (eds.) Applied Bayesian Modeling and Causal Inference from an Incomplete-Data Perspective, pp. 44–56. Wiley, New York (2004)

Imbens, G.W.: Nonparametric estimation of average treatment effects under exogeneity: a review. Rev. Econ. Stat. **86**, 4–29 (2004)

International Conference on Harmonisation.: Final concept paper E9(R1): addendum to statistical principles for clinical trials on choosing appropriate estimands and defining sensitivity analyses in clinical trials. http://www.ich.org/fileadmin/Public_Web_Site/ICH_Products/Guidelines/Efficacy/E9/E9__R1__Final_Concept_Paper_October_23_2014.pdf (2014). Accessed 13 Nov 2017

Li, H., Mukhi, V., Lu, N., Xu, Y., Yue, Q.L.: A note on good practice of objective propensity score design for premarket nonrandomized medical device studies with an example. Stat. Biopharm. Res. **8**(3), 282–286 (2016)

Mehrotra, D.V., Hemmings, R.J., Russek-Cohen, E., ICH E9/R1 Expert Working Group: Seeking harmony: estimands and sensitivity analyses for confirmatory clinical trials. Clin. Trials. **13**(4), 456–458 (2016)

Rosenbaum, P.R.: Observational Studies, 2nd edn. Springer, New York (2002)

Rosenbaum, P.R., Rubin, D.B.: The central role of the propensity score in observational studies for causal effects. Biometrika. **70**(1), 41–55 (1983)

Rosenbaum, P.R., Rubin, D.B.: Reducing bias in observational studies using subclassification on the propensity score. J. Am. Stat. Assoc. **79**, 516–524 (1984)

Rubin, D.B.: Estimating causal effects of treatments in randomized and nonrandomized studies. J. Educ. Psychol. **66**, 688–701 (1974)

Rubin, D.B.: For objective causal inference, design trumps analysis. Ann. Appl. Stat. **2**(3), 808–840 (2008)

Rubin, D.B., Thomas, N.: Matching using estimated propensity scores: relating theory to practice. Biometrics. **52**, 249–264 (1996)

Schafer, J.L., Kang, J.: Average causal effects from nonrandomized studies: a practical guide and simulated example. Psychol. Methods. **13**, 279–313 (2008)

Stuart, E.A.: Developing practical recommendations for the use of propensity scores: discussion of "a critical appraisal of propensity score matching in the medical literature between 1996 and 2003". Stat. Med. **27**(12), 2062–2065 (2008)

Stuart, E.A.: Matching methods for causal inference: a review and a look forward. Stat. Med. **25**, 1–21 (2010)

US Food and Drug Administration.: Use of real-world evidence to support regulatory decision making for medical devices. Guidance for industry and Food and Drug Administration staff. http://www.fda.gov/downloads/MedicalDevices/DeviceRegulationandGuidance/GuidanceDocuments/UCM513027.pdf (2017). Accessed 18 Nov 2017

Yue, L., Campbell, G., Lu, N., Xu, Y., Zuckerman, B.: Utilizing national and international registries to enhance pre-market medical device regulatory evaluation. J. Biopharm. Stat. **26**(6), 1136–1145 (2016)

Yue, L., Lu, N., Xu, Y.: Designing premarket observational comparative studies using existing data as controls: challenges and opportunities. J. Biopharm. Stat. **24**(5), 994–1010 (2014)

Yue, L.Q.: Regulatory considerations in the design of comparative observational studies using propensity scores. J. Biopharm. Stat. **22**(6), 1272–1279 (2012)

Chapter 16
Review of Statistical Issues in Pragmatic Clinical Trials in Current Drug Development Environment

Dingfeng Jiang, Kun Chen, Saurabh Mukhopadhyay, Nareen Katta, and Lanju Zhang

16.1 Background of the Review

Traditionally, regulatory decisions have been based on evidence from randomized controlled trials (RCTs), which have long been the fundamental tool for benefits-risks assessment when studying the efficacy and safety of a new intervention in healthcare (FDA guidance 1998). Common practices of RCTS such as well-defined inclusion and exclusion criteria, blinding, randomization and well-controlled environment, however, limit RCTs' external validity, i.e. the ability to generalize conclusions in an extended population and daily clinical settings (Ware and Hamel 2011). On the other hand, downstream decision makers constantly struggle for the lack of real-world evidence (RWE) of effectiveness and safety in daily practice, where patients are heterogeneous and not well-compliant. Those requests from stakeholders drive the recent rise of 'pragmatism' in clinical trials (Patsopoulos 2011). Both the United States Food and Drug Administration (FDA) and European Medicines Agency (EMA) have spoken publicly about the potential of RWE to inform regulatory decisions. Featured by minimum exclusion, close-to-normal practice environment, open-label and ability to switch, pragmatic clinical trials (PCTs) bridge RCTs and observational studies and are capable of generating high quality RWE of effectiveness and safety.

Although abundant literatures describe applications of PCTs, the methodological and statistical implications of pragmatism are not well discussed; especially when the treatment is an investigational drug or trial results are for regulatory approval.

D. Jiang (✉) · K. Chen · S. Mukhopadhyay · N. Katta
AbbVie Inc., North Chicago, IL, USA
e-mail: dingfeng.jiang@abbvie.com

L. Zhang
Data and Statistical Sciences, AbbVie Inc., North Chicago, IL, USA

L. Zhang et al. (eds.), *Contemporary Biostatistics with Biopharmaceutical Applications*, ICSA Book Series in Statistics,
https://doi.org/10.1007/978-3-030-15310-6_16

This review outlines some well-established statistical principles in RCTs that need to be re-examined to account for the uniqueness of PCTs, together with an example study, the Salford Lung Study. Hopefully this would be the initial step to spark the interest of developing best practice guidelines for PCTs.

The review organizes as follows, Sect. 16.1 provides background, characteristics and definition, and applications of PCTs. Section 16.2 discusses PCTs from design perspective including study population, enrollment, endpoints/outcomes, randomization, blinding, sample size, power and multiplicity. Section 16.3 investigates analysis issues such as analysis population, missing data, heterogeneity of treatment effect and subgroup analysis, etc. The Salford lung study is discussed in Sect. 16.4 as an example of PCT prior to approval. Section 16.5 offers discussion and conclusion to close the review.

16.1.1 History, Characteristics, and Definition of PCTs

The original concept of PCTs dates back to the 1960s when two French statisticians, Schwartz and Lellouch tried to differentiate two types of trials, one to confirm a causal relationship between an administrated intervention and pre-defined outcomes/endpoints ('explanatory'), and the other one to inform clinical and health policy decisions where two or more interventions are involved ('pragmatic') (Schwartz and Lellouch 1967). More recently, a growing number of authors seem to agree that there is a continuum between explanatory RCTs and PCTs rather than a dichotomy between them. PCTs were given a strong boost in the late 2008 and 2009 when a 22-item checklist (Zwarenstein et al. 2008) and the Pragmatic-Explanatory Continuum Indicator Summary (PRECIS) tool were published to help design pragmatic trials (Thorpe et al. 2009). The PRECIS tool and it's improved and validated version PRECIS-2 (Loudon et al. 2015) provided a global visualization tool for assessing a trial's pragmatism.

Tunis et al (Tunis et al. 2003) argued that pragmatism includes four important elements (1) comparison among clinically relevant interventions, (2) a diverse study population, (3) heterogeneous practice environment, and (4) a wide range of health outcomes. Califf et al (Califf and Sugarman 2015) proposed three key attributes of PCTs "(1) an intent to inform decision-makers (patients, clinicians, administrators, and policy-makers), as opposed to elucidating a biological or social mechanism; (2) an intent to enroll a population relevant to the decision in practice and representative of the patients or population and clinical settings for whom the decision is relevant; (3) either an intent to (a) streamline procedures and data collection so that the trial can focus on adequate power for informing the clinical and policy decisions targeted by the trial or (b) measure a broad range of outcomes." They further offered the definition of PCT as a trial "designed for the primary purpose of informing decision-makers regarding the comparative balance of benefits, burdens, and risks of a biomedical or behavioral health intervention at the individual or population level." The definition of PCTs will keep evolving, but the aspects of broader patient population, near-to-normal practice environment, and clinical relevant comparator seems to be well agreed upon.

16.1.2 Applications of PCTs

Early applications of PCTs include "large and simple" trials in the cardiovascular (DIG group 1996) and diabetes areas (DREAM study group 2004). Comparative effectiveness research (CER) is traditionally an important area for PCTs. Both Mullins (Mullins et al. 2010) and Chalkidou (Chalkidou et al. 2012) discussed challenges and solutions of using PCTs in CER. PCTs are also popular tools for post-approval safety study after a new drug becomes widely prescribed. Reynolds (Reynolds et al. 2011) evaluated PCT study designs over other designs in this setting.

Recent interest in PCTs emerges from their potentials for regulatory approval. The twenty-first Century Cures Act requires the FDA to evaluate RWE for supporting approval of a new indication for an already approved drug and for post-approval studies (FDA PDUFA 2016; FDA guidance 2017; Berger et al. 2017). Under this legislation, a draft framework for the evaluation of RWE will be established by the end of 2018 and draft guidance will be issued by the end of 2021. The Adaptive Pathway by EMA (EMA guidance 2016) supports the use of PCTs to generate RWE of effectiveness and safety for approval. Another emerging area is to demonstrate real-world effectiveness to reimbursement agencies or payors. A spotlight example is the GSK's Salford Lung Study (New et al. 2014; Bakerly et al. 2015), the world's first phase 3 PCT to evaluate the effectiveness and safety of an investigational medication prior to regulatory approval and to provide real-world effectiveness for Health Technology Assessment (HTA) agencies. Regardless of the purpose, the real-world effectiveness and safety are the key elements stakeholders seeking from PCTs.

It should be noted that the rising pragmatism should not undermine the importance of RCTs, which is the only source for high-quality efficacy and safety evidence with strong internal validity. PCTs provide a complementary platform for real-world effectiveness and safety with external validity in a broader population (Treweek and Zwarenstein 2009). Both study types are important for evidence generation in drug development and lifecycle management.

16.2 PCTs Design Considerations

Accurate estimation of treatment effectiveness in real-world clinical setting requires many components, addressing confounding and bias being on the top lists. At very least, investigators need to consider the choice of control group and types of design at design stage. Comparing to RCTs, the standard of care (SOC) instead of placebo is more frequently used as a control in PCTs. Non-inferiority, superiority and equivalent study designs are still applicable in PCTs. Other key design issues in a PCT include: (1) a population reflecting patients who currently receive medication in daily care; (2) recruiting and practice environment; (3) clinical meaningful outcomes; (4) blinding; (5) randomization; (6) power, sample size, and multiplicity.

16.2.1 Patient Population: Inclusion and Exclusion Criteria

A common critique of RCTs, particularly on those designed for regulatory approval, is their strict eligibility criteria (Williams et al. 2015), which often seek a homogeneous patient population to prove clinical efficacy. This leads to the exclusion of many patients in need of such care, making it unclear whether the conclusion applies to these excluded patient populations. The reported screen failure rate varies greatly across disease areas. A 20–30% failure rate was reported in prostate and kidney cancer studies (Wong et al. 2016). In other areas, the proportion of excluded patients is substantial. In Alzheimer's diseases, it was suggested that only 10–13% are eligible for clinical trials (Grill and Karlawish 2010). In COPD and asthma, the number could be as low as 3–7% (New et al. 2014). It has been estimated that only about 2–5% of the screened patients were enrolled in early percutaneous transluminal coronary angioplasty (PTCA) vs. coronary artery bypass graft (CABG) studies (Black 1996; Hannan 2008).

On the other hand, how to manage diverse and heterogeneous patient population with limited resource is a common challenge for most healthcare systems. With increasing aging populations in most developed countries, elderly patients, patients with multiple comorbidities and multiple concomitant medications, are very common in practice. However, these patients are often under-represented or even excluded from RCTs (Konrat 2012). Women and ethnic minorities are also under-represented in RCTs (Hoel 2009). Obviously, there is evidence gap for the under-represented populations even for efficacy, not to mention effectiveness.

A PCT is targeted to maximize the generalizability of study findings by enrolling patient cohort that is representative of the real-world patient population. Such effort leads to the minimum exclusion criteria in most PCTs to have a well-represented cohort (Patsopoulos 2011). The exclusion criteria based on ethical and safety considerations, however, should be held up to the same standard as RCTs. It must be noted that broader patient groups will create a larger degree of heterogeneity, an increased variation and a diluted treatment effect.

16.2.2 Recruitment and Practice Environment

For most RCTs, recruitment starts when patients with the study disease seek healthcare service. However, once patients consent, they will be closely monitored with frequent visits. The intensive monitoring plan ensures compliance and good follow-up. However, protocol-driven care often leads to exaggerated treatment effect in RCTs since in real-life practice such protocol-driven care rarely exists (Guisasola 2008). The discrepancy between efficacy and effectiveness, at least to a certain level, is attributed to such practice difference.

PCTs are targeted to estimate effectiveness in real-life practice. Hence reducing intensity of monitoring and/or frequency of visit is a key element for maintaining least intervention to avoid protocol-driven care. However, minimum infrastructure is absolutely needed for a successful study. Hence, balancing close-to-normal practice environment and study logistic necessities becomes an operational challenge in PCTs. Another important factor to consider is the reimbursement of concomitant medication. The principle of PCTs generally requires the study to follow general practice coverage without additional incentives.

In multicenter RCTs, study sites may have very different healthcare systems and SOCs. If the focus of a PCT is to understand the effectiveness and economic implication of a new drug in a particular healthcare system of interest, it is generally challenge to do a multicenter PCTs. Salford Lung Study was done in one city—Salford, England (New et al. 2014; Bakerly et al. 2015). The study fully utilized the existing infrastructure for data capture and patient monitoring. This does not only reduce resource burden of study sponsor but also generates results applicable to real-life practice in that health care system.

16.2.3 Endpoints and Outcomes

Most endpoints in late phase RCTs are clinically focused to quantify efficacy, safety, and quality of life (QoL). In some therapeutic areas, validated surrogate endpoints might be used, e.g. progression-free-survival (PFS) being a surrogate for overall survival (OS) (Saad et al. 2010) in oncology. PCTs target broader stakeholders, besides regulators, often including downstream stakeholders such as health technology assessment agencies (HTA), national/regional or private payers, practitioners or patient advocacy groups. Additional endpoints are desirable to meet their needs. Patient-reported outcomes (PROs) are heavily used in PCTs to bring in patients' perspective. Health resource utilization endpoints are important for payors. For PCTs designed to meet post-market safety surveillance commitment, a broad spectrum of safety endpoints should be collected. Early involvement of stakeholders and incorporation of their input are vital to have a full panel of endpoints. It should be noted that the lack of blinding in most PCTs advocates 'objective endpoints' that is less prone to measurement error and/or subjectivity of study personnel who conduct the assessment.

16.2.4 Blinding

Blinding is intended to reduce known or unknown bias in the conduct and interpretation of a clinical trial due to the knowledge of treatment. For RCTs assessing efficacy, blinding is strongly recommended if ethically and/or operationally feasible (Schulz and Grimes 2002). However, open-label status reflects practice of normal care. From this aspect, open-label is preferred in PCTs, which may explain why most PCTs are open-label in order to estimate real-world effectiveness. On the other hand, in an open-label study, the awareness of treatment assignment is likely to affect patient selection, psychological or physical responses of patients, and assessment of endpoints (Beyer-Westendorf and Buller 2011). Depending on patients and/or physicians' perception, the patients assigned to a new treatment may have a favorable expectation or increased apprehension; those assigned to standard treatment, however, may feel deprived or relieved. So selecting endpoints less prone to subjectivity is a key way to reduce bias introduced by open-label status. In general, reducing bias in an open-label environment remains a challenge task for researchers.

16.2.5 Randomization

A key feature of RCTs is randomization, which allows RCTs to minimize group difference prior to treatment to minimize selection bias. Statistically, randomization permits probability theory to construct a likelihood function so that any differences in outcomes can be primarily attributed to treatment assignment (Bulpitt 1996). For the same reason, randomization is also highly recommended for PCTs. However, special consideration is needed for open label PCTs. At operation level, a centralized randomization process is preferred as the center can control the realization of randomization at subject level per each investigator. New randomization methods such as the cluster randomization (Schulz et al. 2010; Hotopf 2002), and the stepped wedge randomization (Hussey and Hughes 2007; Hemming et al. 2015) may be helpful too. Stratified randomization is preferred when there are signals suggesting different responses among subgroups or subpopulation analysis is of interest. For example, when studying the effectiveness of cardiovascular disease (CVD) preventing agents, it is common to randomize with stratification by gender, since the risk of CVD between male and female are different (Mosca et al. 2011).

16.2.6 Power, Sample Size, and Multiplicity

As mentioned before, PCTs target to enroll diverse patients in real-life clinical practice with minimum monitoring. Hence increased heterogeneity, decreased compliance and increased loss to follow-up will inevitably lead to diluted treatment effect and inflated variance. Statistically speaking, either of the two components (reduced treatment effect and inflated variance) will require a larger sample size to maintain the study power. Under such scenario, the common requirement for RCTs (e.g. 90% of power, 2-sided 5% type I error) will need to be revisited. Can the study power be decreased? Can an inflated type I error rate more than 5% be used in PCTs? Will a smaller effect size be considered clinically meaningful? Whether multiplicity should be addressed to the same level of pivotal RCTs, in which the family-wise type I error rate is strictly controlled for the primary endpoint(s) and the key secondary endpoints. These are relevant questions for the investigator, sponsors, and stakeholders. It is reasonable to expect the same requirement for PCTs, if the results are part of a submission package for regulatory approval. In other cases, input from stakeholders on these issues need to be fully addressed in design stage. In our opinion, PCTs should be adequately powered. Unfortunately, literature reveals that many PCTs are not properly powered, often leading to an inconclusive result and thus a waste of resources.

16.3 PCTs Analysis Considerations

As mentioned above, a continuum exists between RCTs and PCTs. Therefore, the principles of analyzing RCTs are often applicable in PCTs. Special considerations are needed to reflect the uniqueness of PCTs. These analysis considerations should be fully accounted for at the designing stage to increase the chance of a conclusive study.

16.3.1 Analysis Population

Intention-to-treat (ITT) population has long been regarded as the preferred principle to estimate treatment effect (Gupta 2011). ITT analysis is still an important tool in PCTs to assess comparative effectiveness. The ITT principle, however, varies considerably across trials with different levels of adherence to study protocols. Treatment adherence is well managed in most RCTs due to the tight monitoring plan. Nevertheless, in PCTs, patients can switch to any alternative treatments after randomization in order to mimic real-world practice. The classical ITT approach compares patients based on randomized assignment without adjusting for switching. When switching rate is high in PCTs, the validity of ITT principle can be severely compromised. This could contaminate the estimation of true treatment effectiveness. The Per-Protocol (PP) principle, which excludes switched patients from analysis or treat them as censored at the point of the switch, however, may increase selection bias and disrupt covariate balance between treatment groups by randomization.

How to accurately estimate the effectiveness under treatment switch remains the most challenging statistical issue in PCTs. No consensus guidelines currently exist. Fortunately, PCORI (Patient-Centered Outcomes Research Institute) funded projects are in progress to develop such guidelines (PCORI 2018). Some advanced methods are also proposed, including the Inverse probability of censoring weights (IPCW) method (Hernan et al. 2001), the two-stage method (Latimer et al. 2014), the rank preserving structural failure time model (RPSFTM) (Robins and Tsiatis 1991; Mark and Robins 1993), and the iterative parameter estimation (IPE) algorithm (Branson and Whitehead 2002). However, adoptions of these new methods in PCTs are rare (Latimer et al. 2018). An operationally easy approach is to conduct multiple sensitivity analyses using different analysis populations.

16.3.2 Missing Data

Preventing and planning for missing data and describing the methods to address them in a study protocol is an important component of any good research. More planning is needed for PCTs because patients are more likely to drop out as no intensive monitoring planning exists in general. Informing patients about the implications of missing data may help. Tracking all patients and recording not only dropout itself but also the reasons of dropout should be considered as a routine requirement. Other factors such as the censored/diluted/contaminated assessments due to use of recue medications, switch practitioners, and poor treatment compliance only add more complexity to the problem in PCTs. The extent and pattern of missing data must be reported so that the implications are clear to anyone who might make a decision on the results.

All current missing data handling methods rely on missing mechanisms, three common ones being MCAR (missing completely at random), MAR (missing at random), MNAR (missing not at random, also known as non-ignorable nonresponse). To reduce the risk of data-driven selection of approach that could adversely affect either the validity or the relevance of the results, researchers—before seeing the data—should determine how to address them. Multiple imputations (MI) has the potential to produce less biased results than using a single value (such as LOCF or Last Observation Carried Forward) method when missing at random assumption seems reasonable. For MNAR, new approaches such as selection model or pattern mixture model have been adopted. In light of ICH E9 Addendum on estimand, the principles of defining a valid estimand may be more critical for PCTs.

16.3.3 Heterogeneity of Treatment Effect and Subgroup Analysis

Compared to RCTs, PCTs face more heterogeneous population, leading to potentially different responses across subgroups to the intervention of interest. Thus how to assess treatment effectiveness in subgroups becomes more challenging in PCTs. To avoid data-driven approach of multiple post-hoc subgroup analysis, which is prone to false positive, it is recommended that the subgroups need to be pre-specified to ensure valid estimation of stratified treatment effects for testing a limited number of subgroups (with large enough sample sizes).

16.3.4 Analysis of Economic and Humanistic Endpoints

Health economic evaluations are commonly included in PCTs because it is often important to assess cost and cost-effectiveness as well as clinical outcomes to inform policy decisions (Fayers and Hand 1997). Healthcare planners may need information about the total annual budget required to provide a treatment at a particular hospital. An estimate of total cost is often obtained from data in a trial by multiplying the cost in a particular treatment group by the total number of patients to be treated. Such information is not always available in RCTs as the study drug is largely provided by sponsors. In such analysis, the distribution of cost data is frequently skewed because a few patients can produce large costs. For example, some patients receiving the investigational drug can develop rare but costly adverse effects. This can lead to a skewed distribution with a right-tailed skewness generated by the few patients having adverse effects. The difference in the mean costs, therefore, may be inappropriate. Proper data transformation before analysis may be required (Thompson and Barber 2000). Distribution-free approaches, such as the bootstrap method, in particular, may be more appropriate alternatives (Desgagné et al. 1998).

Humanistic endpoints or patient-reported outcomes (PROs) are also frequently used in PCTs, as well as RCTs. It brings patient perspectives into the drug development and is vital to understand the impact on patients' quality of life. The choice of particular PRO instruments needs to be well balanced between stakeholders' need and study feasibility.

16.4 An Example: Salford Lung Study

Concerning patients' representativeness in RCTs, the Salford Lung Study (SLS) is the world's first pragmatic randomized clinical trial of an investigational medication prior to the drug approval (New et al. 2014; Bakerly et al. 2015). It was designed to evaluate the effectiveness and safety of the once-daily combination of inhaled corticosteroids (ICS) fluticasone furoate (FF) and the novel long-acting beta2-agonist (LABA) in a dry powder inhaler (DPI) compared with existing maintenance therapy in a real world population of patients with asthma, and patients with Chronic Obstructive Pulmonary Disease (COPD) in normal care. The primary outcome of COPD is the rate of moderate and/or severe exacerbations, and an improvement in asthma control (Asthma Control Test) in the asthma study. The first patients were enrolled in the COPD study in April 2012 and the asthma study in December 2012.

The SLS COPD is a 12-month, open-label study, with 1:1 randomization to the intervention or the 'usual' care for 12 months by patient's own general practitioner (GP), practice nurse and community pharmacist. The study GPs and pharmacy were instructed to maintain usual normal practice for the study participants in order to preserve the real-world nature. Similarly, the SLS employed minimal exclusion criteria to only exclude patients with an exacerbation within the previous weeks or having chronic oral corticosteroid use. During the conduct of the study, it also maintained a minimum monitoring schedule, with visit 1 for informed consent, visit 2 for randomization, phone contact during visit 3, 4 and 5, and visit 6 for a final assessment of outcomes. Multiple secondary endpoints were used for COPD study, including those of the interest of downstream stakeholders such as health technology assessment agencies and payers. The secondary endpoints include time to first exacerbation and health resource utilization such as COPD-related primary and secondary care contacts. Other endpoints include hospitalization, use of rescue medication, EuroQol-5 Dimensions (EQ-5D) questionnaire. Safety endpoints include death, pneumonia, frequency, and types of serious adverse events (SAEs). Randomization was stratified by the baseline maintenance therapy and by the history of COPD exacerbation in the previous 12 months to ensure treatment balance among subgroups. During the design stage, the SLS intended to achieve 80% power to detect a relative reduction of 12% in the primary endpoint with two-sided 5% significance level. To account for the heterogeneous population, the sample size calculation was based on a negative binomial regression with dispersion rate of 0.7. The primary endpoint was analyzed by the ITT principle. Subgroup analyses were planned using baseline medication, lung function, comorbidity and other factors for the definition of subgroups.

The study was conducted in and around Salford, UK, where a high prevalence of COPD in a community was served by a single hospital. Salford established electronic medical record (EMR) for both primary and secondary care before initiation of SLS, which allowed the study to capture effectiveness and safety data in real time on study participants. A pilot study was conducted to quantify the burden of asthma and COPD, evaluated the outcomes used in the SLS. The SLS's challenge comes from the enrollment of the study participant as a large number of patients needs to be enrolled through GPs and practice nurses, the majority of which had little pre-license clinical research. Hence extensive training in good clinical practice was offered to nurses, pharmacists and GPs in Salford in 2012, as well as a public education campaign in parallel. The existing integrated EMR system that connected the hospital and surrounding primary care practices in real time facilitated the collection of study data; although additional data feed was added to capture access to out-of-hours services, access to health services outside of Salford and deaths.

The results of the SLS COPD study were published in the New England Journal of Medicine in September 2016 (Vestbo et al. 2016). The primary effectiveness analysis population (N = 2269, defined as those who was randomized and have one or more moderate or severe exacerbations in the year before the trial) showed an 8.4% (95% confidence interval 1.1–15.2) lower rate in the FF group compared to the standard of care (p = 0.02). This finding was confirmed in the entire trial population (N = 2799, defined as those who were randomized, and took trail medicine if in the treatment group), with 8.4% reduction in the FF group (95% CI 1.4–14.9, p = 0.02). The incidence of SAE during the treatment was similar in both groups. Although the SLS COPD study meets the primary endpoint, some of the reviewers criticized the study as it did not consider patients' smoking history and spirometric values as subgroup analyses. Rates of serious pneumonia among patients receiving FF as compared with the rate among those who received usual care was also questioned by the reviewers from safety perspectives. Another major concern was the value of pre-specified ITT population as 22% of the patients in the FF group switched back to their previous drugs over the first 3 months (Correspondence 2016).

16.5 Discussion and Conclusion

PCTs are the type of studies bridging RCTs and observational studies and are important sources of RWE for regulatory decision making. How to balance the cost of statistical rigorousness for the gain of PCTs in terms of diversity and pragmatism is a highly individualized choice depending on the nature of the scientific question and stakeholders' need for evidence. FDA is in the process of building infrastructure for PCTs under the twenty-first Century Cures Act and Prescription Drug User Fee Act (PDUFA) VI. Emerging roles of PCTs in drug approval and reimbursement for effectiveness assessment have triggered a new wave of interest to generate high-quality evidence to meet the need of broad stakeholders including patients, clinicians, payers, and policymakers. Potential bias due to open label and treatment switch remains the greatest challenges to PCTs as illustrated in the SLS study. In addition, PCTs face operational hurdles, especially on data collection. PCTs are generally large in scale and involve heterogeneous (realistic) practice sites. This increases the number of data sources and the volume of data that needs to be pulled and standardized for downstream analysis. Once relevant data sources are identified, formats and frequency of data transfers need to be sufficiently evaluated by considering internal infrastructure and scalability. This requires the sponsors of PCTs conduct a careful feasibility analysis beforehand for the potential investment.

To date, limited statistical guidelines exist for investigators as to how to properly design and analyze PCTs. Much of the current practices in PCTs are mimicking those of RCTs. Although the continuum of the explanatory and pragmatic aspects of in RCTs and PCTs makes it reasonable to borrow much of the practice from RCTs, which has long been tested and validated; some unique aspects of PCTs, however, call for new methodology and guidance (Califf 2016). The current review works toward this goal. Many reviewed issues are up for discussion and will be evolving as more exploration is done in this direction. The consensus from industry, academic institutions, and government agencies on the above issues will definitely help to guide the development of PCTs for the benefits of healthcare providers, policymakers and ultimately patients.

References

Bakerly, N., Woodcock, A., New, J., Gibson, M., Wu, W., Leather, D., Vestbo, J.: The Salford lung study protocol, a pragmatic, randomised phase III real-world effectiveness trial in chronic obstructive pulmonary disease. Respir. Res. **16**, 101 (2015)

Berger, M., Daniel, G., Frank, K., Hernandex, A., McClellan, M., Okun, S., Overhage, M., Platt, R., Romine, M., Tunis, S., Wilson, M.: A framework for regulatory use of real world evidence. Duke-Margolis Center for Health Policy. https://healthpolicy.duke.edu/sites/default/files/atoms/files/rwe_white_paper_2017.09.06.pdf (2017). Accessed 15 Aug 2018

Beyer-Westendorf, J., Buller, H.: External and internal validity of open label or double-blind trials in oral anticoagulation, better, worse or just different? J. Thromb. Haemost. **9**, 2153–2158 (2011)

Black, N.: Why we need observational studies to evaluate the effectiveness of health care. BMJ. **312**, 1215–1218 (1996)

Branson, M., Whitehead, J.: Estimating a treatment effect in survival studies in which patients switch treatment. Stat. Med. **21**, 2449–2463 (2002)

Bulpitt, C.: Randomized Controlled Clinical Trials, 2nd edn. Springer, New York (1996)

Califf, R.: Pragmatic clinical trials, emerging challenges and new roles for statisticians. Clini. Trials. **13**(5), 471–477 (2016)

Califf, R., Sugarman, J.: Exploring the ethical and regulatory issues in pragmatic clinical trials. Clin. Trials. **12**(5), 436–441 (2015)

Chalkidou, K., Tunis, D., Whicher, D., Fowler, R., Zwarenstein, M.: The role for pragmatic randomized controlled trials (pRCTs) in comparative effectiveness research. Clin. Trials. **9**, 436–446 (2012)

Correspondence: Effectiveness of fluticasone furoate–vilanterol in COPD. N. Engl. J. Med. **375**(26), 2605–2607 (2016)

Desgagné, A., Castilloux, A., Angers, J., LeLorier, J.: The use of the bootstrap statistical method for the pharmacoeconomic cost analysis of skewed data. PharmacoEconomics. **13**, 487–497 (1998)

EMA.: Guidance for companies considering the adaptive pathways approach (2016)

Fayers, P., Hand, D.: Generalisation from phase III clinical trials, survival, quality of life, and health economics. Lancet. **350**, 1025–1027 (1997)

FDA.: Guidance for industry, an E9 statistical principle for clinical trials (1998)

FDA.: PDUFA (Prescription Drug User Fee Act 2016) VI on enhancing use of real world evidence for use in regulatory decision making. https://www.fda.gov/downloads/forindustry/userfees/prescriptiondruguserfee/ucm511438.pdf (2016). Accessed 15 Aug 2018

FDA.: Use of real-world evidence to support regulatory decision-making for medical devices guidance for industry and Food and Drug Administration Staff (2017)

Grill, J., Karlawish, J.: Addressing the challenges to successful recruitment and retention in Alzheimer's disease clinical trials. Alzheimers Res. Ther. **2**, 34 (2010)

Guisasola, F.: Glycaemic control among patients with type 2 diabetes mellitus in seven European countries, findings from the Real-Life Effectiveness and Care Patterns of Diabetes Management (RECAP-DM) study. Diabetes. Obes. Metab. **10**, 8–15 (2008)

Gupta, S.: Intention-to-treat concept: a review. Perspect. Clin. Res. **2**(3), 109–112 (2011)

Hannan, E.: Randomized clinical trials and observational studies: Guidelines for assessing respective strengths and limitations. J. Am. Coll. Cardiol. Intv. **1**, 211–217 (2008)

Hemming, K., Haines, Y., Chilton, P., Girling, A., Lilford, R.: The stepped wedge cluster randomised trial, rationale, design, analysis, and reporting. BMJ. **350**, h391 (2015)

Hernan, M., Brumback, B., Robins, J.: Marginal structural models to estimate the joint causal effect of nonrandomized treatments. J. Am. Stat. Assoc. **96**, 440–448 (2001)

Hoel, A.: Under-representation of women and ethnic minorities in vascular surgery randomized controlled trials. J. Vasc. Surg. **50**, 349–354 (2009)

Hotopf, M.: The pragmatic randomised controlled trial. Adv. Psychiatr. Treat. **8**, 326–333 (2002)

Hussey, M., Hughes, J.: Design and analysis of stepped wedge cluster randomized trials. Contemp. Clin. Trials. **28**, 182–191 (2007)
Konrat, C.: Underrepresentation of elderly people in randomised controlled trials the example of trials of 4 widely prescribed drugs. PLoS One. **7**(3), 33559 (2012)
Latimer, N., Abrams, K., Lambert, P.: Adjusting for treatment switching in randomised controlled trials - a simulation study and a simplified two-stage method. Stat. Methods Med. Res. **26**, 724–751 (2014)
Latimer, N., Abrams, K., Lambert, P., Morden, J., Crowther, M.: Assessing methods for dealing with treatment switching in clinical trials: a follow-up simulation study. Stat. Methods Med. Res. **27**(3), 765–784 (2018)
Loudon, K., Treweek, S., Sullivan, F., Donnan, P., Thorpe, K., Zwarenstein, M.: The PRECIS-2 tool, designing trials that are fit for purpose. BMJ. **350**, h2147 (2015)
Mark, S., Robins, J.: A method for the analysis of randomized trials with compliance information - an application to the multiple risk factor intervention trial. Control. Clin. Trials. **14**, 79–97 (1993)
Mosca, L., Barrett-Connor, E., Wenger, N.: Sex/gender differences in cardiovascular disease prevention what a difference a decade makes. Circulation. **124**, 2145–2154 (2011)
Mullins, D., Whicher, D., Reese, E., Tunis, S.: Generating evidence for comparative effectiveness research using more pragmatic randomized controlled trials. PharmacoEconomics. **28**(10), 969–976 (2010)
New, J., Bakerly, N., Leather, D., Woodcock, A.: Obtaining real-world evidence, the Salford lung study. Thorax. **69**, 1152–1154 (2014)
Patsopoulos, N.: A pragmatic view on pragmatic trials. Dialogues Clin. Neurosci. **13**, 2 (2011)
PCORI.: https://www.pcori.org/research-results/2015/causal-inference-guidelines-pragmatic-clinical-trials (2018). Accessed 15 Aug 2018
Reynolds, R., Lem, J., Gatto, N., Eng, S.: Is the large simple trial design used for comparative, post-approval safety research? Drug Saf. **34**(10), 799–820 (2011)
Robins, J., Tsiatis, A.: Correcting for noncompliance in randomized trials using rank preserving structural failure time models. Commun. Stat. - Theory Methods. **20**, 2609–2631 (1991)
Saad, E., Katz, A., Hoff, P., Buyse, M.: Progression-free survival as surrogate and as true end point, insights from the breast and colorectal cancer literature. Ann. Oncol. **21**, 7–12 (2010)
Schulz, K., Altman, D., Moher, D., CONSORT Group: CONSORT 2010 statement, updated guidelines for reporting parallel group randomised trials. BMC Med. **8**, 18 (2010)
Schulz, K., Grimes, D.: Blinding in randomised trials, hiding who got what. Lancet. **359**, 696–700 (2002)
Schwartz, D., Lellouch, J.: Explanatory and pragmatic attitudes in therapeutical trials. J. Chronic Dis. **20**, 637–648 (1967)
The Digitalis Investigation Group: Rationale, design, implementation and baseline characteristics of patients in the dig trial, a large simple long-term trial to evaluate the effect of digitalis on mortality in heart failure. Control. Clin. Trials. **17**, 77–97 (1996)
The DREAM Trial Investigators: Rationale, design and recruitment characteristics of a large, simple international trial of diabetes prevention, the DREAM trial. Diabetologia. **47**, 1519–1527 (2004)
Thompson, S., Barber, J.: How should cost data in pragmatic randomised trials be analysed? BMJ. **320**, 1197–1200 (2000)
Thorpe, K., Zwarenstein, M., Oxman, A., Treweek, S., Furberg, C., Altman, D., Tunis, S., Bergel, E., Harvey, I., Magid, D., Chalkidou, K.: A pragmatic-explanatory continuum indicator summary (PRECIS): a tool to help trial designers. J. Clin. Epidemiol. **180**, E47–E57 (2009)
Treweek, S., Zwarenstein, M.: Making trials matter, pragmatic and explanatory trials and the problem of applicability. Trials. **10**, 37 (2009)
Tunis, S., Stryer, D., Clancy, C.: Practical clinical trials, increasing the value of clinical research for decision making in clinical and health policy. JAMA. **290**, 1624–1632 (2003)
Vestbo, J., Leather, D., Bakerly, N., New, J., Gibson, M., McCorkindale, S., Collier, S., Crawford, J., Frith, L., Harvey, C., Svedsater, H., Woodcock, A., Salford Lung Study Investigators:

Effectiveness of fluticasone furoate-vilanterol for COPD in clinical practice. N. Engl. J. Med. **375**, 1253–1260 (2016)

Ware, J., Hamel, M.: Pragmatic trials - Guides to better patient care? N. Engl. J. Med. **364**(18), 1685 (2011)

Williams, H., Burden-The, E., Nunn, A.: What is a pragmatic clinical trial? J. Investig. Dermatol. **135**, e33 (2015)

Wong, S., North, S., Sweeney, C., Stockler, M., Sridhar, S.: Screen failure rates in contemporary phase II/III therapeutic trials in genitourinary malignancies. J. Clin. Oncol. **34**(Suppl 2S), 176 (2016)

Zwarenstein, M., Treweek, S., Gagnier, J., Altman, D., Tunis, S., Haynes, B., Oxman, A., Moher, D., CONSORT and Pragmatic Trials in Healthcare (Practihc) Groups: Improving the reporting of pragmatic trials, an extension of the CONSORT statement. BMJ. **337**, a2390 (2008)

Chapter 17
Evaluating Potential Subpopulations Using Stochastic SIDEScreen in a Cross-Over Trial

Ilya Lipkovich, Bohdana Ratitch, Bridget Martell, Herman Weiss, and Alex Dmitrienko

17.1 Introduction

The continuing developments in the biomedical technology, statistics, and machine learning are leading not only to a growing recognition of the presence of heterogeneity of treatment effects in many clinical conditions but also to the ability of researchers to identify patient characteristics responsible for such heterogeneity. Once established, this knowledge can be used to improve patient outcomes through the practice of precision medicine (also known as personalized medicine) where therapies can be tailored to characteristics of the patients as well as to their environment and lifestyle (Ashley 2015). Recent breakthroughs and decreases in the cost of genome sequencing led to advances in identifying genetic traits that are responsible for variations in disease susceptibility and response to treatments. Such variations

I. Lipkovich
Eli Lilly and Company, Indianapolis, IN, USA
e-mail: ilya.lipkovich@lilly.com

B. Ratitch
Eli Lilly and Company, Montreal, QC, Canada
e-mail: ratitch_bohdana@lilly.com

B. Martell
Yale University School of Medicine, New Haven, CT, USA
e-mail: bridget@bamconsultants.com

H. Weiss
Juniper Pharmaceuticals, Boston, MA, USA
e-mail: hweiss@juniperpharma.com

A. Dmitrienko (✉)
Mediana, Inc, Overland Park, KS, USA
e-mail: admitrienko@medianainc.com

L. Zhang et al. (eds.), *Contemporary Biostatistics with Biopharmaceutical Applications*, ICSA Book Series in Statistics,
https://doi.org/10.1007/978-3-030-15310-6_17

can often be attributed to phenotypic and clinical characteristics of patients (often broadly referred to as "biomarkers") that do not require genetic testing. While in some cases heterogeneity of treatment effects may be suspected or even obvious based on biological considerations, in general, identifying characteristics reliably describing patient profiles with the most beneficial treatment effect or a favorable benefit-risk balance is a challenging task. It often requires advanced analytical approaches which fortunately enjoyed much research and progress in recent years (see, e.g., a review of recent approaches to evaluating treatment effect heterogeneity in clinical trials in Lipkovich et al. 2017a).

There are multiple stakeholders who would benefit from identification of subgroups with enhanced treatment benefits in the patient population. Obviously, patients and their physicians can make use of such knowledge to improve their decision-making and choose the best individualized course of treatment. Clinical study sponsors may face scenarios where a promising therapy fails in the overall study population, yet there is a subgroup of patients with a clinically meaningful treatment effect. In such cases, the sponsors may be able to salvage the experimental treatment by refining the target population and testing the treatment again in a new population. Even when a trial shows a significant effect based on the overall study population, additional analyses may reveal that the observed treatment effect is in fact driven largely by a subset of patients (see, e.g., Basile 2009). In such situations, more targeted indication labeling can lead to considerable health benefits and cost savings.

Numerous advances in statistical methods for subgroup and biomarker identification have been made in the last 10 years by researchers from diverse communities including machine learning, causal inference, and multiple testing (see recent reviews in Lipkovich et al. 2017a, Lamont et al. 2016).

One important distinction can be made between parametric and non-parametric methods. The former typically seek biomarker signatures as smooth functions of biomarkers (e.g. linear combinations that are further thresholded to define meaningful subgroups of patients who may get more benefit from the experimental treatment, control, or none). Non-parametric methods often utilize methods of recursive partitioning to construct biomarker signatures as nodes of regression tree models. Examples include interaction trees (Su et al. 2009) and methods proposed within the GUIDE platform (Loh et al. 2015).

Another group of methods termed *local subgroup modeling* (Lipkovich et al. 2017a) develops biomarker signatures by direct search rather than by estimating a single model on the entire covariate space. Some methods within this group borrow ideas from PRIM (patient rule indication method a.k.a. Bump Hunting, Friedman and Fisher 1999). Examples include (Kehl and Ulm 2006; Chen et al. 2015).

SIDES (Lipkovich et al. 2011) and SIDEScreen methods (Lipkovich and Dmitrienko 2014) also belong in this category. Similarly to the tree-based methods, they construct subgroups by recursive partitioning; however, unlike trees, they generate a collection of overlapping subgroups ("branches") rather than a partitioning on the entire covariate space. Each subgroup is formed by consequent splits of a parent node into two child groups and retaining only one of the two as a candidate for a possible subsequent split.

Methods based on recursive partitioning are notoriously unstable and it is not surprising that various proposals incorporating stochastic elements in subgroup search (e.g. via resampling) were made. For example, Virtual Twins (Foster et al. 2011) uses random forests to obtain a reliable estimate of individual treatment effects; TSDD method (Shen et al. 2015) employs bootstrap for subgroup generation and obtaining reliable estimates of treatment effect; (Huang et al. 2017) use resampling at several stages of biomarker signature development.

Challenges of subgroup identification are accentuated in small data sets (such as the case study presented in this paper). It is well known that small sample size aggravates the instability of methods based on recursive partitioning, so that small changes in the data can lead to important changes in the results. This issue can impact the subgroup search as well as evaluating the magnitude of treatment effects in the identified subgroup, leading to an increased difficulty of replicating and confirming the results in future studies.

In this paper we propose and evaluate an approach for mitigating this issue by combining SIDES with resampling applied at different stages of subgroup search. Resampling is used to help screen out unimportant biomarkers, to reduce the "optimism bias" in the estimated treatment effect in subgroups, and to obtain an indicator of the replicability of the identified subgroup.

This paper describes an application of the SIDES methodology that adapt to specifics of the case study while employing novel strategies described above. The analysis aims at evaluation of potential heterogeneity of treatment effects given available biomarkers while protecting from spurious "findings" which are likely to occur in small data sets.

The paper is organized as follows. In Sect. 17.2 we present an overview of SIDES methodology, show how it can be applied to a cross-over study, and describe several methods for evaluating the predictive ability of candidate biomarkers (variable importance) using resampling techniques that may be especially relevant for small data sets. In Sect. 17.3 we present a Phase 2 cross-over study in females with dysmenorrhea. Section 17.4 presents a simulation study evaluating operating characteristics of two different methods for biomarker selection via variable importance using simulated data that closely mimic our case study. In Sect. 17.5 we analyze the case study from Sect. 17.3 using several SIDES-based methods. Section 17.6 contains some discussion and conclusions.

17.2 Overview of SIDES Methodology

SIDES (Lipkovich et al. 2011), SIDESreen (Lipkovich and Dmitrienko 2014), and Stochastic SIDEScreen (Lipkovich et al. 2017b) are recursive partitioning methods that aim at identifying predictive biomarkers and associated subgroups which are defined in terms of biomarker signatures. The SIDES methodology identifies subgroups with enhanced treatment effect as compared to that in the overall study population. It also provides a significance test for the treatment difference which is adjusted for the number of subgroups that have been investigated, thereby providing a (weak) control of the Type I error rate for all the hypothesis tests potentially conducted as part of subgroup search (based on a single efficacy variable).

Our case study employed a cross-over treatment design. The original SIDES method was developed for randomized studies with parallel design. Nevertheless, as we argue in the following section, subpopulations with enhanced treatment effect can be identified from a cross-over study by applying SIDES to appropriately re-defined outcome and treatment variables.

17.2.1 Applying SIDES Methodology to a Cross-Over Study

The SIDES method was originally developed for use in independent groups of patients receiving different treatments, e.g., in trials with a parallel-arm design. In this paper we adapt the SIDES methodology to a case study with a cross-over design in which each patient receives both treatments, the difference in outcomes for each patient under the two treatment conditions will be analyzed as the outcome of interest. Details of this approach follow.

Let the outcome efficacy variable be a continuous measure Y with larger values indicating poorer clinical outcome; the treatment variable T assumes values "0" for placebo and "1" for experimental treatment. We consider a cross-over trial with two treatment sequences: $A =$ "01" where patients first receive $T = 0$, followed by a period when they receive $T = 1$, and $A =$ "10" where they first receive the active treatment followed by the control. Let U be a derived outcome defined as the difference between the outcome at the end of the second period and the first, $U = Y_2 - Y_1$. The subgroups with a large treatment effect can be identified by applying the SIDES methodology to the outcome variable U.

Let $\mu_{01} = E(U|A = \text{"01"})$ and $\mu_{10} = E(U|A = \text{"10"})$ be the expected means of U for the subpopulations defined by treatment sequences $A =$ "01" and $A =$ "10", respectively. We will argue that a larger treatment contrast $\mu_{10} - \mu_{01}$ can be attributed to a larger differential between the active treatment and placebo in the absence of cross-over effect and other possible biases, where $\mu_{10} - \mu_{01} > 0$ favors the active treatment and $\mu_{10} - \mu_{01} < 0$ favors placebo.

Let $Y(1)$ and $Y(0)$ denote potential outcomes defined as outcomes for a randomly selected patient if treated with T=1 and T=0, respectively. The goal in this clinical trial is to evaluate the true treatment difference $E\{Y(0) - Y(1)\} = \Delta$,where $\Delta > 0$ indicates that the experimental treatment is superior to placebo. Now,

$$\begin{aligned}\mu_{10} - \mu_{01} &= E\left(U|A = \text{"10"}\right) - E\left(U|A = \text{"01"}\right)\\ &= E\left(Y_2 - Y_1|A = \text{"10"}\right) - E\left(Y_2 - Y_1|A = \text{"01"}\right)\\ &= E\left\{Y(0) - Y(1) - [Y(1) - Y(0)]\right\} = 2E\left\{Y(0) - Y(1)\right\} = 2\Delta\end{aligned}$$

Therefore, the estimand associated with the estimated differences in outcome U between sequences $A =$ "10" and $A =$ "01" is twice the true treatment effect. This argument also applies to binary outcomes.

17.2.2 The Base SIDES Method

Let $X_1, X_2, \ldots, X_p$ be candidate biomarkers that can be continuous or categorical. A biomarker signature is defined as a union of elementary subgroups, each defined based on a single biomarker: a range $\{X \le x_0\}$ or $\{X > x_0\}$ if X is continuous (or ordinal), and $X \in S(X)$, where $S(X)$ is a subset of m levels $L_1, \ldots, L_m$ associated with a categorical X.

The SIDES method essentially is a tool for generating (harvesting) multiple promising subgroups by recursively applying to a current parent set the following subgroup generation process.

For each candidate biomarker, the best split is determined by considering all possible splits. A split for a continuous biomarker X forms two child subgroups $\{X \le x_0\}$ versus $\{X > x_0\}$, where x_0 is one of values of X observed in the analysis data set. For a categorical (nominal) biomarker X, splits are formed by dividing the m levels of the biomarker into two mutually exclusive and exhaustive groups. For a categorical variable with m categories, one can form $(2^{m-1} - 1)$ non-trivial splits and the best is selected by evaluating a differential effect splitting criterion as defined below. For example, for a variable with 3 categories L_1, L_2, L_3, the following 3 distinct splits can be formed: L_1 vs. (L_2 or L_3), L_2 vs. (L_1 or L_3), L_3vs. (L_1 or L_2).

The best split for a biomarker X is the one that optimizes the differential effect splitting criterion D (Lipkovich et al. 2011) defined as follows:

$$D = 2\left(1 - \Phi(t)\right), t = \frac{|Z_1 - Z_2|}{\sqrt{2}},$$

where Z_1 and Z_2 are the standardized treatment effect statistics for the two child subgroups resulting from splitting on the biomarker X.

$$Z = \frac{\overline{U}_{10} - \overline{U}_{01}}{s\sqrt{n_{10}^{-1} + n_{01}^{-1}}},$$

where $\overline{U}_{01}$ and $\overline{U}_{10}$ are the mean values of the outcome U and n_{10} and n_{01} are the number of patients in the two treatment sequence groups within the subset, and s is the pooled standard deviation in the subset.

Let $D^*(X)$ be the value of the criterion for a biomarker X associated with the best split. For example, for a continuous biomarker, the value of the criterion is associated with the cutoff value x_0^* resulting in the best (smallest) value of D among all other candidate cutoffs. When optimizing the splitting criterion over all possible splits, selection bias occurs favoring covariates with a larger number of candidate splits. Since splitting criterion $D^*(X)$ is in the form of the p-value associated with the differential test statistic, this problem is addressed by computing a multiplicity-adjusted p-value associated with the best candidate split, $\widetilde{D}^*(X)$. The multiplicity adjustment is based on a version of the Šidák test, which accounts for the correlation among the D-values associated with different cut-offs for a given covariate (for details, see Lipkovich et al. 2011).

The biomarkers are ordered from best to worst so that

$$\widetilde{D}^*(X_{i_1}) \leq \widetilde{D}^*(X_{i_2}) \leq \cdots \leq \widetilde{D}^*(X_{i_p}),$$

where $i_j, j = 1, \ldots, p$ is a permutation of indices $\{1, \ldots, p\}$ corresponding to the ordered values of $\widetilde{D}^*$criterion, and the first M biomarkers are retained. Parameter M is called *width*.

For each biomarker, two child groups based on the optimal split are formed and the biomarker-positive group is selected as the one that produces the larger (positive) value of the standardized treatment effect statistic, Z. As a result, M potential candidate subgroups are generated from a parent group.

The same process is applied recursively to each of the generated M subsets resulting in M^2 terminal subgroups. The recursion is repeated L times resulting in M^L terminal promising subgroups. Typically, L is a small number, $L = 2$ in our example.

The SIDES method includes several tuning parameters, some of which have already been introduced. Here we list them for completeness

- the width (M), the number of covariates to retain as promising at a given splitting level,
- the depth (L), the number of splitting levels (i.e. 1st, 2nd, 3rd) at which the algorithm is to terminate, and
- the complexity parameter (γ) that quantifies the amount of relative improvement in the treatment effect in the child subgroup relative to the parent group required to identify this subgroup as promising.

The relative improvement γ is defined as the ratio of the p-value for the Z statistic (here, testing of U_{10} versus U_{01}) in the child node (i.e. at splitting level l) versus its parent node (i.e. at splitting level $l - 1$). Smaller values of γ impose more stringent conditions for splitting and result in a fewer generated subgroups (more pruning).

The SIDES method controls the Type I error rate associated with the identification of a promising subset by accounting for 3 important sources of multiplicity: (1) multiple candidate covariates, (2) multiple candidate binary splits per covariate, and (3) multiple splitting levels during the recursive search.

Because of the high dimensionality of the subgroup space, the distribution of the "maximally selected" test statistic under the null hypothesis is approximated by using resampling methods. The "null" reference distribution is generated by randomly permuting the treatment labels of patients to create many "null" datasets that retain the relationships among covariates, and relationships between covariates and outcomes, but effectively remove the effects of treatment and treatment-by-covariate interactions. These null datasets are then analyzed using the SIDES method to establish an empirical null distribution of p-values for promising subsets formed by candidate covariates and their associated binary splitting rules. The multiplicity adjusted p-value for each promising subgroup is then computed as the proportion of p-values for the best subgroups from the null sets which are smaller than the observed p-value for the promising subset.

17.2.3 The Adaptive SIDEScreen Method

The Adaptive SIDEScreen method is an extension of base SIDES. It is a 3-stage procedure. At the first stage, base SIDES method is used to generate a large number of subgroups, and a variable importance score *VI*(*X*) is computed for each biomarker (as described below). At the second (screening) stage, a subset of biomarkers is selected by applying to *VI*(*X*) a variable importance threshold computed from the reference (null) distribution. Finally, at the third stage, base SIDES method is applied only to a subset of biomarkers (if any) that passed the screening stage.

The VI score associated with a biomarker X is a measure of its predictive ability, *VI*(*X*), computed as the average contribution of that biomarker across all promising subgroups. Contribution of a biomarker X is set to zero for all subgroups where it is not involved, and as the negative logarithm of the splitting criterion, $\widetilde{D}^*(X)$, if the biomarker was involved in forming the subgroup.

More formally, $VI(X) = K^{-1}\sum_{i=1}^{K}\nu_i$, $\nu_i = -\log \widetilde{D}_i^*(X)$, if the ith subgroup contains biomarker X (in its signature), and $\nu_i = 0$ otherwise. Here, K is the number of identified subgroups, $\widetilde{D}_i^*(X)$ is the splitting criterion evaluated for the biomarker X at the selected (optimal) split and adjusted for multiple splits using the modified Šidák adjustment.

At the screening stage of the Adaptive SIDEScreen method, biomarkers are selected based on a screening rule:

$$VI(X) > \hat{E}_0 + k\sqrt{\hat{V}_0},$$

where $\hat{E}_0$and $\hat{V}_0$ are the mean and variance of the maximal (over all biomarkers) *VI* score under the null distribution obtained by permuting the treatment labels. These mean and variance are estimated from a large number of such samples. The multiplier k is a free parameter that is often calibrated so that $k = \Phi^{-1}(1 - \kappa)$, where κ is interpreted as the probability of selecting at least one noise biomarker in the absence of any predictive biomarkers in the data set.

At the last stage, the base SIDES method is applied only to biomarkers which pass the screening. The final adjusted p-values are computed by replicating the entire thee-stage procedure (i.e., initial subgroup generation, biomarker screening, and final subgroup identification) on a large number of additional null sets. Note that the same screening threshold is applied to each null set, therefore, regardless of the value of the multiplier, the overall Type I error rate of the final subgroup(s) can be controlled at any desired level.

17.2.4 The Stochastic SIDEScreen Method

The key enhancement made in the Stochastic SIDEScreen procedure (Lipkovich et al. 2017b) is that the VI score for a biomarker is computed not from the subgroups "harvested" from the original data but rather from subgroups generated by applying the base SIDES method to multiple (say, $B = 1000$) bootstrap samples from the original data. Each bootstrap sample $\{Y_b^*, A_b^*, X_b^*\}, b = 1, \ldots, B$, of the same size ($N$) as the observed data set is obtained by sampling individual records with replacement N times and is comprised of the outcome variable, treatment indicator (or treatment sequence, as in the setting of our cross-over trial, $A =$ "01" or $A =$ "10"), and a collection of biomarkers. To ensure that bootstrap samples contain the same proportion of patients in each treatment group as in the original data, sampling can be stratified by treatment group, that is, carried out separately for $A =$ "10" and $A =$ "01" and the two samples combined in a single data set. Then the VI scores $VI_b(X)$ for biomarkers X_i, $i = 1, \ldots, p$, are computed from each bootstrap sample by running the base SIDES method on the bootstrap data in the same way that $VI(X)$ is computed by running SIDES on the original data, $\{Y, A, X\}$. The bootstrap distribution of VI scores is generated for each candidate biomarker. It contains useful information that can be utilized in several ways as described below.

The final VI scores are computed by averaging the VI scores from bootstrap samples. Note that the VI scores from the observed data already implement the idea of model averaging, as they are based on the biomarker's contribution to multiple identified subgroups ("models"). In this sense, they are smoothed measures of each biomarker's contribution (the degree of smoothness depends on the scope of the subgroup search controlled by the *width* and *depth* parameters). Stochastic SIDEScreen makes a further improvement in this direction and adds a random component to averaging, the same way as it is done in the *bagging* method (Breiman 1996). It is expected that a greater degree of noise reduction will be achieved by averaging over relatively "independent" VI scores from multiple bootstrap samples. The idea is that strong predictors of treatment effect would consistently manifest themselves across the majority of the bootstrap samples. By contrast, non-informative biomarkers would emerge only in a smaller number of different samples, which will result in cancelation of their importance when averaging over a large ensemble of bootstrap samples.

The VI scores computed from the original data may be fairly unstable when dealing with small sample sizes and/or biomarkers with a large number of potential splits. A bootstrap-based confidence interval of the mean VI score provides an insight into this inherent instability of VI scores. Consequently, this confidence interval can serve as a more reliable tool for biomarker screening compared to the standard VI score computed from the original data, as in the Adaptive SIDEScreen method. This approach provides the foundation for the Stochastic SIDEScreen procedure. The fundamental idea is to construct a more robust biomarker selection rule based on the bootstrap distribution of the VI scores along with the null distribution of the scores.

One approach is to select a candidate biomarker X if $L_\alpha(X) > \hat{E}_0(X)$, where $L_\alpha(X)$ is the lower limit of the $100 \times (1-\alpha)\%$ (say, 80%) bootstrap confidence interval of the VI score associated with the biomarker X, and $\hat{E}_0(X)$ is the mean of the VI score associated with biomarker X under the null distribution obtained by permuting the treatment labels, similarly to how it is done in the Adaptive SIDEScreen method. Several biomarker selection procedures based on different methods of computing $L_\alpha(X)$ can be considered:

1. Percentile method, $L_\alpha(X) = q_{\frac{\alpha}{2}}\,[VI_b(X), b = 1, \ldots, B]$.
2. Normal approximation method with $L_\alpha(X) = \widehat{VI}(X) - z_{1-\alpha/2} \times \sqrt{\hat{V}_B(X)}$, where $\widehat{VI}(X)$ is the variable importance score obtained from the original data set, and $\widehat{V}_B(X)$ is the bootstrap estimate of the variance of $\widehat{VI}(X)$.
3. Normal approximation method with $L_\alpha(X) = \overline{VI}_B(X) - z_{1-\alpha/2} \times \sqrt{\hat{V}_{IJ}(X)}$, where $\overline{VI}_B(X)$ is the *bagging* estimator $\overline{VI}_B(X) = B^{-1} \sum VI_b(X)$and $\hat{V}_{IJ}(X)$ is the variance of the *bagging* estimator $\overline{VI}_B(X)$, computed using the Infinitesimal Jackknife estimator (Efron 2014) or its bias-corrected version (Wager et al. 2014).

Practical implementation of the thresholding method 3 (bagging estimator of VI) would require computing the "smooth" mean of the null distribution $\hat{E}_0(X)$, as the benchmark for variable importance associated with biomarker X which would seem to require performing additional bootstrap resampling for each null set, and computing $\hat{E}_0(X)$ by first averaging variable importance scores $VI^{(b)}_{0,m}(X)$ associated with each predictor X over B_0 samples within the mth null set $(m = 1, \ldots, \mathcal{M})$, followed by averaging over the $\mathcal{M}$ sets (we use B_0 so as not to confuse with B samples applied to the observed data).

That is, first $\overline{VI}_{0,m}(X) = B_0^{-1} \sum_{b=1}^{B_0} VI^{(b)}_{0,m}(X)$ will be computed Then the final average estimate of VI is found by averaging the $\mathcal{M}$ smoothed scores into a single threshold value $\hat{E}_0(X) = \mathcal{M}^{-1} \sum_{m=1}^{\mathcal{M}} \overline{VI}_{0,m}(X)$. However, because smoothing is achieved naturally by averaging over $\mathcal{M}$ null sets, it seems that the smoothing over B_0 samples may be redundant, or B_0 can be taken as a fairly small number, say $B_0 = 10$.

Another approach is to mimic the selection procedure of the Adaptive SIDEScreen but replace the quantities with the counterparts from the bootstrap distribution.

4. Select a biomarker X, if its bagging estimator $\overline{VI}_B(X)$exceeds the threshold based on the null distribution of the bagging estimator, $\overline{VI}_B(X) > \hat{E}_{0,B} + k\sqrt{\hat{V}_{0,B}}$, where $\hat{E}_{0,B}$ and $\hat{V}_{0,B}$ are the null mean and the variance of the maximum of the bagging estimator computed across all p biomarkers in the data set, respectively. Specifically, for each of the $\mathcal{M}$ null data sets, bagging estimators of null VI scores are obtained by averaging across bootstrap samples obtained from each null dataset, $\overline{VI}^m_{0,B}(X_i)\,, i = 1, \ldots, p; m = 1, \ldots, \mathcal{M}$. Then the maximal null VI scores are computed as $\overline{VI}^m_{0,B} = \max_{i=1,..,p}\left\{\overline{VI}^m_{0,B}(X_i)\right\}$ for each $m = 1, \ldots, \mathcal{M}$. The $\hat{E}_{0,B}$ and $\hat{V}_{0,B}$are the mean and variance of $\overline{VI}^m_{0,B}, m = 1, \ldots, \mathcal{M}$, across $\mathcal{M}$ null sets, respectively.

Similarly to the Adaptive SIDEScreen method, the final set of subgroups S_j, $j = 1, \ldots, s$, is then identified by applying the base SIDES method only to the biomarkers in the original data set that passed the screening.

Like with the Adaptive SIDEScreen, the multiplier k in the procedure 4 can be selected so as to ensure a desired probability of selecting at least one biomarker when no true biomarkers exist in the data set. Note that for the Stochastic SIDEScreen procedures 1–3, we could also control this operating characteristic by calibrating the significance level α at which the confidence limit is computed, e.g., by using a conservative Bonferroni type procedure $\alpha^* = \alpha/p$ and applying $L_{\alpha^*}(X)$or using a less conservative approach that takes into account the correlation among the biomarkers and associated VI scores.

The proposed bootstrap-based rule for biomarker selection can be contrasted with the rule used in the Adaptive SIDEScreen procedure: select the biomarker X if

$$VI(X) > \hat{E}_0 + k\sqrt{\hat{V}_0}$$

where k is calibrated to ensure 100α% Type I error rate (to match the rule based on the $100 \times (1 - \alpha)\%$ bootstrap confidence interval), and $\hat{E}_0$ and $\hat{V}_0$ are the expectation and variance of the maximal *VI* score estimated from the null distribution by randomly permuting treatment labels from the original data.

In the remainder of the paper, we will focus on methods 3 and 4, as more relevant for this small data set (see also Lipkovich et al. 2017b).

17.2.5 Obtaining Bias-Corrected Estimates of Treatment Effect in Identified Subgroups

To obtain estimates of treatment effect in each of the identified subgroups S_j, $j = 1, \ldots, s$, that correct for over-optimism inherent in the subgroup search process, we can use bootstrap again.

1. For each bootstrap sample $b = 1, \ldots, B$:

   (a) Select the best subgroup, S_b^*, by applying a simplified version of SIDEScreen:

   - Compute variable importance scores $VI_b(X)$ for each variable and apply the rule: $VI_b(X) - k\sqrt{\hat{V}_{IJ}(X)} > \hat{E}_0(X) <=> VI_b(X) > \hat{E}_0(X) + k\sqrt{\hat{V}_{IJ}(X)}$, where $VI_b(X)$ is the variable importance from the bth sample and $\hat{V}_{IJ}(X)$ and $\hat{E}_0(X)$ are "borrowed" from the variable screening stage (described in the previous section).
   - Apply base SIDES to only those biomarkers that pass the screening rule
   - The best subgroup S_b^* is selected as the one having the smallest (unadjusted) p-value.

   (b) Compute a bias-corrected treatment effect estimate for the best subgroup, $\hat{\Delta}_{BC}\left(S_b^*\right)$ using the Efron's "0.632 estimator" that combines the in-bag and out-of-bag estimate of treatment effect:

$$\hat{\Delta}_{BC}\left(S_b^*\right) = 0.632\,\hat{\Delta}_{OOB}\left(S_b^*\right) + 0.368\hat{\Delta}_{INB}\left(S_b^*\right),$$

   where $\hat{\Delta}_{OOB}\left(S_b^*\right)$ is an estimate of treatment effect based on "out-of-bag" data (data not selected for a given bootstrap sample b) and $\hat{\Delta}_{INB}\left(S_b^*\right)$ is an estimate of treatment effect based on data included in the bootstrap sample b ("in-bag" data).

2. For each patient i in the experimental treatment arm (in the initial data set), compute patient's expected treatment effect:

$$\Delta_i = B^{-1}\sum_{b=1}^{B}\left[\hat{\Delta}_{BC}\left(S_b^*\right)I\left(i \in S_b^*\right) + \hat{\Delta}_{OV}I\left(i \notin S_b^*\right)\right],$$

   where $\hat{\Delta}_{OV}$ is the overall treatment effect from all observed data (not a specific bootstrap sample). Therefore, the bias-adjusted patient-specific estimate of treatment effect is shrunk towards the overall effect when the patient is not present in a given bootstrap-based subgroup.

3. For each subgroup S identified from the original data set, compute the expected treatment effect concentrated in this subgroup:

$$\hat{\Delta}_{BC}(S) = N_1(S)^{-1} \sum_{i=1}^{N} \Delta_i I(i \in S) I(A_i = a),$$

where N is the total sample size, $N_1(S)$ is the number of treated patients in subgroup S, and A_i is the treatment indicator, and a designates the experimental treatment arm (in our example of a cross-over design, A_i is the treatment sequence and $a =$ "10").

As an alternative method for computing bias-corrected estimates of treatment effect in the best subgroup, k-fold cross-validation can be used as follows (see a similar approach in Simon et al. 2011). The data is randomly divided into $k = 10$ sets stratified by the treatment sequence. Then while keeping each set as a test group, Adaptive SIDEScreen is applied to the $k - 1 = 9$ remaining training sets and the best subgroup identified. Based on the descriptor of the best subgroup, patients in the test set are classified as biomarker-positive or negative. After repeating this process for each of the $k = 10$ test sets, all biomarker-positive patients are combined across the k test sets in a single group—mimicking a subgroup that would have been identified in "future data." If the Adaptive SIDEScreen procedure returns no subgroup on any given training set, then all patients in the test set are considered as "biomarker-positive" (the subgroup is considered to be the same as the overall population).

17.2.6 Obtaining Replicability Measures for Identified Subgroups

The same bootstrap samples that were used for obtaining bias-corrected estimates of the treatment effect are also used to obtain replicability indices for each of the subgroups $S_j, j = 1, \ldots, s$ identified in the original set.

The replicability is computed as an average over coefficients of similarity (agreement) for a given subgroup S from the original data set and all best subgroups S_b^* from the bootstrap samples. That is $r(S) = B^{-1} \sum_{b=1}^{B} agree\left(S, S_b^*\right)$.

The similarity between any two subgroups can be evaluated by a variety of measures $agree(S_1, S_2)$, defined for 2 by 2 contingency tables. For example, a popular Jaccard similarly index between the subgroups S_1, S_2 is computed as

$$J(S_1, S_2) = \frac{| S_1 \cap S_2 |}{| S_1 \cup S_2 |},$$

where $|S|$ is the subgroup size.

We used a simple agreement coefficient that is the proportion of patients on the main diagonal of the cross-tabulation:

$$A(S_1, S_2) = \frac{| S_1 \cap S_2 | + | \overline{S}_1 \cap \overline{S}_2 |}{N},$$

where $\overline{S}$ is the complement of S, and N is the total sample size. In other words, this measure represents a proportion of patients for which the classifications by the two subgroups S_1 and S_2 agree (either they are included in both subgroups, or excluded from both subgroups). This measure is preferred over the Jaccard coefficient because it is well defined even when the compared subgroups are both empty. In this case, the index returns a "perfect agreement," which indicates the lack of any subgroups in the data. To make the coefficient more informative, we also computed an *adjusted* (or standardized) *agreement* coefficient defined as

$$A_{adj} = \frac{A - E(A)}{\max(A) - E(A)},$$

where $E(A)$ is the expected value of the index under "independence" between the rows and columns of the contingency table defined by S_1 and S_2 (assuming the sizes of the subgroups S_1, S_2 are fixed at observed values), i.e., an agreement expected by chance. The value $max(A)$ is the maximal value of the index, again assuming fixed margins of the contingency table. It is easy to see that

$$E(A(S_1, S_2)) = \frac{| S_1 \| S_2 | + | \overline{S}_1 \| \overline{S}_2 |}{N^2},$$

$$\max(A\ (S_1, S_2)) = \frac{| S_1 \cap S_2 | + |\overline{S}_1 \cap \overline{S}_2| + 2 \min\left(|S_1 \cap \overline{S}_2|, |\overline{S}_1 \cap S_2|\right)}{N}.$$

The adjusted agreement coefficient is very similar to a familiar Cohen's kappa coefficient, except in the denominator we have $max(A) - E(A)$ instead of $1 - E(A)$.

17.3 Case Study

Primary dysmenorrhea is usually described as cramping pain in the lower abdomen occurring at or near the onset of menstruation in the absence of any other identifiable pelvic pathology. Occurring in more than 50% of menstruating women, dysmenorrhea is by far the most common gynecologic problem reported in this population. It is a significant cause of absenteeism from work or school and loss of productivity in the workplace. Painful menstrual cramping typically occurs 1–2 days each month with greatest pain intensity typically experienced during the first 24–36 h after menses commences (Dawood 2006).

The pain of primary dysmenorrhea is thought to be caused by intense uterine muscle contractions resulting in transient ischemia. Prostaglandins and possibly vasopressin appear to not only initiate these muscle contractions that lead to ischemia but also sensitize nerve endings. Prostaglandin concentrations increase in the myometrium during menses and appear to reach their highest plasma concentrations during the first 2 days of menses (Dawood 2006; Coco 1999).

Nonsteroidal anti-inflammatory drugs (NSAIDs), such as ibuprofen and naproxen, are commonly used to treat primary dysmenorrhea. The current therapeutic approach to dysmenorrhea is treatment rather than prevention. Although there is no recognized preventive treatment for the cramping episodes of dysmenorrhea, a drug capable of blocking or attenuating the intense uterine contractility associated with dysmenorrhea would be expected to be effective in preventing the pain. Lidocaine affords potential as an effective treatment to prevent or alleviate uterine contractions and dysmenorrhea. Lidocaine has been in use for over 60 years as a local anesthetic and intravenously as a modulator of certain cardiac arrhythmias. Lidocaine alters signal conduction in neurons by blocking the fast voltage-gated Na^+ channels in the neuronal cell membrane responsible for signal propagation. With sufficient blockage, the membrane of the postsynaptic neuron will not depolarize and will thus fail to transmit an action potential and hence muscle contractility.

A Phase II, multicenter, randomized, placebo-controlled, double-blind cross-over study to assess the efficacy and safety of 10% (150 mg) lidocaine vaginal gel in women (age 18–40 years) with recurrent dysmenorrhea was conducted. Patients were randomized in a 1:1 ratio to one of two sequences and treated with 10% (150 mg) lidocaine vaginal gel (or placebo), once daily for 4 days, immediately prior to and during menstruation. Patients were treated over the course of two menstrual cycles. The data set for the primary efficacy analysis had $n = 70$ patients with $n_0 = 34$ patients randomized to "10% LIDOCAINE GEL–PLACEBO" (sequence 1) and $n_1 = 36$ to "PLACEBO–10% LIDOCAINE GEL" (sequence 2). Primary efficacy endpoint was time-weighted average pain intensity (TWAPI) collected via a 4-point scale and assessed using an analysis of covariance (ANCOVA) model. The sample size was computed so as to ensure about 85% power with a 2-sided significance level of 0.05, based on the assumption that the standardized treatment difference (effect size) $\delta = \frac{\mu_{10}-\mu_{01}}{\sigma} = \frac{2\Delta}{\sigma} = 0.74$, which corresponds to the absolute treatment difference parameter, $\Delta = \frac{0.74}{2}\sigma = 0.167$ (assuming the error standard deviation $\sigma = 0.45$). The primary analysis was based on an ANOVA model for a cross-over design, including terms for treatment sequence, subject within sequence, treatment, treatment cycle and pooled study center.

The set of candidate biomarkers includes 13 biomarkers (see Table 17.3 in the Appendix). These biomarkers are labeled in the analysis data set as AGE, BMI, ALB, ALKPH, CDCO2, BILTOT, EOSIN, HEMOG, LAC_DEH, LYMPH, NEU_ABS, BMI_cat, RACE. The candidate biomarkers were chosen based on what clinical and phenotypic characteristics were known to be associated with recurrent dysmenorrhea and widely discussed in the published literature. We stress that pre-selection of biomarkers should be done prior to study unblinding, so as to prevent selection bias. Eliminating irrelevant biomarkers based on clinical considerations is important as including them in the analysis set would automatically expand the "search space" resulting in a larger multiplicity burden. Including a larger number of irrelevant variables obviously makes detection of true predictive biomarkers problematic, which is especially true for Phase 2 studies with relatively small data sets.

For the primary efficacy variable, TWAPI over the 4 treatment days using the 4-point categorical scale in the ITT Analysis Set, mean [SD] values were 0.91 [0.635] and 0.92 [0.552], with the 10% lidocaine bioadhesive gel treatment and the placebo gel treatment, respectively. The least squares (LS) mean difference between the two groups was not statistically significant ($p = 0.905$). Analysis of this data by the SIDES methodology will be reported in Sect. 17.5.

17.4 Simulation Study

The simulation study aimed at assessing potential benefits of variable screening using Stochastic SIDEScreen compared to Adaptive SIDEScreen when applied to a small data set that is very similar to our case study. The simulated data sets mimic the real study in the following features

- The same 13 covariates as in the original set of 70 patients were included in each data set resampled with replacement so that each generated data set inherits similar covariates with their correlations
- The outcome variable was generated by a version of parametric bootstrap based on a very general non-parametric regression model, Random forest (Breiman 2001) fitted to the original data where the outcome was replaced with residuals from the treatment arm means. Therefore, the simulated data by construction exhibit no predictive biomarkers (as treatment variable was not used in the regression model), while retaining correlations among covariates and dependency of outcome on covariates.
- An overall treatment effect of the same size as observed in the original data was added to the data.

A subgroup effect was added to the data for a specific subgroup based on the covariate Age: S = {Age ≤ 33}. This subgroup had the size ns = 52 in the original data. Formally, the simulation can be described as follows.

1. Fit a random forest to study data $\left\{\widetilde{Y}, \boldsymbol{X}\right\}$, where $\widetilde{Y}$ is the outcome variable, expressed as a deviation from the treatment mean (here, the mean for the treatment sequence of a given patient, see Sect. 17.3) and $\boldsymbol{X}$ is the matrix with columns corresponding to 13 continuous biomarkers in the original data set measured on 70 patients. Obtain estimated function $\hat{g}_{RF}(X)$ and associated residual error variance, $\hat{\sigma}^2_{RF}$
2. Simulate K sets as follows. For $j = 1, \ldots, K$,

    (a) obtain a random permutation of rows of $X_j^* = perm(\boldsymbol{X})$

(b) generate patient-level outcomes ($i = 1, \ldots, 70$) using random forest from step 1, $Y^*_{ij} = \hat{g}_{RF}(X^*_{ij}) + \mu_{OV} \times A_i + \mu_{SU} \times A_i \times I\left(age_i \leq 33\right) + e_{ij}$, $e_{ij} \sim N\left(0, \hat{\sigma}^2_{RF}\right)$, where μ_{OV} and μ_{SU} are the overall and subgroup effects, respectively, and the sequence indicator $A_i = \{+1, -1\}$ for sequences '10' and '01', respectively.

In this simulation, we used $\mu_{OV} = 0.0089$, and μ_{SU} was set to 0, 0.2, 0.25, 0.3, 0.35, 0.4, 0.45. Because of the computational burden of Stochastic SIDEScreen, only $K = 500$ data sets were analyzed.

The goal of the simulation was to evaluate the two methods, Adaptive SIDEScreen and Stochastic SIDEScreen, for their ability to detect the correct biomarker (Age) in this very simple setting with a single predictive covariate. We used Adaptive SIDEScreen with multiplier $k = \Phi^{-1}(0.9) = 1.28$. To make the two algorithms comparable we use the approach 4 from Sect. 17.2 with the same k. The null mean and variance for variable importance in Adaptive SIDEsreen was assessed using 100 permutations. For Stochastic SIDEScreen we performed $B = 100$ bootstrap samples to compute the bagging estimator on the simulated data. For each simulated data set, we evaluated the null distribution using $M = 100$ permutations and within each permutation performed 10 bootstrap re-samples. The parameters of the base SIDES procedure used to generate subgroups in both procedures were

- *min_subgroup_size* = 20
- *criterion_type* = "differential effect"
- *depth* = 2
- *width* = 5
- $\gamma = 1$ (at both levels 1 and 2)

For each method we evaluated the proportion of simulated sets when (1) only incorrect biomarkers passed the threshold and (2) for scenarios with $\mu_{SU} > 0$,the proportion of times when the correct biomarker "Age" was included in the set of biomarkers that passed the threshold. The estimated probabilities (1) and (2) are plotted in Fig. 17.1 against each subgroup's effect size, i.e., the mean treatment difference divided by the standard deviation of outcome, which was computed as the square root of the pooled within-treatment variance across $K = 1000$ sets, $\hat{\sigma} = 0.45$.

Several observations can be made from the simulation results:

- For $\mu_{SU} = 0$ (null scenario), both procedures control the nominal probability of 0.1 of falsely classifying at least one biomarker as predictive.
- For $\mu_{SU} > 0$ (alternative scenarios), both methods select the correct biomarker (Age) with the same probability (the two dashed lines in Fig. 17.1).
- Stochastic SIDEScreen, on the other hand, has a better ability to filter out irrelevant biomarkers compared with the Adaptive SIDEScreen. For example, when the subgroup effect is 0.8 and "Age" is not selected, there is about 18.5% chance that a noise biomarker will be selected by the Stochastic SIDEScreen, while for Adaptive SIDEScreen this probability is 26%.

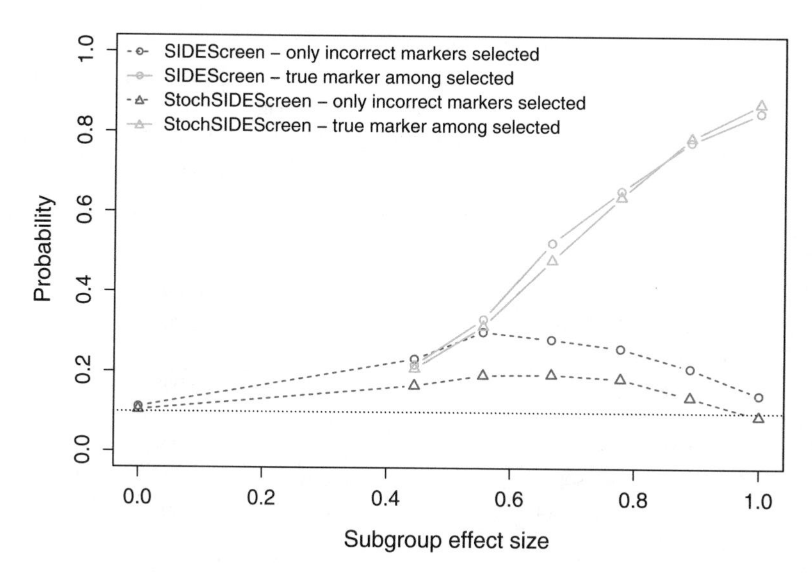

Fig. 17.1 The dashed lines represent the proportion of simulation runs when only incorrect biomarker(s) were falsely identified as predictive by Adaptive SIDEScreen (open circles) and Stochastic SIDEScreen (triangles). The solid lines represent the proportion of times the correct marker (age) was among those that passed the variable importance cutoff: The horizontal dotted line is drawn at $\alpha = 0.1$

We note that the power for such a small data set is fairly low and the probability of detecting correct biomarkers becomes meaningful only for effect sizes >0.8.

Clearly, conducting such simulation studies would be beneficial at the design stage of confirmatory Phase 3 trials, when some information on biomarkers and their relationship with outcomes has been collected from earlier Phase 2 studies. The outcomes can be simulated under various scenarios and the simulations can help obtain estimates of the power for effect sizes that may be clinically relevant and plausible.

17.5 Analysis of the Case Study Using SIDES and Related Methods

In order to examine the performance of the Adaptive and Stochastic SIDEScreen methods using the data from the case study, we first start by generating candidate subgroups using base SIDES with the following parameters:

- *min_subgroup_size* = 10
- *criterion_type* = "differential effect"
- *depth* = 2
- *width* = 5
- *gamma* = 1 at both levels 1 and 2, $\gamma = (1, 1)$

Table 17.1 Subgroups identified using base SIDES method

Subgroup	Size	Splitting criterion	Apparent treatment effect	Treatment effect P-value	Multiplicity adjusted P-value
Overall population	70	NA	0.0089	0.452	NA
Using $\gamma = (1, 1)$					
BILTOT > 0.6	10	3.01	0.52	0.0005	0.55
AGE > 20	59	2.85	0.05	0.2803	1.00
AGE > 20 and ALB > 4.5	25	3.30	0.34	0.0006	0.57
AGE > 20 and HEMOG > 13.8	24	3.02	0.32	0.0009	0.62
AGE > 20 and LYMPH ≤ 27	13	2.95	0.52	0.0004	0.51
AGE > 20 and EOSIN ≤ 2.4	32	2.36	0.22	0.0222	0.95
ALB > 4.5	31	2.56	0.22	0.0171	0.94
ALB > 4.5 and BMI_cat > 2	11	2.94	0.71	<0.00001	0.13
ALB > 4.5 and NEU_ABS > 4.02	21	2.08	0.34	0.0027	0.77
ALB > 4.5 and LAC_DEH > 139	20	1.56	0.27	0.0047	0.83
ALB > 4.5 and LYMPH ≤ 26.1	10	1.54	0.53	0.0021	0.74
LYMPH ≤ 27	15	2.77	0.41	0.0032	0.79
ALKPH > 70	16	2.31	0.32	0.0088	0.89
Using more stringent $\gamma = (0.1, 0.1)$					
BILTOT > 0.6	10	3.01	0.52	0.0005	0.54
ALB > 4.5	31	2.56	0.22	0.0171	0.83
ALB > 4.5 and BMI_cat > 2	11	2.94	0.71	<0.00001	0.28
LYMPH ≤ 27	15	2.77	0.41	0.0032	0.69
ALKPH > 70	16	2.31	0.32	0.0088	0.77

The subgroups reported by base SIDES are given in Table 17.1. For comparison we also included the results of base SIDES with the complexity parameter (γ) set to a more stringent value of 0.1 at both levels (i.e. requiring that a child subgroup's p-value is at least 0.1 times the p-value for the parent group), resulting in much smaller set of subgroups. The multiplicity-adjusted p-values in the last column reflect the proportion of null sets (10,000) where the smallest subgroup p-value was more significant than the p-value from the original data (base SIDES was applied with the same parameters to the null data sets, obtained by permutation of the treatment sequence variable). As we can see, none of the p-values remain significant after the multiplicity adjustment was applied. Clearly, we cannot rely on unadjusted p-values or on apparent treatment effects in the subgroups.

To better evaluate potentially predictive biomarkers we then used more powerful procedures based on variable importance screening. First, we used Adaptive SIDEScreen with multiplier $k = 0.5$ and 1.0 (Table 17.2). With the Adaptive SIDEScreen, we obtained much smaller multiplicity adjusted p-values compared to those for the base SIDES.

This decrease in multiplicity adjusted p-values is expected and reflects the trade-off between the complexity of the search space and the multiplicity burden

Table 17.2 Subgroups identified using Adaptive SIDES screen method with the threshold = 1.35855, based on $E0 + 0.5\sqrt{V_0}$ and threshold = 1.62394, based on $E0 + 1.0\sqrt{V_0}$

Subgroup	Size	Splitting criterion	Apparent treatment effect	Treatment effect P-value	Multiplicity adjusted P-value	Adjusted treatment effect (bootstrap)	Replicability index (A)	Adjusted replicability index (A_{adj})
Overall population	70	NA	0.0089	0.452	NA			
Using threshold = 1.35855 (multiplier $k = 0.5$)								
AGE > 20	59	2.85	0.049	0.280	0.2158	0.041	0.559	0.168
AGE > 20 and ALB > 4.5	25	3.30	0.345	0.000601	0.0348	0.028	0.527	0.327
ALB > 4.5	31	2.56	0.221	0.0171	0.1148	0.031	0.540	0.283
ALB > 4.5 and AGE > 29	10	1.14	0.350	0.00390	0.0699	0.007	0.437	−0.173
Using threshold = 1.62394 (multiplier $k = 1.0$)								
ALB > 4.5	31	2.55878	0.221	0.0171	0.0668	0.031	0.518	0.254

The thresholds are computed from 1000 null sets. Adjusted P-values are computed based on additional 10,000 null sets associated with the search. By effectively reducing the search space via variable importance screening, Adaptive SIDEScreen reduces the need for multiplicity adjustment after subgroup identification while gaining more power because of filtering out noise covariates. For example, when using a multiplier equal to 0.5, the adjusted p-value for subgroup $S = \{\text{ALB} > 4.5\}$ was 0.1148, versus a p-value of 0.94 when using base SIDES with a liberal setting $\gamma = (1, 1)$. Using a more stringent setting $\gamma = (0.1, 0.1)$ of course pruned out many subgroups, however did not have a substantial effect on the adjusted p-values.

With a more stringent multiplier $k = 1$, Adaptive SIDEScreen returned a single subgroup, $S = \{\text{ALB} > 4.5\}$ with a smaller p-value of 0.068. We can see that while for the base SIDES, the ratio of the adjusted to unadjusted p-value for this subgroup was $0.94/0.0171 = 55$, for the Adaptive SIDEScreen the ratio was much smaller, $0.0668/0.0171 = 3.9$.

Although a reduced multiplicity adjustment burden with the Adaptive SIDEScreen is expected, the procedure does not guarantee an "improved" significance over the base SIDES method. For example, in the fairly common case when no biomarker would have passed the pre-specified threshold (no subgroup selected), there will simply be no p-value to adjust. Hence, there is no free lunch.

The graphical representation of screening in Adaptive SIDEScreen is shown in the left panel of Fig. 17.2. The dashed line corresponds to the multiplier of 0.5 and the dotted line to the multiplier of 1.0 (meaning one standard deviation above the noise level). Clearly, the variable importance scores barely reach both levels of the threshold and the results do not appear very robust.

As mentioned earlier, one concern with the Adaptive SIDEScreen procedure may be that it relies on variable importance measures which may be highly variable in a data set of such a small size. To evaluate the robustness of findings under data resampling, we applied the strategies from the Stochastic SIDEScreen procedure (Lipkovich et al. 2017b) and extended them further so that to achieve the following 3 objectives (via bootstrapping):

- Screening biomarkers using bagged estimates of variable importance
- Evaluating treatment effect in the identified subgroups
- Assessing replicability of identified subgroups

Biomarker screening was performed by comparing the lower limits of the bootstrap 80% confidence intervals using the percentile method (shown with solid circles) and, typically less conservative, Efron's formula for the variance of the bagging estimator (shown with open circles) with variable-specific thresholds. The thresholds are obtained as the means of the null distribution for each respective biomarker. The results are such that none of the lower limits exceeded the threshold.

For subgroups identified by the Adaptive SIDEScreen, the bootstrap-based estimates of treatment effect are shown in Table 17.2. Using a tenfold cross-validation, estimates of treatment effect were much less optimistic (negative values): $\text{TE(CV)} = -0.029$ when using a threshold based on the multiplier $= 0.5$, and $\text{TE(CV)} = -0.035$ when using a threshold based on the multiplier $= 1.0$.

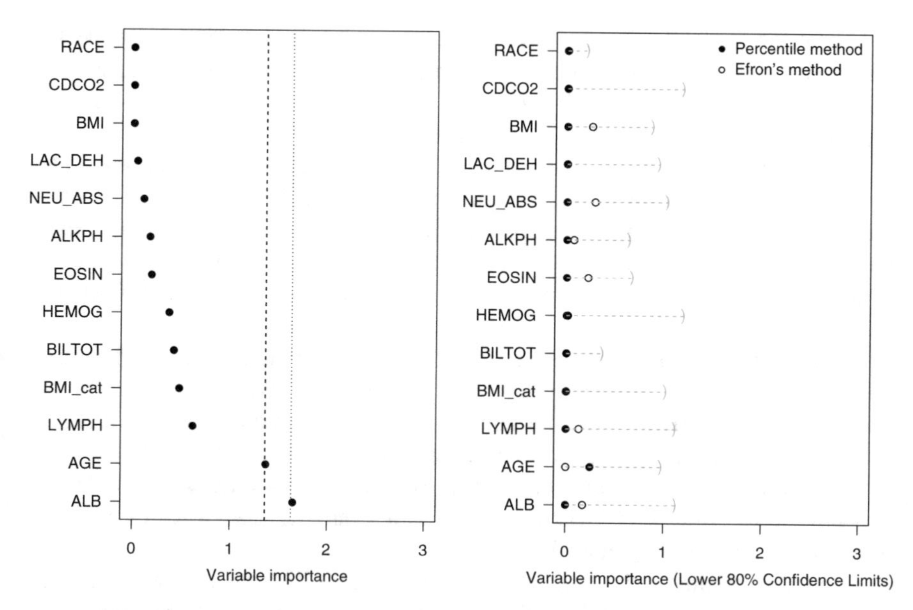

Fig. 17.2 The left panel shows the variable importance generated by base SIDES. The dashed line indicates the threshold based on $E_0 + 0.5\sqrt{V_0}$ and the dotted line is for $E_0 + 1.0\sqrt{V_0}$. The right panel illustrates identifying biomarkers using Stochastic SIDEScreen method. The solid circles represent the tenth percentiles of bootstrap distribution of variable importance, and the open circles—based on Efron's variance estimator. The horizontal dotted lines extend to the means of null distributions for each variable. To pass the screening, the circles must be on the right-hand side of the null means

The cross-validation estimates are more conservative than the bootstrap estimates which could be explained by the fact that they target somewhat different estimands and mimic "future" samples using different resampling mechanisms. The performance of these methods could be evaluated using simulation studies under different scenarios. Note, however, that converting treatment differences into effect sizes (dividing by the standard deviation of $\hat{\sigma} = 0.45$) results in rather small positive effects based on the bootstrap procedure, all of them are less than 0.1.

The assessment of replicability using the adjusted agreement A_{adj} (in the last column of Table 17.2) suggests that the small subgroups listed in this table are unlikely to be replicated in future trials. For example, the replicability of the subgroup {ALB > 4.5 and AGE > 29} with only 10 patients is negative. This agrees with zero treatment effect from the bootstrap estimates.

17.6 Discussion and Conclusions

In this paper we provide a compressive analysis of a data set from a failed Phase 2 cross-over study using SIDES methodology. We showed that the machinery of SIDES can be extended to a cross-over design, by using the treatment sequence as a new "treatment" variable and the difference between outcomes of the two treatment periods as a new outcome variable. We emphasize the need to account for uncertainty associated with subgroup search when evaluating the subgroup identification findings, especially from a small trial.

Extensions of base SIDES, namely the 3-stage procedures based on the Adaptive biomarker screening allow us, first of all, to identify potentially important biomarkers. This may be more relevant for small studies than determination of exact biomarker signatures. The adaptive screening can be enhanced by bootstrapping which takes into account the uncertainty in the subgroup search as reflected in variable importance scores.

Additionally, bootstrap can be used to obtain more realistic estimates of treatment effects in subgroups identified from the original data. We found that for our case study, these estimates did not agree with those based on the cross-validation procedure (a conservative benchmark). Although this result is expected, more research is needed to understand the properties of various approaches to constructing bias-adjusted estimates of treatment effect (for example, see Foster et al. 2011, Simon et al. 2011, Huang et al. 2017).

Simulation studies mimicking empirical associations in a real data set (e.g., using random forest or other "black box" methods to simulate potential outcomes) is a useful tool to realistically evaluate the probability of discovering a true subgroup that the sponsors should use in the design of Phase 3 studies where subgroups are planned to be evaluated.

The SIDES methodology was applied to a case study in recurrent Dysmenorrhea with somewhat ambiguous results. There is a weak evidence, as suggested by evaluating observed variable importance scores with Adaptive SIDEScreen that patients with albumin over 4.5 g/d and age over 20 years may have greatest treatment benefit from lidocaine bioadhesive gel as a treatment for pain associated with primary dysmenorrhea. This hypothesis may have some clinical merit, given that increasing age is inversely correlated to disease severity and higher albumin production is inversely correlated with proinflammatory disease, thus suggesting that the lidocaine bioadhesive gel may be primarily effective in patients with a less severe or a less proinflammatory disease. However, when taking into account sampling variability of variable importance via the Stochastic SIDEScreen method, these results appear spurious. While, as is always the case, additional studies would be an ultimate test for the validity of the hypotheses generated from Phase 2 data, sponsors need to carefully weigh in various risks when deciding on conducting future studies tailoring subpopulation, especially with a rather weak support from the data.

When evaluating results of a Phase 2 study, the sponsor is often tempted to ignore the issues of multiplicity altogether, arguing that these analyses are "merely for internal decision making/signal detection." We emphasize that, when evaluating predictive biomarkers from a Phase 2 study, control of the overall type I error rate (or the false discovery rate, FDR) is still critical to prevent incorrectly "detecting" a noise biomarker for the signal and carrying it further into a Phase 3 program (as is often the case in drug development practice). However, one obviously does not have to adhere to the proverbial 5% error rate, but rather aim at a reasonable balance between the Type I and II errors rates. An attractive option when choosing biomarkers based on Variable Importance may be to control the proportion of selected noise biomarkers (FDR) rather than the probability of selecting at least one noise biomarker. This will be explored in the context of the SIDEScreen method in future work.

Acknowledgements This work is dedicated to the memory of James (Chip) Hackett.

A.1 Appendix

Table 17.3 List of candidate covariates and potential confounders

Covariate	Label in data set	Variable type; pre-specified cut-offs	Mean(SD) [min, max] or % for categorical
Age (years)	AGE	continuous	64.3%, 35.7%
Race	RACE	"White Non-Hispanic" (59) vs "All others" (11)	84.3%, 15.7%
Body mass index (kg/m^2)	BMI, BMI_CAT	Continuous and ordinal variable with levels: ≤20; (20–25); >25	8.6%, 48.6%, 42.7%
ALBUMIN	ALB	Continuous	4.5 (0.25) [3.8–5.0]
ALKALINE_PHOSPHATASE	ALKPH	Continuous	61.5 (14.2) [39–102]
CARBON_DIOXIDE_CO2	CDCO2	Continuous	23.1 (1.96) [18–28]
BILIRUBIN_TOTAL	BILTOT	Continuous	0.47 (0.21) [0.2–1.3]
EOSINOPHILS	EOSIN	Continuous	2.4 (1.8) [0.3–11.2]
HEMOGLOBIN	HEMOG	Continuous	13.6 (0.92) [11.2–15]
LACTIC_DEHYDROGENASE	LAC_DEH	Continuous	147.4 (23.6) [99–212]
LYMPHOCYTES	LYMPH	Continuous	31.5 (7.7) [6.2–54.2]
NEUTROPHILS_ABSOLUTE	NEU_ABS	Continuous	4.3 (1.5) [2.1–10.9]

References

Ashley, E.A.: The precision medicine initiative: a new national effort. JAMA. **313**, 2119–2120 (2015)

Basile, J.: Blood pressure responder rates versus goal rates: which metric matters? Ther. Adv. Cardiovasc. Dis. **3**, 157–174 (2009)

Breiman, L.: Bagging predictors. Mach. Learn. **24**, 123–140 (1996)

Breiman, L.: Random forests. Mach. Learn. **45**, 5–32 (2001)

Chen, G., Zhong, H., Belousov, A., Viswanath, D.: PRIM approach to predictive-signature development for patient stratification. Stat. Med. **34**, 317–342 (2015)

Coco, A.S.: Primary dysmenorrhea. Am. Fam. Physician. **60**, 489–496 (1999)
Dawood, M.Y.: Primary dysmenorrhea: advances in pathogenesis and management. Obstet. Gynecol. **108**, 428–441 (2006)
Efron, B.: Estimation and accuracy after model selection. J. of Am. Stat. Assoc. **109**, 991–1007 (2014)
Foster, J.C., Taylor, J.M.C., Ruberg, S.J.: Subgroup identification from randomized clinical trial data. Stat Med. **30**, 2867–2880 (2011)
Friedman, J.H., Fisher, N.I.: Bump hunting in high-dimensional data. Stat. Comput. **9**, 123–143 (1999)
Huang, X., Sun, Y., Trow, P., Chatterjee, S., Chakravatty, A., Tian, L., Devanarayan, V.: Patient subgroup identification for clinical drug development. Stat. Med. **36**, 1414–1428 (2017)
Kehl, V., Ulm, K.: Responder identification in clinical trials with censored data. Comput. Statist. Data Anal. **50**, 1338–1355 (2006)
Lamont, A., Lyons, M.D., Jaki, T., Stuart, E., Feaster, D.J., Tharmaratnam, K., Oberski, D., Ishwaran, H., Wilson, D.K., Horn, M.L.W.: Identification of predicted individual treatment effects in randomized clinical trials. Stat. Methods Med. Res. **27**, 142–157 (2016)
Lipkovich, I., Dmitrienko, A., Denne, J., Enas, G.: Subgroup identification based on differential effect search (SIDES): a recursive partitioning method for establishing response to treatment in patient subpopulations. Stat. Med. **30**, 2601–2621 (2011)
Lipkovich, I., Dmitrienko, A.: Strategies for identifying predictive biomarkers and subgroups with enhanced treatment effect in clinical trials using SIDES. J. Biopharm. Statist. **24**, 130–153 (2014)
Lipkovich, I., Dmitrienko, A., D'Agostino, R.B.: Tutorial in biostatistics: data-driven subgroup identification and analysis in clinical trials. Stat. Med. **36**, 136–196 (2017a)
Lipkovich, I., Dmitrienko, A., Patra, K., Ratitch, B., Pulkstenis, E.: Subgroup identification in clinical trials by stochastic SIDEScreen methods. Stat. Biopharm. Res. **9**, 368–378 (2017b)
Loh, W.Y., He, X., Man, M.: A regression tree approach to identifying subgroups with differential treatment effects. Stat. Med. **34**, 1818–1833 (2015)
Simon, R.M., Subramanian, J., Li, M.C., Menezes, S.: Using cross validation to evaluate the predictive accuracy of survival risk classifiers based on high dimensional data. Brief. Bioinform. **12**, 2013–2214 (2011)
Shen, L., Ding, Y., Battioui, C.: A framework for statistical methods for identification of subgroups with differential treatment effect in randomized trials. In: Chen, Z., Liu, A., Qu, Y., Tang, L., Ting, N., Tsong, Y. (eds.) Applied Statistics in Biomedicine and Clinical Trials Design. Springer, New York (2015)
Su, X., Tsai, C.L., Wang, H., Nickerson, D.M., Li, B.: Subgroup analysis via recursive partitioning. J. Mach. Learn. Res. **10**, 141–158 (2009)
Wager, S., Hastie, T., Efron, B.: Intervals for random forests: the jackknife and the infinitesimal jackknife. J. Mach. Learn. Res. **15**, 1625–1651 (2014)

Chapter 18
What Is the Right Comparison? ROC Curve and Trade-Off Between Key Diagnostic Test Errors (ROCKE)

Norberto Pantoja-Galicia and Gene Pennello

18.1 Introduction

Diagnostic tests can help to establish clinical management of the patient, support the use of therapeutic products, and guide therapy selection with the aim to maximize good outcomes or benefits and minimize adverse outcomes or risks (Pennello et al. 2016). In the evaluation of the performance of a new diagnostic test, errors and consequences can be more easily assessed if the new diagnostic test is compared with an already established alternative. Different errors have different consequences (Evans et al. 2016). The ROC curve provides key information regarding the implicit and/or explicit trade-off between false positive (FP) and false negative (FN) test errors that can be employed to establish adequate comparison between the new test and the established alternative.

18.2 Test Performance and Diagnostic Errors

We present the ideas and concepts in the context of assuming a new screening diagnostic test (N) for the detection of colorectal cancer. The test provides a binary outcome and a positive result may indicate the presence of colorectal cancer and should be followed by colonoscopy. Different errors carry different consequences, for example a FP result can lead to unnecessary diagnostic follow-up or treatment with its associated risks, suboptimal allocation of resources (for

N. Pantoja-Galicia (✉) · G. Pennello
U.S. Food and Drug Administration/Center for Devices and Radiological Health, Silver Spring, MD, USA
e-mail: Norberto.Pantoja-Galicia@fda.hhs.gov

L. Zhang et al. (eds.), *Contemporary Biostatistics with Biopharmaceutical Applications*, ICSA Book Series in Statistics,
https://doi.org/10.1007/978-3-030-15310-6_18

example, colonoscopy for patients who do not need it), and unnecessary stress. On the other hand, a patient with a FN result might not receive the necessary treatment and the disease may progress unattended.

Sensitivity (or true positive fraction) and specificity (true negative fraction or 1 − the false positive fraction) are metrics that are commonly used to assess performance of a diagnostic test. Let N be a new test which is to be compared to a standard test (S). A non-inferiority criterion implies that N should be better than merely adding a random test to an approved test in a believe-the-positive or believe-the-negative combination of the results (a believe-the-positive (negative) combination means that if test 1 OR (AND) test 2 are positive, then the result of the combination is considered positive. See Pepe 2003, p. 268). This non-inferiority criterion implies that the relative change in true positive fraction (ratio of sensitivities) between test N and an approved test is greater than the relative change in false positive fraction (ratio of 1 − specificities), or that the relative change in false negative fraction (1 − sensitivity) is less than the relative change in true negative fraction. See Biggerstaff (2000). Please note that the Biggerstaff diagram below (Fig. 18.1) plots the true and false positive rate coordinates (sensitivity, 1 − specificity) for theoretical tests N and S. In this case, compared to the Test S coordinate, performance of Test N may be unacceptable because neither Positive

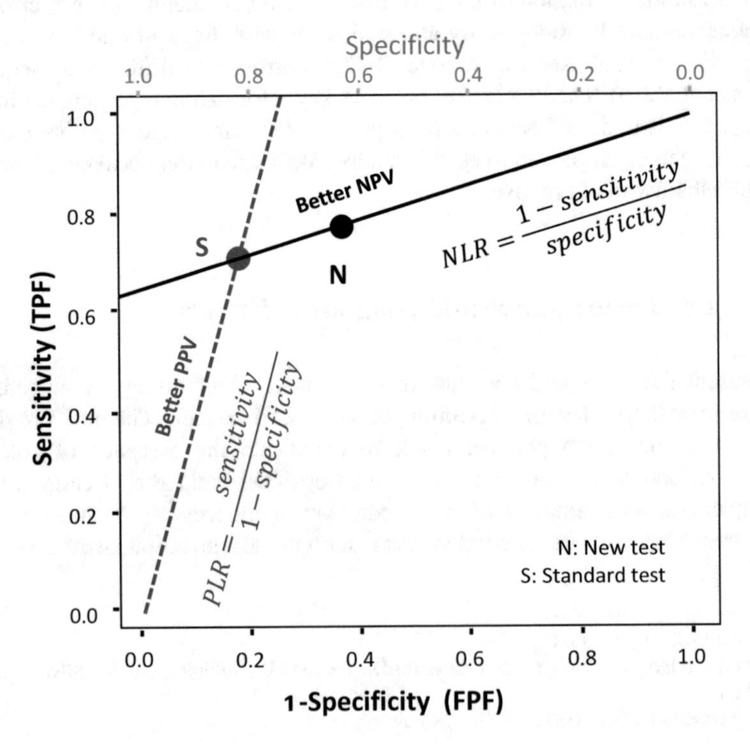

Fig. 18.1 Biggerstaff plot new test N vs. standard test S

Predictive Value (PPV) or Negative Predictive Value (NPV) is better for N. In general, a test is inferior to Test S if its coordinate falls in the region that is below the solid black and dotted blue lines passing through the test S coordinate, as these lines have slopes equal to the positive and negative likelihood ratios (PLR, NLR), respectively.

18.3 ROC Curve Information

Let us assume that each test, S and N, has an underlying numeric result with a corresponding cutoff that defines the positivity threshold for the respective binary outcome. Nonparametric ROC curves can then be estimated by plotting sensitivity vs. specificity for all possible positivity thresholds or cutoffs c. These are shown in Fig. 18.2 respectively for tests S (blue) and N (red).

For a diagnostic test, let C_{FP} denote the seriousness or cost (not necessarily financial cost, but it could be cost to public health, for example) associated with making a False Positive (FP) error and C_{FN} is that for a False Negative (FN) error. At a given operating point, the slope m of the tangent line to the ROC plot confers the implicit trade-off between FP and FN diagnostic test errors that is being made. This slope is $m = (C_{FP}/C_{FN})(1 - \rho)/\rho$, where ρ = prevalence. This derivation follows the results of Zweig and Campbell (1993), and Pepe (2003, p.72, Result 4.5) and is summarized in the Appendix.

In Fig. 18.2, the slope of the tangent line at the operating point of S is shown in blue and is steeper compared to that of N shown in red. This implies that the seriousness of a FP error relative to a FN error is implied to be greater for S at its operating point than for N at its operating point. N and S are operating at points that attribute a different trade-off between FP and FN errors, even though the screening population is the same.

If S were to operate at the same implied trade-off as N, the slope of the tangent line would be the same (at a different operating point S_* along the ROC curve. See Fig. 18.3). From the ROC plots in Fig. 18.3, S_* would have a better PPV (greater PLR) than N, with better NPV (lower NLR).

In addition, if N were to operate at the same implied trade-off as S, the slope of the tangent line would be the same (at a different operating point N_* along the ROC curve towards the origin). Then, N_* would have worse PPV (lower PLR) than S, and worse NPV (greater NLR).

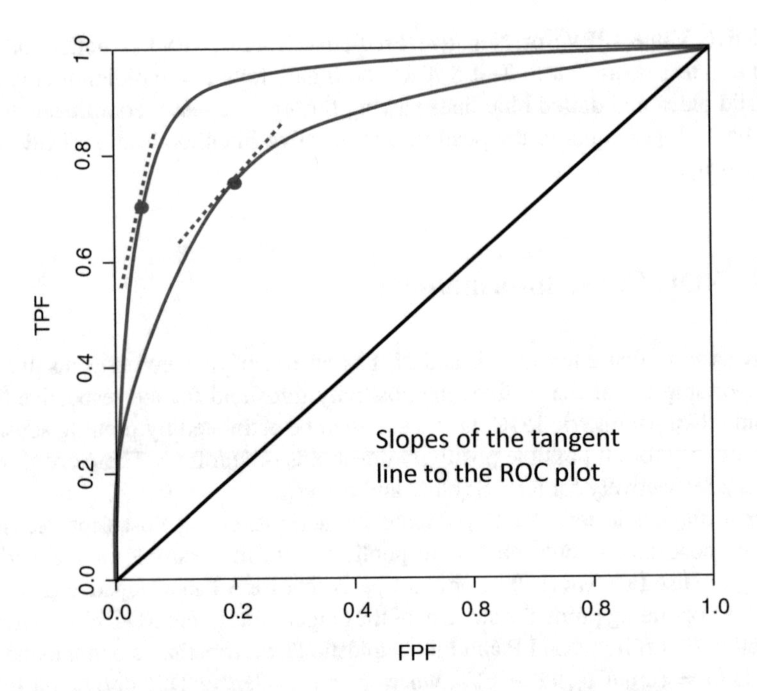

Fig. 18.2 Slopes of the tangent line to the ROC plots at operating points N and S

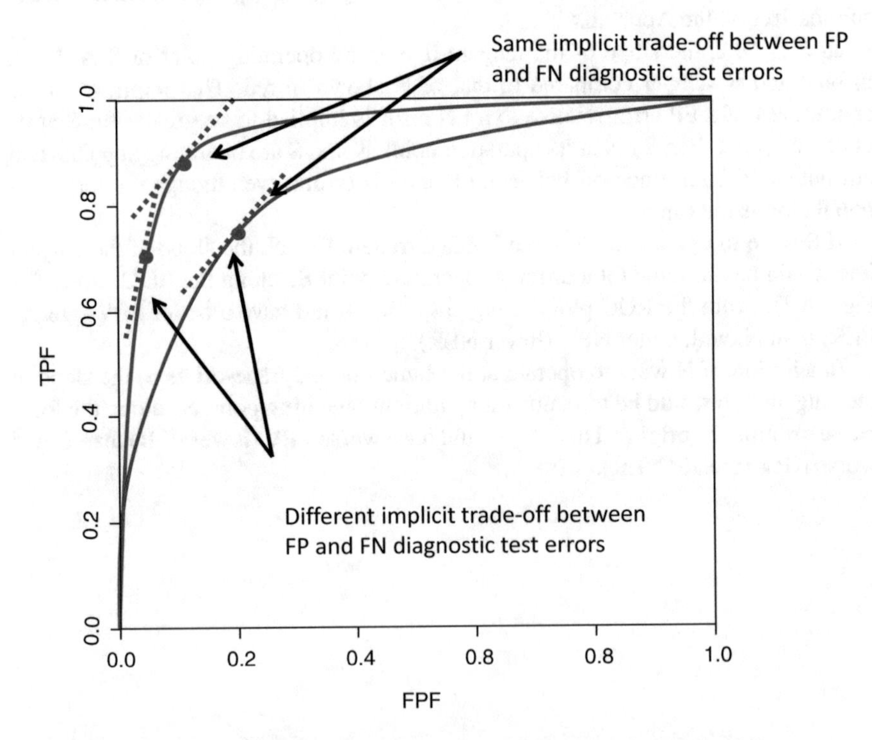

Fig. 18.3 Slopes of the tangent line to the ROC plots at operating points N, S and S_*

18.4 Discussion

At its operating point, a new test N may not be comparable with a standard test S because the corresponding slopes, i.e., the seriousness or importance of a FP error relative to a FN error, are different. If N were to operate at the same implied trade-off between FP and FN errors as S, the slope of the tangent line would be the same and it would correspond to a different operating point N_* in the ROC curve that would make the slopes comparable. In our example, the new test Test N does not seem to be comparable with test S at their corresponding operating points. Conversely, If S were to operate at the same implied trade-off between FP and FN errors as N, the slope of the tangent line would be the same and it would correspond to a different operating point S_* in the ROC curve that would make the slopes comparable and a comparison using the Biggerstaff plot could be conducted as in Fig. 18.4. This comparison would render test N as inferior since it has worse PPV and NPV compared to a test with an operating point at the coordinate S_*.

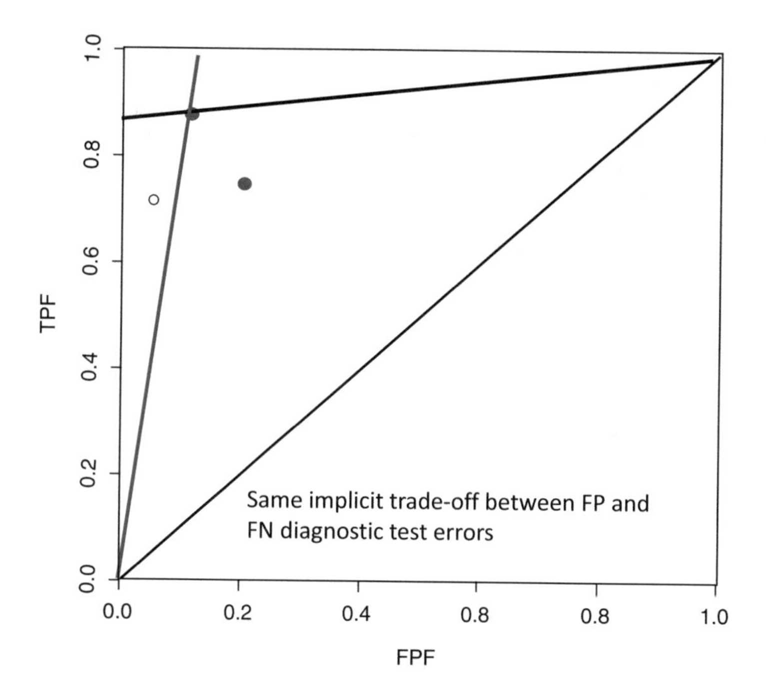

Fig. 18.4 Biggerstaff comparison of the tests at operating points N and S_*

A.1 Appendix

Cost may be expressed as: cost = C_{FN} (1 − sensitivity) ρ + C_{FP} (1 − sensitivity) (1 − ρ).

Therefore, when operating the test at False Positive Fraction or (1 − specificity) = t, the cost is

$\text{cost}(t) = C_{FN}\,(1 - \text{ROC}(t))\,\rho + C_{FP}\,(t)\,(1 - \rho)$.

Solving $\partial \text{cost}(t)/\partial t = 0$ provides the result for the slope $m = \partial \text{ROC}(t)/\partial t = (C_{FP}/C_{FN})(1 - \rho)/\rho$.

References

Biggerstaff, B.J.: Comparing diagnostic tests: a simple graphic using likelihood ratios. Stat. Med. **19**, 649–663 (2000)

Evans, S.R., Pennello, G., Pantoja-Galicia, N., Jiang, H., Hujer, A.M., Hujer, K.M., Manca, C., Hill, C., Jacobs, M.R., Chen, L., Patel, R., Kreiswirth, B.N., Bonomo, R.A.: Benefit: risk evaluation for diagnostics: a framework (BED-FRAME). Clin. Infect. Dis. **63**, 812–817 (2016). https://doi.org/10.1093/cid/ciw329

Pennello, G., Pantoja-Galicia, N., Evans, S.R.: Comparing diagnostic tests on benefit-risk. J. Biopharm. Stat. **26**, 1083–1097 (2016). https://doi.org/10.1080/10543406.2016.1226335

Pepe, M.S.: The Statistical Evaluation of Medical Tests for Classification and Prediction. Oxford University Press, Oxford (2003)

Zweig, M.H., Campbell, G.: Receiver-operating characteristic (ROC) plots: A fundamental evaluation tool in clinical medicine. Clin. Chem. **39**, 561–577 (1993)

Index

L. Zhang et al. (eds.), *Contemporary Biostatistics with Biopharmaceutical Applications*, ICSA Book Series in Statistics,
https://doi.org/10.1007/978-3-030-15310-6

Printed in the United States
By Bookmasters